POSTHUMOUS LIFE

CRITICAL LIFE STUDIES

CRITICAL LIFE STUDIES

Jami Weinstein, Claire Colebrook, and Myra J. Hird, series editors

The core concept of critical life studies strikes at the heart of the dilemma that contemporary critical theory has been circling around: namely, the negotiation of the human, its residues, a priori configurations, the persistence of humanism in structures of thought, and the figure of life as a constitutive focus for ethical, political, ontological, and epistemological questions. Despite attempts to move quickly through humanism (and organicism) to more adequate theoretical concepts, such haste has impeded the analysis of how the humanist concept of life is preconfigured or immanent to the supposedly new conceptual leap. The Critical Life Studies series thus aims to destabilize critical theory's central figure, life—no longer should we rely upon it as the horizon of all constitutive meaning but instead begin with life as the problematic of critical theory and its reconceptualization as the condition of possibility for thought. By reframing the notion of life critically—outside the orbit and primacy of the human and subversive to its organic forms—the series aims to foster a more expansive, less parochial engagement with critical theory.

Luce Irigaray and Michael Marder, *Through Vegetal Being: Two Philosophical Perspectives* (2016)

POSTHUMOUS LIFE

THEORIZING BEYOND
THE POSTHUMAN

EDITED BY JAMI WEINSTEIN
AND CLAIRE COLEBROOK

Columbia University Press
New York

Columbia University Press
Publishers Since 1893
New York   Chichester, West Sussex
cup.columbia.edu

Library of Congress Cataloging-in-Publication Data
Names: Weinstein, Jami, editor.
Title: Posthumous life : theorizing beyond the posthuman /
edited by Jami Weinstein and Claire Colebrook.
Description: New York : Columbia University Press, 2017. |
Series: Critical life studies | Includes bibliographical references and index.
Identifiers: LCCN 2016041871 (print) | LCCN 2017000837 (ebook) |
ISBN 9780231172141 (cloth : alk. paper) | ISBN 9780231172158 (pbk. : alk. paper) |
ISBN 9780231544320 (e-book)
Subjects: LCSH: Philosophical anthropology. | Human beings—Forecasting. |
Evolution (Biology)—Forecasting.
Classification: LCC BD450 .P5857 2017 (print) | LCC BD450 (e-book) |
DDC 128—dc23
LC record available at https://lccn.loc.gov/2016041871

Columbia University Press books are printed on permanent
and durable acid-free paper.
Printed in the United States of America

CONTENTS

III. INORGANIC RITES

7. AFTER NATURE: THE DYNAMIC AUTOMATION OF TECHNICAL OBJECTS

LUCIANA PARISI

155

8. NONPERSONS

ALASTAIR HUNT

179

9. SUPRA- AND SUBPERSONAL REGISTERS OF POLITICAL PHYSIOLOGY

JOHN PROTEVI

211

10. GEOPHILOSOPHY, GEOCOMMUNISM: IS THERE LIFE AFTER MAN?

ARUN SALDANHA

225

IV. POSTHUMOUS LIFE

11. PROLIFERATION, EXTINCTION, AND AN ANTHROPOCENE AESTHETIC

MYRA J. HIRD

251

PREFACE: POSTSCRIPT ON
THE POSTHUMAN

CLAIRE COLEBROOK AND JAMI WEINSTEIN

T here is a very real sense in which the posthuman should be the last question we pose for our times—a postscript to the story of the human. Given that the human species is now beginning to change the scale of thinking by reference to a framework of deep geological time—to a time and place before human beings existed and began to scar the earth permanently—and, for the first time in its brief history, starting to imagine a future in which it ceases to exist, we must wonder: What questions would a being who arrives after humans have wanted us to pose? Is it not the height of hubris, myopia, narcissism, megalomania, and denial to talk about the various senses in which the human has been surpassed (whether by way of the end of human exceptionalism or by enhancing and exceeding the limits of human life) precisely when humans have been given various prognoses regarding the sixth great extinction, the end of the human species, or at the very least the end of liberal personhood and "man's" current age of affluence and favorable conditions (Mulgan 2011)? Let us imagine, as geologists are beginning to do, that if there were something or someone able to view the planet Earth after human extinction, it would be able to discern that a species event occurred of such magnitude that the planet ceased to be a stable living system. This thought experiment of the Anthropocene, imagining not just a world without us but also a readable geological archive that testifies to our once forceful existence, opens a new mode of historical reflection that is literally *after* humans while simultaneously reinforcing the sense that there is something identifiably and inescapably human: human environmental and geological impact. We might be post-human in no longer having exceptional moral value or even an essential species distinction, but, as advocates of the Anthropocene argue, we are well and truly distinct in our earth-destructive capacity. For any event, we might imagine various possible narrations and contestations, but something entirely new has occurred with the posing of the Anthropocene, and this is the possibility of human life and *human history after humans*: Humans will be readable in the scar they left upon the earth.

Given the uncritical turn to life that has often accompanied current trends of Anthropocene thinking, we argue that theorists must instead take up a *critical* study of life (critique defined in the Kantian sense). In addition to the questions humans pose to life, critical life studies considers the ways in which certain appearances of life (and its possible nonexistence) pose questions to the human that would be critical or that would raise a sense of human limits and require thinking beyond the human conditions of existence. In that guise, it is essential to revisit the question of the posthuman anew and to do so from several angles. The first sense of the posthuman we would like to pose is one that is rarely, if ever, used: There will be a *post*human time, a world without and after humans. Such a sense renders many of the other senses of the posthuman enigmatic, if not null and void. Before the twenty-first century's increasing preoccupation with the end of human life (as evidenced in the proliferation of postapocalyptic cinema and literature), the posthuman was more often than not a liberating (and perhaps trans- or super-human) motif: We would no longer imagine ourselves to be exceptional, no longer imprisoned in the identity of "man." The "superman," to use Gilles Deleuze's terminology, is an expansion of living force beyond the human species:

> The forces within man enter into a relation with the outside, those of silicon which supersedes carbon, or genetic components which supersede the organism, or agrammaticalities which supersede the signifier. In each case we must study the operation of the superfold, of which the "double helix" is the best-known example. What is the superman? It is the formal compound of the forces within man and these new forces.
>
> (Deleuze 1988b, 131–132)

Articulated at a time when the "end of man" was a conceptual and philosophically liberating gesture, such dreams of a life beyond life seem to be at odds with today's increasing sense of the contraction of life to mere survival for an increasing number of the planet's humans. Deleuze's imagined "superman" will, one assumes, not be readable from the literally posthuman future imagined by the thought of the Anthropocene. In its positing of silicon's surpassing of carbon-based life forms, the idea of the superman is at odds with the increasing "no exit" warnings of today's climate scientists and geologists. If the posthuman gestured at one time to a world of cyborgs, supermen, and "inorganic" life, that past sense of a life posed beyond life has been countered by a future of a post-Holocene world where the earth as a living system is beginning to fail and instead promises a future inhospitable to all living forms. Such a future will differ from Deleuze's "superfold," in which life is freed from the organic survival mechanisms of the human animal; it will be distinctly *post*human not simply because humans will have ceased to exist but (more importantly) because our having existed will have irrevocably altered the planet's composition and forces.

This highly literal sense of the posthuman allows us to return to the problem of last questions; these are not just questions about how we plan the end and what we will leave for the future. The multiple senses of the posthuman—ranging from a world without life at all to a world in which "man" has freed himself from the limits of life—are and must always be in play and in tension with one another. This is not because the human and life are prima facie values that must absorb our concerted efforts in what may well be the final moments of Holocene stability but rather because whatever the human is or will be cannot be distinguished from the sense of an end. Such an "end" of man ranges from Western philosophy's commitment to a telos beyond mere life, to Amazonian culture's capacity to think of human life as a fleeting contraction of an ensouled life that takes on bodily distinction (Viveiros de Castro 2009), and to Australian Aboriginal conceptions of a time and space populated by spirit beyond human dwelling (Povinelli 2002). The human (if there is such a thing) has, more often than not, imagined itself by way of what surpasses the human (whether universal rationality, life, or spirit)—even if in its Western mode, "man" is that distinct being who has no end other than the end he sets for himself. When Jacques Derrida refers to the "ends of man," he draws upon an always-doubled sense of the end as both limit and goal: "man" has always imagined that his end (his way of choosing to be and the limits he sets for himself) is something not yet given, and he has always tried to put an end to any definitive conception that would limit his future. In this respect it has always been the nature of "man" to be posthuman, not accepting any supposed end or function that may define him as one more thing within the world (Derrida 1969). Even the early Karl Marx, referring to our "species-being" (*Gattungswesen*), nevertheless sees free conscious activity as definitive; for any human to accept his position within labor and history is a form of alienation:

> It is therefore in his fashioning of the objective that man really proves himself to be a species-being. Such production is his active species-life. Through it nature appears as his work and his reality. The object of labor is therefore the objectification of the species-life of man: for man reproduces himself not only intellectually, in his consciousness, but actively and actually, and he can therefore contemplate himself in a world he himself has created.
>
> (Marx 1974, 329)

For Marx, then, the very nature of "man" is that he defines his own nature by transforming his world. Today, however, the very sense of this perennially self-surpassing and always posthuman being of "man" is rendered unintelligible by the thought of a future without humans. Will we still be human in a world of such depleted conditions that our supposedly essential rights and freedoms are no longer available? There will be, perhaps sooner rather than later, not merely a speculative imagination of a time without humans but an actual experience of ending. This experience does not coincide with mainstream postapocalyptic, "end-of-the-world" disaster tropes—where "man" vanquishes some virus or threat that makes his humanity precarious or finds refuge on another planet—since they largely intensify an ultrahumanist posthumanism where even zombie viruses, catastrophic floods, alien invasions, and resource depletion are no match for a species that can either hunt down the clues of a mother nature who has always been a serial killer (e.g., the film *World War Z*)[1] or populate another galaxy and replicate earthly existence in a higher form (e.g., the films *Elysium* and *Interstellar*). If we start to think of ourselves as the last humans, as the beings capable of thinking beyond the time of humanity, what questions does the future pose to us? This is not the question of how "we" will survive but of how—given our eventual nonsurvival—we might be inscribed (including the literal climactic inscription on the earth). This mode of questioning the present's continued existence has perhaps already started to be felt in various forms of generational resentment: The next generations of humans will inherit debt, resource depletion, high levels of environmental toxicity, a warmed and warming planet, and acidified oceans, along with embedded structures of nepotism, finance, and bureaucracy that will exacerbate the effects of that inheritance. There will not be the conditions to sustain the consumption and pollution, not to mention the rights, which have defined the human. This generation, with the sense of an ending, might be entitled to ask: What questions should the last humans have posed, and would these questions be posthuman?

On the latter question, we might stop to consider two dominant conceptions of the posthuman. Nick Bostrom distinguishes his version of the posthuman from what he deems to be its main competitor. For Bostrom, posthumanism is a predicate or attribute that would, by and large, be a beneficial supplement for humans (Bostrom 2008). To be posthuman in this sense, which in other contexts is also referred to as transhuman, is to conquer death, and maybe cognitive and moral deficiency. Indeed, extending or enhancing humans is something that almost has the urgency of a moral imperative (and, for Bostrom, is intertwined with the concomitant demand to save humanity from the existential risks that come from runaway artificial intelligence and all other threats to the species). The only reason we have not regarded posthumanism as a good is because of weakness and a lack of expertise (Savulescu and Bostrom 2010). We should face up to the horrors of human death and limits and pour all our resources into extending and enhancing human life beyond the natural limits of the human. The other sense of the posthuman, which Bostrom distinguishes from his own, is that of N. Katherine Hayles. Here, being posthuman is not about supplementing the natural human body but occurs as an awareness and theorization of all the ways in which what we once took to be definitively human—a mind organized according to core principles of reason and progressively skilled in arts of attention, analysis, and sustained argument—has now been dispersed (Hayles 2007). Hayles argues for an entirely new conception of the human archive, including digital texts, computer-generated sequences, and, more importantly, new speeds of composition that return humans to what once dominated their prehumanist past: hyperattention. Years of reading and culture, as well as the required leisure and safety of civilization, allowed humans to develop deep attention, exemplified by the sustained close reading of novels and the articulation of complex philosophical arguments. Now, not only is the brain supplemented by the memories and skills of computers, but it has also reverted to the hyperattention (or fleeting and multi-task-oriented modes of captivation) that would once have served highly threatened hunter-gatherers. For Hayles, like Bostrom, posthumanism simultaneously is a new event enabled by novel technologies and is not so much a theory of the human as it is a negotiation of future possibilities in a post-human era marked by new technologies.

If it is this sense of the posthuman that occupies us—either as the celebration of overcoming human mortality or as a resignation to a new era of humans coupled with technology—then it is hard to imagine that the last humans (the future humans) would be grateful that we were occupied with extending and maximizing ourselves. If we—as we are indeed beginning to do—start to imagine that we are the last humans and that a really posthuman world might begin as early as 2030, then posthumanism perhaps might have the sense not of the means by which the human extends, surpasses, and supplements itself but of a different mode of the "post-" that destroys the ultrahumanism that has always been a posthumanism. In addition to the three senses already described—the literal disappearance of humans, enhancement of humans (Bostrom), and technological coupling (Hayles)—Cary Wolfe deploys the posthuman as a method for destabilizing the human species' claims for distinction. Rather than assuming that poststructuralism sweeps away human exceptionalism or that "we" are somehow locked within a prison house of human language, Wolfe draws from deconstruction in order to question how we think about what the human is and its supposed superhuman or posthuman mode. If we cease to define language as a privileged domain of meaning but instead regard all life as composed of differential systems, then we can neither collapse the human back into some self-present prelinguistic life nor abandon ethical concerns regarding nonhumans. Wolfe's posthumanism does not destroy the rights, sentience, and finitude of the human but questions the extent to which we grant that humanity to the human species alone. Wolfe writes of a "bond between human and nonhuman animals as beings who not only live and die as embodied beings, but also communicate with each other in and through a second form of finitude that encompasses the human/animal difference, forming a bond that is all the more powerful because it is 'unthinking' and in a fundamental sense unthinkable" (Wolfe 2009, 123). Wolfe criticizes versions of what he refers to as "humanist posthumanism" for simply granting rights or attention to animals while keeping the category of the human coherent and distinct. This strategy of adding animals to the already humanist paradigm— whereby magnanimous man grants the putatively lesser animal some moral value—does little to advance the human beyond itself and in that sense is not really "post" anything. Along with Wolfe's critique, we suggest a strategy of extending and contesting a wider range of ostensibly posthumanist claims to exit from the human, precisely because whatever has defined itself as human has always done so by distancing itself from any determined or specified humanity. This is so at least for what has called itself and questioned itself as human within the Western tradition. Likewise, Eduardo Viveiros de Castro—far from romanticizing non-Western humanisms—aims to decolonize the thinking that casts the human as the only being blessed with anthropocentrism. Every being perceives its own world in utterly personal and spiritual terms, personifying and humanizing according to its encounters. If Western humans have thought that they could step outside their own myopia by appealing to some special otherness of the other, then this is only because they are unaware of all the ways other humans and "nonhuman humans" experience the world as composed of persons, all of whom have their own world. Perhaps, then, the desire for the posthuman intensifies a human/nonhuman binary that—far from being something we ought to dissolve—should be multiplied, such that every being in the world unfolds its own distinctions and divisions, never settling into "a" difference or "no difference at all" but a proliferating *indifference*, a single plane from which there are countless claims and dismissals as to what counts as a human or person:

> The point of contesting the question, "what is (proper to) Man?", then, is absolutely not to say that 'Man' has no essence, that his existence precedes his essence, that the being of Man is freedom and indetermination, but to say that the question has become, for all-too obvious historical reasons,

one that is impossible to respond to without dissimulation, without . . . continuing to repeat that the chief property of Man is to have no final properties, which apparently earns Man unlimited rights to the properties of the other. This response from 'our' intellectual tradition, which justifies anthropocentrism on the basis of this human 'impropriety,' is that absence, finitude and lack of being [manqué-a-etre] are the distinctions that the species is doomed to bear, to the benefit (as some would have us believe) of the rest of the living. The burden of man is to be the universal animal, he for whom there exists a universe. . . . As for non-Occidental humans, something quietly leads us to suspect that where the world is concerned, they end up reduced to its smallest part. We and we alone, the Europeans, would be the realized humans, or, if you prefer, the grandiosely unrealized, the millionaires, accumulators, and configurors of worlds. Western metaphysics is truly the fons et origio of every colonialism.

In the event that the problem changes, so too will the response. Against the great dividers, a minor anthropology would make small differences proliferate—not the narcissism of small differences but the anti-narcissism of continuous variations; against all the finished-and-done humanisms, an 'interminable humanism' that constantly challenges the constitution of humanity into a separate order. . . . It is not a question of erasing the contours but of folding and thickening them, diffracting them and rendering them iridescent.

(Viveiros de Castro 2014, 44–45)

We would like to articulate this residual humanism in posthumanism in three interrelated claims: First, whatever might have passed as humanism has always been a form of posthumanism. The human, or man, has never been one being or animal among others but is precisely the being who believes himself to be in command of his own open, self-creative, and purely potential being. This is why Plato directs the proper gaze of humans—and their discovery of their own good—toward an interior exterior or a truth, logic, and ideality that cannot be reduced to humans as a worldly, embodied, or sensibility-confined species. This is why Agamben (2003), supposedly directing himself against the entire history of human propriety and the expulsion of animality, nevertheless presents the task of the future as one in which humans finally own up to their own lack of essence, lack of propriety, and (therefore) exposed and open being. If every other being in the world simply *is*, then it is the "nature" of humans not to have a nature, either because they are so self-cultivating as to have lost all traces of determinacy or because they have the capacity to vanquish nature and self-consciousness and become one again with a general mindful life or global brain. We might say that many of the anti-Cartesian attempts to think an "extended mind" (Menary 2010) or "mind in life" (Thompson 2007) are reaction formations that insist on humanity's closeness to nature precisely at the point where our destructive detachment from nature is all too evident.

Second, we might distinguish distinct modes of the posthuman that are not so much sublations of humanism as they are ultrahumanisms. If posthumanism is the refusal of any determined human essence, with an insistence on the human as nothing more than self-creating potentiality, then such posthumanisms intensify the human quality par excellence: self-fabrication. The supposedly radical mode of this posthumanism that defines itself against essentialism—where humans are self-negating rather than self-fabricating, insofar as they are nothing other than the encounters or "becomings" they experience—is also a dynamic intensification of the human as set apart from a world of beings. There is a form of anti-Cartesianism that is hyper-Cartesian: We would reject man as thinking substance—man as a special being—in the name of man as nothing at all, as not a thing, as pure performance, event, or act. If Descartes defined thinking substance against extended substance, then the claim that humans are not substance but pure becoming is an ultra-Cartesianism and hyperhumanism. It is, moreover, a shrill and frenzied proclamation of man's capacity for being radically futural precisely at that point in his history where the human species seems to be threatened with its actual nonbeing.

Third, this means that ostensible modes of the posthuman—such as extending humans through technology, or denying any human essence, or being in favor of becoming—operate to occlude the simpler but rarely confronted literal sense of the posthuman. There will be a time after humans. And if this occurs sooner rather than later—as those who call themselves "humanity" accelerate the sixth great extinction event—it may be because one of the main reasons for accelerated climate change is the mindset that "man" does not exist and is nothing other than his own pure becoming and action. What often passes for posthumanism—in all its modes of celebrating becoming, potentiality, and the dynamism of the singularity of a life of which humans are an inessential component—will contribute to the literal end of man and do more to avoid, rather than confront, the posthuman sensibility that has always situated the human species apart from its determined and human-all-too-human being. The more stridently we insist upon the radical openness of the future, the less capable we become of considering the virtual futures that our present harbors. The virtual, here, is not absolutely any future whatsoever—a future where one might imagine that anything that is now might also not be. Such a future would be of the type that Quentin Meillassoux (2010) imagines and would follow from the always possible thought of the absolute: Whatever is can always be thought of—without contradiction—as not being. Such a future is possible, but it is possible only while erasing the potentiality of the world's virtual futures. Every moment of the present harbors unfolding multiple eternities. Some of them might see all life and matter erased, and others might see the generation of new forms of life—possibly inorganic or non-carbon-based. But the claim that there is, in the present, a potentiality for "man" to extend and flourish as he is, is a logical but empty possibility. That is to say that while it is logically possible, it is *not at all probable*; it is imaginable, but only if the imagination occludes all the potentialities of the present, all the timelines that unfold from an actuality that is all the richer for harboring virtual futures that are not those of what is merely logically possible. If the posthuman is the simple negation of the human that imagines surpassing human limits and allowing man to live on despite what he has been and done, then the posthuman is nothing more than what the human has always already been: an impoverished actuality that holds on, only, to man as he is. If, however, we think of "man" as he might have been, where other modes of being human—including those histories and possible futures of peoples who do not "enjoy" the trajectory of hyperconsumption, hyperproduction, and universalizing globalism—then the posthuman might not be "man" imagining his end or redemption. Instead, what has known and always surpassed itself as "man" might think beyond its parochial notion of "the" human.

The posthuman is not merely a polysemous term (like humanism, postmodernism, and almost any other "-ism"). The different senses and tendencies of the term are symptomatic of a problem that is neither simply linguistic nor a question of taxonomy and periodization. We can begin with the very problem of humanism as being always a posthumanism. It is undeniable that there is something like a human species and that the distinct nature of this species lies partly in the extent to which it denies its being as a species; that is, "humanism" has always been a way of refusing to see humanity as a biological event within life and has (at least in its Western metaphysical mode) always seen the human as a rational, sentimental, technical, spiritual, cultural, or historical means of surpassing life. It is also the case that just as the specificity of the human takes on its greatest evidential force—as we begin to see humans as having caused climate change and geological stratifications, united by way of impact if not by way of essence—that the countercry of the posthuman reaches fevered pitch. Why now, when human exceptionalism and specificity are beginning to display their damaging power, do at least two conflicting senses of the posthuman (the desire for extended or enhanced life *and* the vision of extinct humanity's having permanently scarred the earth) become ever more insistent? Ours is the age of the Anthropocene *and* of the "singularity," of technological maturity and preliminary mourning. Recapitulating, then, we can chart a series of posthuman projects: the posthuman as a form of cognitive, moral, lifespan, or emotional enhancement; the posthuman as technological extension (ranging from smartphones and computer software to new archival and mnemonic methods); the posthuman as a philosophical claim regarding the end of humanism insofar as we abandon the sense of our exceptional worth; and the posthuman as an imagined historical epoch in which there may well be organic life and a planet but there will be no humans. All these senses are framed by another tendency entirely, the *inhuman*. As already suggested, the very motif of the "post-" is inextricably tied to what has marked itself as human; there is a species, supposedly, that may always surpass any reading it has of its own determination. This is why posthumanism in all its senses—of surpassing, extending, overcoming, and even self-extinguishing—is tied to a humanism that has always defined itself dynamically and dialectically. What if the human and the post- (or a term and its problematization) were possible only because of something outside the terrain of human history altogether? Such inhuman forces—be they technological, geological, physical, mnemonic, or virtual—would not only surpass the species but might also extend beyond specification.

The problem of the posthuman, then, is composed of a series of over-lapping conundrums. The notion that there is something unique, essential, and universal about "man" cannot be sustained, given that all the qualities that allegedly set humanity apart (such as language, reason, morality, art, altruism, technology) occur in the nonhuman world. Yet despite this sense of the posthuman as the destruction of human exceptionalism, the current era also evidences a distinction of the human. Humans cannot simply will away their existence; ecological devastation and anthropogenic climate change have prompted the concept of an Anthropocene epoch, such that even after humans have become extinct *Homo sapiens* will have marked the earth in a manner quite distinct from any other organic life form. Second, posthumanism can appear as a historical event: There were, once upon a time, essential and species-bound humans, but technology has created the conditions for humans to extend themselves into prosthetics, artificial intelligence, cloning, and perhaps even immortality (thus dispelling the notion that humans are marked by the finitude of death). Yet one could also claim that humans have always been cyborgs, always been supplemented by technology and prosthetics, and that even the earliest forms of humanism were already forms of posthumanism (in that "man" is never capable of being captured by definition—a result of the self-constituting nature of the human). Finally, the twenty-first century is marked at once by a quite literal *post*-humanism (with cinema, literature, and television producing an increasing number of thought experiments that imagine a life without or after humans). Yet while the "posthuman" can refer to a world in which humans no longer exist, there have also been a series of moves in philosophy that claim to break with the philosophical tradition by adhering to new forms of epistemic realism about the world beyond the grasp of the human. What has come to be known as "speculative realism" is an insistence that there can be thinking and knowledge that is not limited by human finitude. Yet this deflation of the human is also an inflation regarding what we might claim to know (Bryant, Srnicek, and Harman 2011). We might say that it is precisely when humans face actual extinction (or their physical nonexistence) that they also claim an ideal or virtual capacity to be released from time, capable of finally knowing the world as it is in itself, the world beyond the limits of human representation.

What the foregoing exploration into the posthuman tells us is that the posthuman is not merely a multivalent concept but moreover a *problem* in the specific Deleuzo-Guattarian sense (Deleuze 2004, 198; Deleuze and Guattari 1994, 16). Deleuze and Guattari contend that "all concepts are connected to problems without which they would have no meaning and which can themselves only be isolated or understood as their solution emerges" (1994, 16). The concern is, thus, not whether we are "pro-" or "anti-" posthuman. Problems are not solved by choosing one or the other side of an already existing opposition. Nor are problems solved by definitions, as though we might find some accurate capturing of what the posthuman is. Problems are disruptions of an actualized field. Problems require a redistribution of forces to the point that new questions are created. Problems are by no means confined to philosophy, theory, academia, human life, or even life itself. Indeed, we might say that a problem requires the redistribution of fields and the creation of a new plane. Questions—such as whether drugs should be legalized or whether there should be international intervention in human rights violations—are only possible if *problems* are not composed. What might it be to question the very being of drugs and the notion of the proper human body and its external supplements? What might it be to ask how it is that something like a human right could act as a weapon in international war and politics? Questions that seem to have ready answers—yes or no, pro- or anti- —are only possible because of previous problems that have now lost their tension. We can think of the emergence of organic life as the stabilization of a problem: There is an interaction of forces, the creation of instability, and then the emergence of single-celled organisms. Only once the problem has yielded a certain stability can questions emerge. So, once a relation among forces allows organic life to emerge, certain questions—such as how the organism is to survive—may be formed. Some of our most vexed questions today (concerning rights to life, rights to defense, rights to privacy) are the results of a problem posed, such as the problem of what counts as a political unit—or what counts as a life, a human, or a person—or whether there are any entities that precede political formation. The question posed by Nick Bostrom, whether we should become posthuman, whether we should enhance and extend human capacities, is the outcome of a resolved problem—where we have preestablished the human as a reasoning, feeling, and valuable individual. But there are questions today that stretch beyond the posthuman in Bostrom's sense, questions that cannot be answered within the current posthuman/human terrain and must instead be answered by a new problem, such as the *critical* study of life. These questions require the formation of such new problems, precisely because the usual questions—"Is there any such thing as the human?"; "Is the posthuman good or bad?"; "Is the posthuman sufficiently political?"; "How useful is the term?"—are not at all adequate to answering the current state of disturbance. What is required are not answers to questions but a reconfiguration of the forces from which questions emerge. Such a reconfiguration cannot simply be an act of will, such that we simply decide to form a new lexicon or new method. On the contrary, this volume is united by a commitment to the idea that the problem of the posthuman is both posthuman and inhuman (in the sense of being outside the domain of our own making and in the sense of still being a problem, still being locked within a human-all-too-human field that we are struggling to rethink). Although the question of the ethical nonexistence of humans may occasionally be posed, it is too frequently articulated as a question and not a problem. That is, we are not called upon to think about, decide, and judge the worth and existence of humanity; instead the question is posed in a simple human versus non-human form. Do we want to live or die, survive or become extinct; are we worthy or not?

And the reason the posthuman and nonhuman have thus far remained questions and not problems is that they are largely posed within the current forms and genres of contemporary narrative. The current genre of the postapocalyptic is composed of the standard ultrahuman dramatis personae (lovers, families, villains) along with narrative trajectories structured around inhuman others (zombies, aliens, terrorists) that pose a threat to certain favored humans. Perhaps more than any other genre, what has come to be known as the postapocalyptic provides a way for humanity to view itself, find itself threatened by a nonhuman other, and then refind itself by reaffirming its proper mode. The structure of the postapocalyptic finds "man" almost extinguished because of an accidental occurrence—a mistaken experiment that allows a virus to run free, an overuse of the planet that presents humanity with a dire end, or an invasion by an antihuman species—and then it is man's proper mode that triumphs: man as consumer, destroyer, and dominator is redeemed by man as ecologically attuned life. *Avatar*, cited by Bruno Latour in his "Compositionist Manifesto," opposes a rapacious military man of aggressive capitalism to a utopian humanity—a humanity we might once have been and could still be (Latour 2010). This new humanity is also an ideal and lost humanity that is not separated from the world, not human by way of distinction, but posthuman and ultrahuman by way of being fully attuned to life. The idea of humanity as intrinsically or perpetually inhuman is at once presented and circumvented by dividing humanity into a proper humanity at war with a rogue and violent nonhumanity; the war between the properly human and the accidentally inhuman (zombies, invaders, terrorists, clones) allows the small redeemed essence of humanity to emerge.

This genre of condemning humanity as it is, toying with humanity's seemingly just vanquishing, and then concluding with a new and proper humanity is not at all confined to contemporary cinema (or even the contemporary). Mary Shelley's *Frankenstein* can be read as a cautionary tale that opposes a hyperhumanism of self-production and scientific overreaching to a promised or "monstrous" humanity that is oriented to nature, the archive, and others. Shelley's monster emerges into the world not as a subject but as what Deleuze, after Bergson, refers to as "pure perception" (Deleuze 1988a, 22): "A strange multiplicity of sensations seized me, and I saw, felt, heard, and smelt, at the same time; and it was, indeed, a long time before I learned to distinguish between the operations of my various senses" (Shelley 1999, 128). And it is precisely the trope of monstrous pure perception (also affective, impersonal, and prepersonal) that we signify in our shift to the *posthumous* from the posthuman. This modification not only indicates the posthuman "death of man" but also includes a repudiation of the vestiges of humanism lingering in the notion of life by going beyond it. The posthumous does not boast any form of redemptive humanistic telos as the culmination of undergoing a process of temporary dehumanization brought on by threat (either to the species or individual life). Rather, the posthumous remains a disturbance and a vibration orienting around the chaotic intensities that swirl in the absence of a concept of life as a controllable, containable, namable force attributed to humans (and, occasionally, and in lesser ways, other organic creatures). It is life both as excess and privation but decidedly not post*human*.

One of the few motifs shared by both Jacques Derrida and Luce Iriga-ray is that "man," or at least man as subject, has always defined himself as other than any simply given actuality; the subject can always, at will, destroy and remake himself—always become other than, and elevated above and beyond, any appearance (Irigaray 1985). In this respect, some forms of proclaimed posthumanism or transhumanism—such as the belief that "we" can simply abandon the human—are ultrahumanisms and as such do not pose the genuine problem of the human. Indeed, one way we might think about the overly broad concept of poststructural-ism is as a rendering problematic of the posthuman: The human is both necessary and impossible, for we are at once (historically, culturally, fig-urally, genetically, structurally) bound to a human finitude that it would be naïvely "humanist" to deny, and, at the same time, every attempt to grasp or annul that humanity repeats the most tired gestures of a mythic human freedom of pure self-creation. The posthuman as a problem, therefore, should not be a proclamation that man is dead and that we can now live happily ever after in symbiosis with plants and animals. Nor should it allow us to indulge in a crisis culture of lamentation in the face of the loss of reason, the loss of "our" ecology, the loss of the political (or any other of the supposedly human excellences that have been placed on an endangered list). Nor should it be a negation, denial, or demysti-fication of "man" (a claim that there is no such thing and that humanity is too complex and varied an event to be captured by a single term or "-ism")—for the claim that "man" always transcends any essence is all too human. Finally, a critical posthumanism should not allow for a joyous embrace of technoscience, where we happily transform ourselves into cyborgs or proclaim proudly that we have always been cyborgs (Clark 2003). Rather, the posthuman as a problem consists of a series of inter-secting impasses that stall questions as they are currently formulated and require a new terrain. The posthuman includes and excludes cognitive and moral enhancement: Writers such as Kurzweil (2005) and Bostrom (2008) pose necessary questions regarding what intensified technology might enable in terms of human expansion, precisely at the moment in the species' history when the injustices, violence, and limits of technol-ogy distribution come to the fore. The posthuman as a problem marks an affinity with nonhumans and nonlife—evidenced in the intelligence of everything from animals to bacteria and machines—just as claims are

being made that humans are exceptional by way of anthropogenic geological transformation.

At core, the posthuman problem is composed of conceptual elements
mined from a series of proper names that include Friedrich Nietzsche,
Martin Heidegger, Charles Darwin, Gilles Deleuze, Jacques Derrida,
Donna Haraway, Simone Weil, and Michel Foucault, all of whom are bivalent. Nietzsche is the philosopher of pure forces, forces that are outside
"man" and also outside anything that might be recognized as having a
stable being beyond the illusion of stability. But Nietzsche is also, with
his *amor fati* (1974, 233), the affirmer of living life's fate and contingency
as if one must choose this life and this time. Nietzsche is also the proper
name associated with nihilism, but he is paradoxically the late nineteenth
century's most affirmative and joyous promulgator of life as the aesthetic
self-creation of the *Übermench* (1982, 124). Heidegger is the staunch critic
of *das Man*, metaphysics, and humanism (1993): "The 'who' is not this one,
not that one, not oneself, not some people, and not the sum of them all.
The 'who' is the neuter, the 'they'" (1962, 164). Yet in an all-too-metaphysical-
humanist move, Heidegger also exalts the German people to the status of
Ideal Humans, and the Jewish people are cast scathingly as an in- or posthuman disturbance of the human: "The question of the role of world Jewry
is not a racial question, but the metaphysical question about the kind of
humanity that, without any restraints, can take over the uprooting of all
beings from being as its world-historical 'task'" (quoted in Seigel 2015).
Perhaps the most stunning example of human self-annihilation as human
affirmation arrives in the form of Darwinism, which places the human
as an emergent effect within life yet has also enabled a twenty-first century of neo-Darwinisms and evolutionary psychologies that allow us to
read our history as one grand "just-so" story. Darwin (1859) becomes the
thinker who annihilated human essentialism while nevertheless allowing
contemporary neo-Darwinists to read all our behaviors and technologies
as having an ultimately readable reason of life (Nagel 2012). Closer to our
own time, we might read the most posthuman of thinkers as also haunted
by the figures of the human: the Deleuze of the virtual (2004) versus the
Deleuze of vitalism (1988a), who allows us to refer the world back to one
grand life and becoming. Other contemporary posthuman thinkers are
also not exempt from such bivalent logics. There is the Derrida of writing
(1977) versus the Derrida of the gift of death (1995), which opens toward

the radically singular ipseity of the other; the Haraway of cyborgs (1991) versus the Haraway of companion species (2003); the Weil (1962) with contempt for the person and rights versus the Weil who asserts the grace that attaches to human life (2002); the Foucault of the anonymous murmur (1972) versus the Foucault of creating the self as a work of art (1985). And so we could go on, with every seeming destruction of humanism nevertheless allowing for a reaffirmation of all that had hitherto never been quite human enough. However, these doubled figures are not just two accidental readings or misreadings that we might separate and adjudicate but are tendencies that mark any attempt to depart from, redefine, destroy, or sanctify the human. Along with the Janus-faced nature of proper names, we can examine the *events* that mark the supposed posthuman and cannot be adequately addressed in the mode of our current questions. September 11 might signal a historical threshold that forces us to ask new questions about the borders of the political at the same time that it allowed for a series of sentimental, racist, and reactive "security" measures to protect a supposedly intrinsic human freedom from outside threats. What has now been referred to as the Anthropocene seems to chasten humanity by noting a destructive impact that reaches geological intensity, but the Anthropocene also invites projects of geoengineering and uncritical uses of a once multiple but now reunified humanity. The 2008 global financial crisis destroyed notions not only of mastery but also of capitalism as some master plot, and it paradoxically allowed for the ongoing legitimacy of old-fashioned human tropes, such as the rights of individuals over corporations.

If the human has increasingly become less manageable at a theoretical level, it has nevertheless pressed itself forward with a practical and aesthetic urgency. This is not at all to say that arcane questions from the humanities should be set aside while we defer to questions of technoscientific urgency—such as saving our species from the ecological catastrophe it has caused—for those scientific questions that assume a "we" and assume a human reason capable of managing its own desires and imagination have perhaps rested too easily on the plane of the human/posthuman dichotomy. Instead of abandoning or saving the humanities, we argue that we should harness the current state of disturbance of the human in the service of framing new problems. With increased fervor, we must put an end to the attempts to respond to the *questions* of the posthuman and supplant them with the goal of reconfiguring the forces and intensities from which they originated—because these questions require nothing less than the formation of novel problems, not answers. And it is to that purpose that this volume is directed.

NOTE

1. "Mother Nature is a serial killer. No one's better. Or more creative. Like all serial killers, she can't help the urge to want to get caught. What good are all those brilliant crimes if no one takes the credit? So she leaves crumbs. Now the hard part, why you spend a decade in school, is seeing the crumbs. But the clue's there" (*World War Z* 2013, performed by Elyes Gabel).

WORKS CITED

Agamben, Giorgio. 2003. *The Open: Man and Animal.* Trans. Kevin Attell. Stanford, Calif.: Stanford University Press.

Bostrom, Nick. 2008. "Why I Want to Be a Posthuman When I Grow Up." In *Medical Enhancement and Posthumanity*, ed. Bert Gordijn and Ruth Chadwick, 107–137. Dordrecht: Springer.

Bryant, Levi, Nick Srnicek, and Graham Harman, eds. 2011. *The Speculative Turn: Continental Materialism and Realism.* Melbourne: re:press.

Clark, Andy. 2003. *Natural-Born Cyborgs: Minds, Technologies, and the Future of Human Intelligence.* Oxford: Oxford University Press.

Darwin, Charles. 1859. *On the Origin of Species by Means of Natural Selection, or the Preservation of Favoured Races in the Struggle for Life.* London: John Murray.

Deleuze, Gilles. 1988a. *Bergsonism.* Trans. Hugh Tomlinson and Barbara Habberjam. New York: Zone.

——. 1988b. *Foucault.* Trans. Seán Hand. Minneapolis: University of Minnesota Press.

——. 2004. *Difference and Repetition.* Trans. by Paul Patton. London: Continuum.

Deleuze, Gilles, and Félix Guattari. 1987. *A Thousand Plateaus: Capitalism and Schizophrenia.* Trans. Brian Massumi. Minneapolis: University of Minnesota Press.

——. 1994. *What Is Philosophy?* Trans. Hugh Tomlinson and Graham Burchell. New York: Columbia University Press.

Derrida, Jacques. 1969. "Ends of Man." *Philosophy and Phenomenological Research* 30 (1): 31–57.

——. 1977. *Limited, Inc.* Trans. Samuel Weber. Evanston, Ill.: Northwestern University Press.

——. 1995. *The Gift of Death.* Trans. David Wills. Chicago: University of Chicago Press.

Foucault, Michel. 1972. *The Archaeology of Knowledge and the Discourse on Language.* Trans. A. M. Sheridan Smith. New York: Pantheon.

——. 1985. "What Is Enlightenment?" In *The Foucault Reader.* Trans. Catherine Porter. Edited by Paul Rabinow. New York: Pantheon.

Haraway, Donna. 1991. "A Cyborg Manifesto: Science, Technology, and Socialist-Feminism in the Late Twentieth Century." In *Simians, Cyborgs, and Women: The Reinvention of Nature.* New York: Routledge.

——. 2003. *The Companion Species Manifesto: Dogs, People, and Significant Otherness.* Chicago: Prickly Paradigm.

Hayles, N. Katherine. 2007. "Hyper and Deep Attention: The Generational Divide in Cognitive Modes." *Profession*: 87–199.

Heidegger, Martin. 1962. *Being and Time.* Trans. J. Macquarrie and E. Robinson. Oxford: Basil Blackwell.

Heidegger, Martin. 1993. "Letter on Humanism." Trans. F. A. Capuzzi and J. Glenn Gray. In *Martin Heidegger: Basic Writings*, ed. D. F. Krell, 217–265. London: Routledge.

Irigaray, Luce. 1985. *Speculum of the Other Woman.* Trans. Gillian C. Gill. Ithaca, N.Y.: Cornell University Press.

Kurzweil, Ray. 2005. *The Singularity Is Near: When Humans Transcend Biology.* New York: Penguin.

Latour, Bruno. 2010. "An Attempt at a Compositionist Manifesto." *New Literary History* 41 (3): 471–490.

Marx, Karl. 1974. *Early Writings.* Ed. Q. Hoare. New York: Vintage.

Meillassoux, Quentin. 2010. *After Finitude: An Essay on the Necessity of Contingency.* Trans. Ray Brassier. London: Bloomsbury Academic.

Menary, Richard., ed. 2010. *The Extended Mind.* Cambridge, Mass.: MIT Press.

Mulgan, Tim. 2011. *Ethics for a Broken World.* London: Acumen.

Nagel, Thomas. 2012. *Mind and Cosmos: Why the Materialist Neo-Darwinian Conception of Nature Is Almost Certainly False.* Oxford: Oxford University Press.

Nietzsche, Friedrich. 1974. *The Gay Science.* Trans. Walter Kaufmann. New York: Vintage.

——. 1982. *The Antichrist* and *Thus Spoke Zarathustra.* In *The Portable Nietzsche.* Trans. Walter Kaufmann. New York: Viking Penguin.

Povinelli, Elizabeth. 2002. *The Cunning of Recognition: Indigenous Alterities and the Making of Australian Multiculturalism.* Durham, N.C.: Duke University Press.

Savulescu, Julian, and Nick Bostrom. 2010. *Human Enhancement.* Oxford: Oxford University Press.

Seigel, Zachary. 2015. "Seven New Translated Excerpts on Heidegger's Anti-Semitism." *Critical Theory* (February 23). http://www.critical-theory.com/7-new-translated-excerpts-on-heideggers-anti-semitism/.

Shelley, Mary. 1999. *Frankenstein.* 2nd ed. Ed. D. L. Macdonland and K. Scherf. London: Broadview.

Thompson, Evan. 2007. *Mind in Life: Biology, Phenomenology, and the Sciences of Mind.* Cambridge, Mass.: Belknap Press of Harvard University Press.

Viveiros de Castro, Eduardo. 2009. *Métaphysiques cannibales.* Paris: PUF.

——. 2014. *Cannibal Metaphysics.* Trans. Peter Skafish. Minneapolis: Univocal.

Weil, Simone. 1962. "Human Personality." In *Selected Essays, 1934–43*, 9–34. Oxford: Oxford University Press.

——. 2002. *Gravity and Grace*. Trans. Emma Crawford and Mario von der Ruhr. London: Routledge.

Wolfe, Cary. 2009. *What Is Posthumanism?* Minneapolis: University of Minnesota Press.

FILMS

Elysium. Directed by Niell Blomkamp. 2013. Culver City, Calif.: Tristar Pictures.

World War Z. Directed by Marc Forster. 2013. Hollywood, Calif.: Paramount Pictures.

Interstellar. Directed by Christopher Nolan. 2014. Hollywood, Calif.: Paramount Pictures.

POSTHUMOUS LIFE

INTRODUCTION

Critical Life Studies and the Problems of Inhuman Rites and Posthumous Life

JAMI WEINSTEIN AND CLAIRE COLEBROOK

This book belongs to the very few. Perhaps none of them is even living yet. . . . Only the day after tomorrow belongs to me. Some are born posthumously. . . . The conditions under which I am understood, and then of necessity—I know them only too well. . . . One must have become indifferent; one must never ask if the truth is useful or if it may prove our undoing. The predilection of strength for the questions for which no one today has the courage; the courage for the forbidden; *the predestination for the labyrinth.*

—NIETZSCHE, *THE ANTICHRIST*

This volume opens onto the problem of posthumous life captured in this epigraph. As Friedrich Nietzsche insists, to live posthumously, and to understand and be understood while amid that praxis, one must remain both indifferent—to the uncritical "truths" that sway those who endeavor to locate their use-value—and fearless in posing the sorts of questions often rendered off-limits by those who dictate the scripts of what can and cannot be asked. By shifting the terrain semantically to the inhuman and the posthumous, we hope to transform the questions of the posthuman (discussed in the preface) into problems that can generate the kinds of concepts and questions that have heretofore been forbidden.

Paramount to this collection, thus, is a metamorphosis, or a becoming, toward what we refer to as *critical life studies*. Critical life studies figures as the framing problematic in which the related problems of the posthuman, the inhuman, and posthumous life posed in the volume are located. By using the term "critical" we are at once invoking a post-Kantian tradition of critique that is wary of positing *any* given term or substance as a straightforward epistemic ground, while being mindful of counter-Kantian ambitions that would move beyond the distancing practices of critique in order to consider the emergence of thought, theory, conditions, and everything that we take to be human and subjective. Neither an uncritical literalism that posits *life as such* nor an unbridled speculation that simply accepts life as a foundational given, critical life studies begins with the force of a problem: One cannot simply dismiss the problem of life, precisely because life presents the most immediate of political imperatives, given the major threats to life in all its organic forms that mark the twenty-first century. Yet the concept of "life" has been used to humanize, racialize, gender, pathologize, and manage human and nonhuman bodies. Emergency measures to save, maximize, and sustain life have marked what has become known as biopolitics, which might prompt us to be critical of any use of the concept *life*. At the same time, theory, philosophy, and politics bear the responsibility of being open to the ways in which what we know as life cannot be abandoned. Any attention to language, concepts, history, culture, or other forms of mediation, far from leaving life out of play, requires a thoroughgoing interrogation of "life" that places it in relation to the various knowledge and social practices through which life is thought and formed.

We invoke the term "critical," thus, as a specific orientation toward questioning, rendering problematic, and examining conditions and assumptions, and also with some sense of critique as having run out of steam. The very concept of critique in its Kantian sense implies a necessary distance from life (Latour 2004)—we only know things relationally and as synthesized, never as they are in themselves—and for that very reason any appeal to life as a ground would be *un*critical and would also enable biopolitics: the subjection of political decisions and disputes to supposed facts that can be grasped and managed by expertise (Campbell 2011). *Life* needs to be considered critically insofar as it increasingly seems to provide an uncritical ground—whether in political imperatives to maximize and save life (as in both pro-life and pro-choice rhetorics) or in theoretical work that strives to return to or give voice to life under the guise that there exist vital norms that must be met. There is a long philosophical tradition, going back to Aristotle at least, of thinking that both politics and ethics emerge from the potentiality of life and insisting that there is no *life* as such but rather a range of capacities, including *reasoning life*, which gives itself its own end rather than simply extending metabolic and perceptive life. In addition to the thorny question of the knowability of life, and whether there is such a singular entity of definition one could locate, there is also the profound problem of life's history: It may be, as Foucault suggests, that "life did not exist" until the eighteenth century (Foucault 1970, 139). And it may also be that we emerge as human only by positing some animality that is other than, or distant from, our properly human being—though today we have seemingly abandoned this distance while nevertheless still thinking of animality (including our animality) as inhuman (Agamben 2003). Added to the problem of our knowledge of life and our own historically varied experience of our own life, it has also been suggested that whatever our humanity may have dictated in the past in terms of life's limits, these can now be overcome (Kurzweil 2005). In addition to the "becoming posthuman" or "becoming-superhuman" of human life, there is the further possibility of life beyond life or at least beyond organisms. David Toomey, following a study by the National Research Council to evaluate the possibility of nonstandard biochemistry supporting life in the solar system and conceivable extrasolar environments, lists a series of contenders for this radically posthuman future of "weird life" (NRC 2007): "beta life, hypothetical life, nonstandard life, nonterran life, unfamiliar life, life as we do *not* know it, alternative biology, and (you knew this was coming) Life 2.0" (Toomey 2014, xvii). Moreover, we can think about life before life, pre- or protolife, or a kind of wild, savage life before it was reined in by the constraints of humanism. This "wild life" we argue, is a condition of possibility of life in its humanistic forms, even if theorists are only now starting to imagine what life without the human could mean. While various attacks have been made on the enterprise of critique, ranging from Bruno Latour's insistence on the composition of reality from embedded and intertwined practices to François Laruelle's project of subtracting the human and reflective viewpoint from philosophy (Laruelle 2013), we aim to sustain a notion of critical *problems* while not granting any force (such as life or the human) the privilege of being the source of critical distance.

In this new milieu of twenty-first-century life sciences, life philosophies, and life politics, critical life studies does not propose a novel ontology, epistemology, or theoretical turn. Neither is it about solidifying alliances or positions in an ongoing debate: We do not argue *for* or *against* life. Rather, diagnosing the state of the humanities, sciences, and social sciences with regard to the various turns and studies, we have determined a certain inability of going further separately. For each of these fields meets an insurmountable hurdle when faced with the problematic of life. Critical life studies, we suggest, is inclusive of the gamut of recent theoretical "turns" and "studies" (including among them: the affective turn; new vitalism; new materialism; the ontological turn; anti-, in-, and posthumanisms; critical climate change; speculative realism; and feminist, trans, queer, critical race, postcolonial, animal, technoscience, and Anthropocene studies) insofar as they all, to varying degrees, theorize around several intertwined concerns: the continual modulations of the epistemology, ontology, and resituation of the status of the human (now seen as a living rather than a knowing being) and the various beings included in/excluded from it; the reconsideration of embodiment, matter, and materiality; the enigmatic question of what constitutes life; and the fraught determination of whose lives matter. Grouping these multiple projects together under the rubric of critical life studies might help inspire new modes of thought by way of a critical reevaluation of the human, bodies, materiality, and the humanities more generally. Although posthumanism, biopolitics, and vitalism have been topics within theory for the past few decades, this volume seeks to place various strands of theory into conversation, in part as provocation to that very recent history.

It is in this respect that we see life as our framing problem; new modes of knowing life—ranging from epigenetics, virology, neuroscience, and nanotechnology to geology, astrobiology, and cosmology—present critical theory with the task of creating concepts. Life is not one more thing in the world, for ways of thinking about, knowing, and transforming life dramatically change what might count as living and the epistemic and ontological status of life itself. In short, between a vitalism that proclaims life as an absolute value or generative power and a critical discourse that focuses on the uses and abuses of the concept of life, this volume situates life at the hinge between realism and mediation—delving into the inchoate labyrinth of life and posing forbidden questions. As such, within the framework of a critical relation toward life, and from a tangle of approaches, this volume seeks to recapture the tension that the posthuman problem has ceased to sustain—in the guise of posthumous life and the inhuman rites associated with it.

With respect to the latter, we submit that the inhuman is the terrain within which the figure of the human (and its posthuman others, which would include animals, bacteria, viruses, plants, technology, and virtuality) takes place. How did something like the human as a being that could imagine its own end emerge, and what does the complexity of the "post-" enable and delimit for a future that can no longer easily be imagined as "ours?" Inhuman, thus, is meant as a departure from the posthuman. Sidestepping the *post-* and the inevitable reuptake of the human vestiges that the post- is meant to surpass, *inhuman* orients us to all that is not human, not just that which comes after the human. It also pushes us to scales beyond the human—temporalities and spatialities both deep and astronomical. Inhuman also contains murmurs of a humanist ethical upending, in that it denotes a lack of the ostensibly human qualities of compassion and mercy as well as "cruelty, savageness, and barbarism." Savageness and barbarism themselves overlap with commonplace racist and colonialist discourses often used to sketch the perimeters of what and who gets counted as human, implying that adopting the position of the inhuman serves to rearrange these pernicious discourses. And when inhuman is coupled with the more affectively haunting *rites*—which emerge from fear, suspicion, desire, risk, ritual, and habit—we shift away from discourses around (human) *rights* and their bond to modernity, reason, and universality. If, today, we are wrestling with rights—what humans can demand and what can be demanded of them—this is because of a prior plane of rite, as an event and encounter that establishes an ongoing series of practices, relations, differences, and forces, the history of which can never be fully rationalized, solidified, or settled. This shift allows a parallel retreat from a kind of ethicopolitical worldview spawned from notions of the equality and sameness typified by ontologically flat (DeLanda 2002, 47) theories and toward one founded on the irreducible, infinite multiplicity of pure difference that is revealed in *indifference* (Weinstein 2016). The inhuman is also a gesture toward what Deleuze and Guattari describe as "archaism": The human emerges from a series of events and habituations that constantly mutate but are also repeated and cannot simply be expunged or overcome (Deleuze and Guattari 1983, 232; 1987, 16). The "rites" of the inhuman—the forces and repetitions that have allowed for something like the human to appear and have a history—are precultural (in the sense of *human* culture) but no less cultivated. Finally, with echoes of funeral rites or last rites in the background we once again return to the opening suggestion of our preface that the posthuman should be the last question of our time—as if the inhuman rites we are performing in this volume were precisely the last rites of the posthuman.

In regard to the former problematic (also the title of the volume), posthumous life, we argue that in light of the vestiges of humanism still present in the concept of life and of the Anthropocene possibility that imagines not only the absence of human (and organic) life but also the likelihood of human impacts being legible (written after life), there needs to be a necessary but impossible theoretical transformation from the post*human* to the post*humous*. Whereas the posthuman is imbricated only in the event of the "death of Man" (or human) and remains a human question, the concept of posthumous goes a level deeper by indicating that the remnants of humanism present in our conventional notions of life, too, must be transcended—signaling the "death of life" and the problem of the posthumous. Multivalent, *posthumous* resonates in a variety of registers, all of which have some bearing on essays contained within this volume. The first, derived from the Latin *humus*—or "ground," "soil," "earth"—carries, in a unifying gesture, the sense of being after the earth, after conceptions of humans as emerging from the earth, and after all the notions of the earth as home or ours. Posthumous life interpreted as such is, thus, life not of this earth and not human. A second layer of significance codes as "arising, occurring, or continuing after one's death." In this case, posthumous life is life after death, the afterlife, ongoing, and the troubling of the very categories of life and death. A third spin indicates "being born after the death of the father, originator, or author." As overdetermined as this might be, it is hard not to read this sense of the posthumous through psychoanalysis or poststructuralism—both with attendant references to the oedipal or name of the father (superego, phallus, symbolic order) and the generation of identity and with the "death of the author" (think Anthropocene, human extinction, and the legible scar that will endure), which engenders a subsequent opening to critical, interpretative abandon (Barthes 1967). We argue that this move toward the posthumous or the afterlife—located in the temporality of the future aftermath—is vital if we wish to refigure our ethicopolitical frameworks to account for contemporary global crises.

One might be tempted to claim that what we are claiming as critical life studies is nothing less than a scientific revolution or paradigm shift—which Thomas Kuhn chronicled as entailing not merely a turn to another theory but an entire transformation of worldview such that we see the same objects in a completely new way. The catalyst for a paradigmatic transformation, to Kuhn, must always be a crisis (Kuhn 1962, §VI–X). However, Kuhn's underlying presupposition is that there is an observable common world or life that different paradigms refocus, such that new entities can thereby come to life. With the type of metamorphosis critical life studies proposes in terms of our understanding of life and of time, this commonly observable world ceases to exist *as such* or as we know it. Accordingly, the crisis here cuts far deeper than even a revolutionary paradigm shift could capture, as it eviscerates the world as such of its observability. This volume and critical life studies in general is consequently not a reflection about how we know a world but rather about how worlds—or different modes of time and life—generate knowing and not knowing.

The essays in this volume, though differing widely in style, method, and disciplinary base and not at all unified by a single claim, are nevertheless focused on what we insist is a decidedly singular problem, *life*—notably as it is reflected through the tropes of the posthuman, inhuman, and posthumous. The volume is thus divided into four intertwined sections, each taking on one dimension of the problem that life opens up. The first section, "Posthuman Vestiges," addresses the problem of posthumanism in its waning state of sway in order to set the stage for the impossible but necessary theorizing beyond the human, beyond the organic, and beyond life itself. Nicole Anderson leads off by tackling the issue from the position that we at once have *always already* and *have never* been posthuman. Likewise, she argues that the human as such, or the anthro-ontological human of humanism, was only ever an ontological reality (if at all) for a select elite few. Thus the human and, in turn, the posthuman is an entity not only restricted to certain bodies but also temporally located in multiple temporal registers—both futurity and history simultaneously.

As Anderson puts it: "If there is a 'post-'humanism or a 'posthumanism,' it is neither a one-directional leap into the future, nor is it a one-directional unfolding of time from past to future (and thus a perpetuation of an evolutionary continuum that 'ends' in an ideal of human perfectibility); rather . . . it is to move simultaneously forward and back to the future and the past. In this way the human species has never/has always been 'post/humanimal.'" Taking the question of the posthuman one step backward, Frida Beckman, following Deleuze and Guattari and others, claims that *we have never even been human*—and the very narrative that we have been is evidence that perhaps we are idiots in at least two senses: First, we have failed to understand that the human has an origin and history, and second, philosophers have taken the capacity to think as a given. To avoid this idiocy, Beckman wonders whether it might not be wiser to question the importance of narratives and the sort of thinking that shaped both the concept of the human and the posthuman. She declares: "The understanding of the posthuman as occupying a place in a historical progression, that is, the ontological conception . . . delimits our possibilities of addressing the question as it, even when it manages to let go of humanist presumptions about the subject, nonetheless continues to think the posthuman in relation to the human."

Susan Hekman, in turn, negotiates the space between materialist and constructivist accounts of the subject by choosing the unlikely Judith Butler to argue her case. While Butler has been critiqued for being a wholesale social constructivist and eliding materiality, Hekman finds that "the perspective of the 'new materialism' that attempts to incorporate the material without abandoning social construction [via its intra-action between the material and discursive]" is "even more problematic." In an attempt to arrive at a defensible formulation of a posthuman (woman) subject, Hekman turns to Butler's more nuanced recent work and to Andrew Pickering's concept of "the mangle" for answers.

The second and third sections—"Organic Rites" and "Inorganic Rites"—together comprise the more general theme of *inhuman* rites. "Organic Rites" investigates the organic others that are deemed to be not human and, therefore, elided under the humanist framework in order to position the posthuman interrogation of the human "somewhere else"— somewhere outside the bounds of anthropocentrism. For one of the many ways in which the human has sought either to surpass or erase itself has been by way of a recognition of animality, either through companion-ship, ecological dependence, common intelligence, or the recognition of a continuum—rather than divide—between humanity and its multiple complex animal others. This section also extends that now relatively common human-animal discussion to plants. First, Akira Lippit questions the possible meaning or epistemology of autobiography, the defining characteristic of the human, from the vantage point of the nonhuman animal in order to explore whether such epistemology must be put under erasure when not attached to a human perspective. He concludes that "the very difference of life itself appears to open up across the divide of human and animal being. If the animal—signifier, figure, and living being—has served to establish the existential lines of a primary, primal difference, a primal scene of differentiation from which difference is instituted, then it is perhaps in the field of autobiography that this difference is deconstructed." For Jeff Nealon, this rapprochement with animality is constitutive rather than disruptive of modern "man." It was only by way of discovering *life*—common to animals and humans—that man became the modern biopolitical being that he is today, at once defined by but distant from the life he shares with animal others. Man's radical other is not the animal but the plant, the being whose life does not allow man self-reflection and recognition. And, in Cary Wolfe's analysis, animals are not other beings with whom we share a common world; drawing upon Jacques Derrida's figure of the island, Wolfe argues for a radical disjunction among worlds. Like Nealon, Wolfe argues that there is nothing comforting or simply ecological about the recognition of animals; there is no animal in general whose common life unites us in one happy *oikos*.

The "Inorganic Rites" section takes up the inhuman from a non-carbon-chauvinist perspective—meeting with the inorganic, the technological, corporations and nonpersons, and the supra- and subpersonal. This section opens with Luciana Parisi's chapter on the automated, algorithmic human and the ways in which computation has transformed the characteristically human faculty of reason into an inhuman force. Following Simondon and Alfred North Whitehead (and others), Parisi argues that rather than rejecting this inhuman reason tout court, we should understand the ways in which it circumscribes the limit of the problem of the posthuman. For it reveals the incapacity of the human as a rational species to be reconciled with the living, since thought—in its most artificial dimension (the computational ordering of information)—is a speculative functioning "that goes beyond the ontological grounding of being in the energetic impulse of living." Alastair Hunt, for his part, explores what happens when we take the collapse of these humanistic binaries—between the human and nonhuman—too far. To reveal this, Hunt weaves together two cases: the political debacle of the U.S. Supreme Court decision in *Citizens United*, after which corporations effectively became persons under the law, and a cotemporaneous European Court of Human Rights decision not to grant a chimpanzee the status of personhood. Through a reading of Hannah Arendt, and under the pretext of both thinking the inhuman and the person together and questioning the epistemology of personhood overall, he endeavors to reveal "that the inhuman, indeed impersonal formal dimensions of the figure of the person suggest that its reduction to the human is less an ontological necessity than it is an ideological effect." In a similar vein, John Protevi seeks both to dig beneath and rise above the surface of the constituted subject through an inhuman political physiology, which, he maintains, is the "production of bodies politic" and "the way in which the human subject is formed by linking the social and the somatic." Concluding with "a hydro-solar-bio-techno-political multiplicity" as an affective political assemblage, Protevi manages to find a way through the humanist political hierarchical stasis.

One of the most common self-narrations of new materialism and new vitalism is that we have suffered too long from the myopia of linguistic, cultural, social, and textual construction, which systematically and regrettably ignores the analysis of life and matter. The constructivist turn, after all, was born in part as a response to the history of using biological essentialism as a weapon of oppression; thereby thinking about race, gender, sexuality, or politics in terms of bodies and matter was a risky endeavor. Arun Saldanha's criticism and defense of Marxist materialism, framed as *geocommunism*, seeks to chart a path between a turn to "life" that fails to pay heed to the ways in which life and materiality have been channeled, forced, and organized by the history of capital and a Marxism that neglects to acknowledge that capitalist flows cannot be fully understood in terms of humans, labor, resources, or any of the organic categories of traditional political theory. His claim for geocommunism is, therefore, not only an ethical claim that would take heed of nonhuman forces but a new mode of theory that would consider "life" or "matter" *differentially*.

Neither life nor matter are substrate terms from which the polity can be explained or redeemed; while there are always lives and matters that are not necessarily bound to organic forms, there are also flows and forces of money, text, image, and figure. If one thinks of materialism *neither* as an attention to some underlying substance (like matter) *nor* as the history of a thinking subject who through time comes to awareness of its own material being (as in dialectical materialism), one might think provocatively of the risk of, for example, race in the present.

If theory turned away from life, biology, bodies, and matter, it was because the specter of essentialism could not but recall eugenics, sexism, racism, homophobia, and fascism. But today, in response to the insistence that "black lives matter," one hears its echoes in the specious retort that "*all* lives matter." The liberal, egalitarian, and humanist protestation that race should not matter *differently* or *exceptionally* presupposes a flawed, essentialist metaphysics of life: Deep down we are all equal, and every life is precious. Moreover, this retort exposes the bourgeois frustration of the privileged, white subject that making a claim for black lives mattering is bringing race back after decades of trying to achieve a postracial world— forgetting all the while that this postracial world, if such a world could exist, has yet to be realized. If one were to focus instead on the richer sense of "black lives matter," one would grasp the critical nature of the concept of life: Rather than a simple argument for the *value* of black lives (that they matter and do so differently), there is an additional claim that blackness is a question *of matter* and, therefore, *materiality*—of a kind located in the interstice between neovitalist/neomaterialist and historical dialectical accounts.

Thus, talking about matter in this sense is neither to make a claim for race (which is already an organization of matter along certain geohistorical narratives, whether one claims one single humanity or distinct races) nor a claim that "we" are postracial—rather it is a recognition both of the distributions of matter among material bodies and the ways in which race *blunts* the question of difference. And this modulation, following Saldanha, could be generalized in ways that might reveal posthumous life in these Anthropocene times. As he puts it: "To be a materialist, one must respond to the world. If man is dead, one has to reinvent life after him."

In the last section of this collection, "Posthumous Life," we find theorists who are striving to push the envelope further by presenting creative responses to life after life. Myra Hird's chapter explores the ramifications of certain visualization practices—in particular extinction and proliferation—endemic to the Anthropocene aesthetic and the antiaesthetic of waste. She investigates the ways in which these operate in advanced capitalist societies to define who and what is, and which lives are, deemed worthy of saving. As she articulates it: "The microorganisms metabolizing our detritus may well be creating new life forms that may equally escape taxonomic containment . . . [which reveals] the imprescriptibility of anticipating, calculating, or controlling either proliferations or extinctions at a planetary scale." Tim Morton also approaches the question of ethical value via an exploration of ecology—a figure, he holds, that is so often deployed to return humans to life and to grant life a vital worthiness. He contends that ecology needs to be considered less as a mode of interconnectedness and more as spectral and uncanny. What we know as "life" and "nature" is always given in multiple, fleeting, partial, haunting, and disturbing encounters. And as such, he claims, "a convocation of specters will aid us in imagining something like an ecocommunism, a communism of humans and nonhumans alike." Next, we encounter Eugene Thacker's concept of "darklife." Thinking through paradoxical creatures that have come to be known as "extremophiles," those microorganisms who have adapted and "actually flourish under conditions of extreme heat, cold, acidity, pressure, radioactivity, and darkness," living creatures that are contrary to what we understand and have come to accept as life itself (particularly in the life sciences), Thacker opens up the philosophical question of "what is life?" anew. Reading through Schopenhauer's theories on life, Thacker concludes his study with the claim that "all life is dark life, and thus . . . [shifts] from life-in-itself as a regional problem of epistemology to a fundamental fissure within ontology. Its limit is . . . life-as-nothing, life thought in terms of negation." This ultimately leads Thacker to advocate what he calls an *affirmative meontology* (nothingness) of life. Finally, this section and the volume itself conclude with a previously untranslated chapter by Isabelle Stengers. In it, she argues that the problem of life has changed. In a close reading of the works of Deleuze and Deleuze and Guattari, alongside Raymond Ruyer and Henri Bergson, Stengers traces the trajectory from the dualism between knowing and life itself to the creative unfolding of life. Her claim is that there is neither a primordial totality called "life" that preexisted analysis nor that there is an ultimate totality we one day might discover. It is, she insists, the very thinking about life that simultaneously brings thought to its end—in both senses of the word—and changes the problem of life. For, life (following Deleuze and Guattari) is inorganic, germinal, and intensive, not continuous and representational. And any hasty move to think life otherwise results in a return to a morality of salvation and choice. Life instead must be thought as ever "reconstituting its stakes" and against logics of control, progress, belonging, and salvation. Citing Deleuze and Guattari, she concludes: "Never believe that a smooth space will suffice to save us" (1987, 500).

What we must keep in mind while reading these explorations into posthumous and inhuman territories is that thinking beyond and outside the habit of the human, let alone life, is a relentless struggle—it is the challenge of trying to carve out a "something else" that ultimately might never be identified. However, in spite of the incommensurability of post-humous life, untangling and theorizing these inhuman impossibilities is necessary in order to provide avenues of escape from recalcitrant, contemporary praxes of life and to open up new possibilities for ethicopolitical living. This approach must be creative and experimental, expansive and self-overcoming, insofar as an analysis of life that utilizes traditional methods and concepts risks an unwitting return to the predictable, universal, habitual, and hegemonic. For, once again, we might turn to Nietzsche's exhortation that *indifference* is needed, since:

> Nothing could be more wrongheaded than to want to wait and see what science will one day determine once and for all concerning the first and last things and until then continue to think (and especially to believe!) in the customary fashion—as we are so often advised to do. . . . We have absolutely no need of these certainties regarding the furthest horizon to live a full and excellent human life: just as the ant has no need of them to be a good ant.
>
> (Nietzsche 1996, 308)

Thus, in being "critical" in the senses outlined above, these varied and creative studies of life eschew sanitization and the safety of the familiar in order to remain contaminated and disquieting, engaged in the disharmony and endless insecurity, untaxonomizable, and continually render problematic and question the premises and assumptions that constitute our current habitual praxis of life.

WORKS CITED

Agamben, Giorgio. 2003. *The Open: Man and Animal*. Trans. Kevin Attell. Stanford, Calif.: Stanford University Press.

Barthes, Roland. 1967. "Death of the Author." Trans. Richard Howard. In *Aspen* no. 5+6, item 3. http://www.ubu.com/aspen/aspen5and6/threeEssays.html#barthes.

Campbell, Timothy. 2011. *Improper Life: Technology and Biopolitics from Heidegger to Agamben*. Minneapolis: University of Minnesota Press.

DeLanda, Manuel. 2002. *Intensive Science and Virtual Philosophy*. London: Bloomsbury.

Deleuze, Gilles. 2004. *Difference and Repetition*. Trans. Paul Patton. London: Continuum.

Deleuze, Gilles, and Félix Guattari. 1983. *Anti-Oedipus: Capitalism and Schizophrenia*. Trans. Robert Hurley, Mark Seem, and Helen R. Lane. Minneapolis: University of Minnesota Press.

——. 1987. *A Thousand Plateaus: Capitalism and Schizophrenia*. Trans. Brian Massumi. Minneapolis: University of Minnesota Press.

Foucault, Michel. 1970. *The Order of Things: An Archaeology of the Human Sciences*. London: Routledge.

Kuhn, Thomas. 1962. *The Structure of Scientific Revolutions*. Chicago: University of Chicago Press.

Kurzweil, Ray. 2005. *The Singularity Is Near: When Humans Transcend Biology*. New York: Penguin.

Laruelle, François. 2013. *Principles of Non-Philosophy*. Trans. N. Rubczak and A. P. Smith. New York: Bloomsbury Academic.

Latour, Bruno. 2004. "Why Has Critique Run Out of Steam." *Critical Inquiry* 30 (2): 225–248.

National Research Council (NRC). 2007. *The Limits of Organic Life in Planetary Systems*. Washington, D.C.: National Academies Press.

Nietzsche, Friedrich. 1982. *The Antichrist*. In *The Portable Nietzsche*. Trans. Walter Kaufmann. New York: Viking Penguin.

——. 1996. *Human, All Too Human: A Book for Free Spirits*. Trans. R. J. Hollingdale. Cambridge: Cambridge University Press.

Toomey, David. 2014. *Weird Life: The Search for Life That Is Very, Very Different from Our Own*. New York: Norton.

Weinstein, Jami. 2016. "Vital Ethics: On Life and Indifference." In *Against Life*, ed. Alastair Hunt and Stephanie Youngblood. Evanston, Ill.: Northwestern University Press.

I

POSTHUMAN VESTIGES

1

PRE- AND POSTHUMAN ANIMALS

The Limits and Possibilities of Animal-Human Relations

NICOLE ANDERSON

I n his book *We Have Never Been Modern*, Bruno Latour (1993) argues: "Modernity comes in many versions as there are thinkers or journalists, yet all its definitions point, in one way or another, to the passage of time" (10). So too for "posthuman/ism,"[1] a referential term that assumes not only an "origin," and hence an eschatological-teleological progression from pre- to posthuman, from ape to ultrahuman (in an organic evolutionary trajectory), but also simultaneously refers to a human future beyond the organic. In attempting to break with the past by pointing to the future and denying the past by moving beyond the organic, posthuman/ism is a periodizing concept. Yet ironically, this eschatoteleological process, encapsulated by the referential "post-," perpetuates an ipseity[2] founded on humanist values and ideals, and thus inevitably the *word* "posthuman/ism" embodies this ipseity and its attendant humanistic dichotomous consequences: an all-too-modernist trait.

Without suggesting that we should or indeed can abandon this teleological process, we need, as Cary Wolfe (2010) argues, a new thinking: "We must take yet another step, another post, and realize that the nature of thought itself must change if it is to be posthumanist." In other words, for Wolfe, this means "when we talk about posthumanism we are not just talking about a thematic of the decentering of the human in relation to either evolutionary, ecological, or technological coordinates . . . we are also talking about *how* thinking confronts that thematic, what thought has to become in the face of those thematics" (xvi). Yet how do we begin to seek and embody a new mode of thinking not founded on traditional knowledge and reason when we continue to use a language (in this case the *word* "posthuman" or "posthumanism") that always already constitutes the humanistic effects of ipseity that distinguish us from nonhuman animals, while at the same time denying our past, which connects us to our own animality? In other words, how do we, as a species, think differently when we continue to use a word (a language) that references selectively, which can be seen, as Wolfe points out, in the most popular versions of posthumanism, when they reinstitute and perpetuate the "ideals of human perfectibility, rationality, and agency" (xiii) by which we define ourselves as a species and that carry all the hallmarks of the effects of ipseity and dichotomous logic?

Posthumanism can be defined as a set of competing, at times contradictory, theories about what constitutes and characterizes the "posthuman." Consequently, the posthuman is defined in various ways, but generally the posthuman is described as either a being that functions representationally to reconceive and/or challenge current humanistic ideologies about what it means to be human or as a future being, one whose "capacities radically exceed those of present humans as to be no longer unambiguously human by our current standards" (Bostrom 2003, 5). It is this latter definition that characterizes the most popular versions of the posthuman Wolfe alludes to above. Within this view of the posthuman as a "future being" there are two further, at times overlapping, conceptions of how and when this future being will become reality.

The first popular conception (which I will refer to as "transcendent posthuman") views cybernetics, intelligent machines, and artificial intelligence as the means by which a break *will* take place with the purely organic evolution of human beings, therefore human history, and hence modernity. In this popular conception technology will enable the "best human traits," such as rationality, intelligence, and agency, to be retained (Kurzweil 2005, 9). The most famous representatives of this view include Ray Kurzweil and Hans Moravec, who believe that digital technology in various forms will emancipate humans from their bodies by merging human consciousness (in particular rational intelligence) with machine software (robots, for instance), thus enabling humans to transcend the limitations of their bodies. While evolution began with biological bodies, we have "usher[ed] in another form of evolution: technology" (Kurzweil 2005, 487). In other words, there has been an "increasing intimate collaboration between our biological heritage" and technology, to the extent that evolution will continue in and through technology: What originated in biology, reason, and rationality will continue without the flesh. This break will come with what Kurzweil calls the "Singularity," a "future period during which the pace of technological change will be so rapid, its impact so deep that human life will be irreversibly transformed" (7). At which point, "there will be no distinction, post-Singularity, between human and machine or between physical and virtual realities" (9). We will have transcended the limitations of our bodies and our biological roots:

> The Singularity will allow us to transcend these limitations of our biological bodies and brains. We will gain power over our fates. Our mortality will be in our own hands. We will fully understand human thinking and will vastly extend and expand its reach. By the end of the century, the nonbiological portion of our intelligence will be trillions of trillions of times more powerful than unaided human intelligence.
>
> (9)

By finally enabling the Cartesian mind/body split and creating a metaphysical foundation that is "technicity," not only will the human transcend itself (its matter) while retaining what Kurzweil argues are the "best human traits"—intelligence, rationality, presence, that is, ipseity—but we will have reached our ontotheological end. We will have become gods.

Instead of human minds being uploaded into machines, making the human body redundant, in the second popular version of the posthuman (what is known as "transhumanism"), there is still a belief in the continuation of the carbon-based materiality of the human body. However, in this version the human body is believed to be capable of transformation through enhancing or augmenting technologies, such as tissue engineering, therapeutic cloning, regenerative and reproductive medicines, nanotechnology, genetic engineering, psychopharmacology, life-extension therapies, neural interfaces, memory-enhancing drugs, and so on. Representative of this version is Nick Bostrom, famous for setting up the World Transhumanist Association and the most well-known advocate for enhancement technologies. He argues that "enhancement interventions aim to improve the state of an organism beyond its normal healthy state" (2008, 1). In other words, transhumanists hold

> that current human nature is improvable through the use of applied sci-
> ence and other rational methods, which make it possible to increase con-
> trol over our own mental states and moods. . . . Transhumanists believe
> that . . . human enhancement technologies will offer enormous potential
> for deeply valuable and humanly beneficial uses. Ultimately, it is possible
> that such enhancements may make us, or our descendants, "posthuman,"
> beings who may have indefinite health-spans, much greater intellectual
> faculties than any current human being—and perhaps entirely new sensi-
> bilities or modalities—as well as the ability to control their own emotions.
>
> (Bostrom 2005, 202–214)

Despite their differences, these versions have much in common. For both versions, nonhumans are not part of their discourses or narratives. Furthermore, they both uphold and perpetuate humanistic values. For instance, they believe that reason, autonomy, intelligence, sentience, self-presence, etc. (that is, ipseity) are universal attributes that will remain constant not only through time but through human transformations away from human materiality and bodies. In other words, as Eugene Thacker (2005) critically observes, the "human can be transformed and still remain 'human'" (76). Or, as Katherine Hayles (1999) questions in relation to Moravec's and Kurzweil's beliefs that human consciousness can be downloaded into computer machines, how is it "possible . . . to believe that mind could be separated from body? Even assuming that separation was possible, how could anyone think that consciousness in an entirely different medium would remain unchanged, as if it had no connection with embodiment?" (1). In the same vein, Thacker (2005) also points out that in the use of enhancement technologies, biotechnology "is utilized to redefine biological materiality" (89) with the consequence that

> nature remains natural, the biological remains biological, plus the natural
> and biological can now be altered without altering their essential proper-
> ties (growth, replication, biochemistry, cellular metabolism, and so on).
> The capacity of these technologies, and their aforementioned invisibility,
> enables researchers to conceive of a body that is not a body—a kind of

lateral transcendence. The technologies of therapeutic cloning, tissue engineering, and stem cell research all point toward a notion of the body
that is purified of undesirable elements (the markers of mortality, disease,
instability, unpredictability), but that nevertheless still remains a body
(a functioning organic-material substrate).

(93)

Finally, what both these versions also have in common is an optimistic
belief in the future of humanity to transcend itself while still remaining
the same (Hauskeller 2014, 1–2), which is why from now on I will refer to
both these versions, for efficiency, as *utopic*.

Given this, I argue that despite their attempts either to break with or
perfect the organic, the utopic versions nevertheless refer to a future that
still encapsulates humanistic values. Consequently, the effects of ipseity
are either resituated into machines themselves, or machines serve to perpetuate the humanist ideal. The "post-" of "human/ism," in other words,
refers to a future whereby an "ultra"-human/ism[3] is conveyed either in the
transformed perfected human body or embodied away from the purely
organic into machine form. Unlike Haraway (1985), then, who argues
that the positive aspect of the posthuman is that it ideologically questions
the "dichotomies between mind and body, animal and human, organism
and machine, public and private, nature and culture, men and women,
primitive and civilized" (205), in what follows I argue instead that despite
this ideological questioning, these dichotomies or oppositions (even if
reversed) continue to be instituted precisely because humans (in whatever
form) continue to dominate and attempt to maintain, as we can see in
these utopic versions, "something essential of itself" (Thacker 2005, 75).
I will return to this issue shortly.

Critical versions of posthumanism challenge and critique the humanist ideals encapsulated in and perpetuated by these utopic versions. Ideologically, critical posthumanists put into question the value placed on
the "human" and its ipseity over and above nonhumans. And supporting the critical posthumanist position here, I would argue that the utopic
versions, in valuing and privileging the human as distinct and categorically "different" from the nonhuman, seem to make no allowance for the
varying and multiple differences in kinds and degrees within and between
humans and nonhuman animals. As Halberstam and Livingston (1995)

put it, the utopic versions "only end up absolutiz[ing] difference" (10), thereby widening if not creating an unbridgeable gulf between only two homogenized forms of life: the human and nonhuman. Critical posthumanists do not, however, privilege the human. Instead they argue that humans are a negotiated position within a system and therefore heteronymous, fluid, and hybrid, despite the perceived and experienced "effects" of ipseity. In other words, if there is ipseity, it is a result of heteronomy. Thus, in a deconstructive move critical posthumanism argues that the future cannot be dominated by humanistic values or by "humans" because it is uncontrollable and always open to the unknown and the unpredictable. For critical versions, then, "posthuman/ism" is reinscribed and redescribed away from its popular utopic context and representation (and therefore as a periodizing concept), in an attempt to move it away from the humanistic values it generally encapsulates. For example, Catherine Waldby (2000) views posthumanism as "a general critical space in which the techno-cultural forces which both produce and undermine the stability of the categories of 'human' and 'non-human,' can be investigated" (43). And Katherine Hayles (2003) attempts to deconstruct the "radical breaks" with human history that the utopic versions convey by arguing that posthumanism "exists in a relation of overlapping innovation and replication," what Hayles calls "serieation" (2003, 134; see also Hayles 1999, 13–17).

However, this does not necessarily mean that all critical posthumanists reject technology. For the most part they argue that our bodies are always already technologized; however, they counter the political and ideological utopic positions by demonstrating that the increased incorporation of technology into our bodies and lives confuses and blurs the oppositions between human and nonhuman (machine and animal) as well as gender identities, thus enabling a deconstruction and revision of current ideas of what it means to be human (Taylor 2012, 1–2; Hauskeller 2014, 5). As Hauskeller (2014) puts it, "posthumanism is clearly a liberationist ideal: the hoped for redistribution of difference and identity is ultimately a redistribution of power . . . it is a way of undermining existing structures of domination" (7).

While I admire and agree with the redescription and reinscription of the word "posthumanism" away from its humanistic tendencies as put forth by Wolfe, Hayles, Waldby, Halberstam, and Livingston, among

others, what both the utopic and the critical versions have in common is
the continual use of that word, which operates as another form of "naming" or labeling that—despite the attempted and deliberate blurring of
the boundaries between human and nonhuman that the posthuman
represents—continues to distinguish the human from the other (nonhuman). Or as Neil Badmington (2003) puts it, "the 'post-' is forever tied up
with what it is 'post-ing'" (20) and therefore "does not . . . mark or make
an absolute break from the legacy of humanism" (21).

Given this, the aim of this chapter is to attempt to demonstrate that
regardless of how we might reinscribe and redescribe posthumanism, its
referentiality will continue to perpetuate a humanist hierarchical opposition between nonhuman animals and humans, thus consigning the former to a preposthuman and the latter to a posthumanist time and space.
In other words, the "post-" explicitly refers to what comes after the human
(the future) and thereby implicitly refers to something *before* the "post- "
and this is what I am calling the "preposthuman" (the past-present).
Because the future is idealized, as we can see in the writings of many
utopic writers such as Ray Kuzweil, Hans Moravec, and Nick Bostrom, a
binary opposition is constructed between pre- and post- where the latter
is privileged (because it is generally associated with the progress of the
human, not the nonhuman). If we follow the logic encapsulated in the
referentiality of the term and confirmed by utopic and even some posthumanist discourses, we can define pre- and posthuman as follows: The
posthuman time and space can be defined broadly as hypertechnologized
human life (robotics, cybernetics, informational networks, enhancement
technologies, etc.). Preposthuman time and space is where the effects of
this hypertechnology may indeed be felt (for instance the impact of technology on habitats and species) but without those humans or nonhuman
animals in those particular times and places having control over, or access
to, the technology that creates the effects (either positively or negatively).
(And I say nonhuman animals because often it is domesticated animals
that through their human owners do have access to and benefit from
technologies). The critical and utopic versions, then, not only create and
perpetuate a hierarchical opposition between pre- and post- but continue
to privilege the human over the animal, thereby reinforcing the "(onto)
theological determination, which is present in every humanism" (Balibar
2009, 66).

While these oversimplified definitions and opposition (what I have been labeling pre- and posthuman time and space) are implicitly and explicitly conveyed in the utopic *and* critical versions, throughout this chapter I will not only be referring to "pre- and posthuman time and space" but also to "pre- and posthuman *realms*." When I refer to the former I do so to reveal and expose the opposition instantiated by the referentiality of the words posthuman and posthumanism. This opposition perpetuates a teleological progression to an ideal future, while the past is always something to move beyond and to overcome in and through technology. Consequently, there is the time and space of the "pre-" (which is left behind) and the time and space of the "post-," which is privileged and valued. That is, the opposition serves to categorize all differences between humans and animals and nonhumans, animals and animals, humans and humans, into overly simplistic, thus depoliticized and overarching, categories (that of pre- and post-) that further work implicitly to render invisible the hierarchical operations of this opposition where the posthuman (and hence "Human") is valued over the preposthuman (hence "Animal").[4]

Unlike "pre- and posthuman time and space," the phrase "pre- and posthuman *realms*" I use deconstructively. Bringing together both the work of Pierre Bourdieu and Jacques Derrida, my notion of "realms" is similar to Bourdieu's notion of "field," where various modes of living and being in the world can coexist. As Bourdieu (1993) defines it, "a field is a separate social universe having its own laws of functioning independent of those of politics and the economy" (162). Leaving aside the question whether any field is independent of politics and the economy, I would go beyond this and extend it to include nonhuman animals by bringing it together with Derrida's (2008) notion that there is a "heterogeneous multiplicity of the living" (31). Thus "realms" attempts to convey that there are worlds within worlds that exist sometimes independently, sometimes codependently, but simultaneously and with different modes of organization and different political and ethical effects. To elaborate further, by "realms" I am referring to a time and place or space that is not necessarily past, although it can be that too, and one that is not homogenous because it exists simultaneously with the present, such that modern technologized realms can coexist alongside and at the same time and place with untechnologized realms. In this way, the notion of "realms" attempts to destabilize conventional notions of time and space and thus any idea that there is either some radical leap or evolutionary continuation into a posthumanist era that has a clear demarcation from the humanist or human era (which is conveyed in and through the phrase "pre- and post-human time and space"). For example, there are human beings who live, in varying degrees, without the traditional definition of technology at the same time, and in the same countries, as those humans that do. Or, there are animal realms that exist but are invisible to some forms of human life. This simultaneous existence of realms puts time "out of joint," as Derrida would say, so that the past haunts the present and future, and vice versa, in continuous and noncontinuous ways: Time is always already *différance*, not necessarily or only the gradual unfurling of history as past, present, and future.

So throughout this chapter when I refer to the phrase "pre- and post-human realms" (in the plural), keep in mind I do so as a means of revealing that there is not just one posthumanist or one preposthumanist time and space. To put it another way, the word "realms" operates not only to highlight the overly simplistic and therefore homogenizing effects of this binary opposition inherent in the referentiality of the word "posthuman/ism" but, as I mentioned earlier, also to deconstruct that hierarchy. By that, I do not mean to argue that the opposition doesn't exist; it does. Nor do I want to blur the opposition so that it ceases to exist (if that is at all possible, especially given that it will continue to exist so long as the word posthuman/ism does) because this would only reduce difference to the same. Furthermore, to attempt to blur this opposition would be to attempt to obliterate potential political-ethical action and differences and the positive effects of pluralistic realms. That is, the aim of the word "realms" is to attempt to reveal the instability of this pre- and post- opposition in order to expose further the simultaneous existence of a plurality of pre- and posthuman realms and thus reveal the endless differentiations between humans and animals that both proliferate from and undermine this initial opposition. The word "realms," then, functions to demonstrate the prescriptive, humancentric, and one-sided ethical effect that the term "post-human/ism" has on animal-human relations. My hope is that it will also enable us potentially to rethink those relations away from their humanistic tendencies.

Having laid the foundation for how and why on occasion I will use the phrase "pre- and posthuman realms," I want to proceed by unraveling in more detail the referential operations and implications of the word "post-human/ism" for the nonhuman animal. This will enable me to demonstrate, immediately following, that the consignment of the nonhuman animal (as a general category) to a preposthuman time and space is one where human ethics either does not or cannot apply or is applied anthropomorphically, so that the coconstitution of the human and nonhuman animal (both morally and socially) is denied or is acknowledged only conditionally and within the bounds of humanist discourse. In this sense there is potential danger not only for the human and nonhuman animal to end up without relation to each other but for all differences between animal species to be reduced to the same, so that the domination of the human over the animal prevails.

❦

The traditional eschatoteleological thinking on time is captured in the reference to at least two meanings or forms of the "post-" in posthumanism. On the one hand, the "post-" refers to a humanist conception of time (and space) as the gradual unfurling of human history as past, present, and future, and this conception is supported and conveyed by a scientific method of building on previous knowledge and experience in a developmental continuum (Bergson 1928, xiv; Anderson 2010a, 136). This continuum from pre- to posthuman is supported by a notion of human development as conceived by Darwinian evolution as gradual variation.[5] Although the notion of time as teleological tends to be conflated with Darwinism, this is not to say that evolution is not a true account or that it always unfolds in a developmental continuum. In regards to the latter, the biologist Robert Wesson (1991) argues that there are in fact various versions of evolution from gradual variation to "rapid" or "stasis punctuation" (148), so that while "evolution . . . carries on in one direction that has been adaptive of the past" (192), it does not proceed toward a goal (in some eschatoteleological uniform process), and thus there is no final or overall meaning (to the human), which a traditional humanistic conception of time encapsulates. This is why Derrida argues that this eschatoteleological notion of time actually reinforces a belief in the origin of "man" or the human (with a particular beginning, say as ape, in the past) and the "end of man" (that is, pointing and moving through time, to an endpoint; an ideal, utopian, or uberman, in the future). On the other hand, the "post-" refers to a break from a version of Darwinian evolution defined as gradual variation and instead suggests a radical leap. It refers to something "beyond" or after the human: a transformation of the organic human into a postorganic future (via technology) where the human transcends its animal (preposthuman) origins.

If these references to time appear different (one version conveys time as a continuum, as represented in the transhumanism of Bostrom; the other as a radical break, as represented by the "transcendent posthuman" version of Moravec and Kurzweil), what they both have in common, as alluded to earlier, is their encapsulation in a humanistic eschatoteleological notion of the human. Or, as Derrida (1986) phrases it:

> The end of man (as a factual anthropological limit) is announced to thought from the vantage of the end of man (as determined opening or the infinity of a *telos*). Man is that which is in relation to his end, in the fundamentally equivocal sense of the word. Since always. The transcendental end can appear to itself and be unfolded only on the condition of mortality, of a relation to finitude as the origin of ideality. The name of man . . . has meaning only in this eschato-teleological situation.
>
> (123)

What Derrida is arguing here is that the thinking of and on the end of man *is* humanistic. That is, in a humanistic thinking there is nothing but the end, so even if man doesn't know or cannot say what that end will be, or even if there is no end and thus no "determining limit or essence of "human nature," as Claire Colebrook (2010) argues, man "gives himself his own end" (119).

While the "post-" of posthumanism explicitly references eschatoteleological time, *at the same time*, the "human/ism" in posthuman/ism implicitly (because of its absence) references the animal (there is no human without a notion of the animal as other). As the reference to the animal is implicit and therefore not brought into presence, the animal is given no place in either this eschatoteleological continuum or radical leap and is left aside in preposthumanist time and place. The consequence of this is that the word posthuman/ism (along with the utopic versions), to appropriate Derrida's (2008) words here, "lends support to a version of humanism that saw man put himself at the top of the animal hierarchy domesticating and dominating and subjecting everything below the hierarchy to man's authority" (16). As Derrida goes on to tell us, this thinking of the limit between humans and animals not only leads to animals' subjection but to the homogeneity of animals under the label "Animal" (with a capital A), as if there were only one limit, border, or edge between Animal and "Man" (with a capital M). So the implicit (but absent) reference to the animal in the word posthuman/ism serves to perpetuate a simplistic distinction not only between preposthuman and posthuman/ism but between human and animal. It is a distinction that hierarchizes that relation (and we are familiar with that story from Aristotle through Descartes to Heidegger, Levinas, and so on), thus further enabling a justification for the profits of what Derrida (2008) calls "biological continuism and geneticism" (30). However, this homogenizing of "Animal" and "Human" fails to acknowledge that

> beyond the edge of the so-called human, beyond it but by no means on a single opposing side, rather than "the Animal," or "Animal Life" there is already a heterogeneous multiplicity of the living, or more precisely . . . a multiplicity of organisation or lack of organisation among realms that are more and more difficult to dissociate by means of the figures of the organic and inorganic, or life and/or death.[6]
>
> (31)

The word posthuman/ism (especially its utopic versions), therefore, either references an eschatoteleological time and space—one that moves from origin to end in a progressive motion—or it references that which is beyond the organic through a radical epochal leap. Conventionally what is defined or conceived of as the "past," then, is that which is less developed than the present, while the future is projected as an Ideal end eventually attained. In this humanist conception of time-space where man "gives himself his own end" (Colebrook 2010, 119), only the human is privileged as that which belongs to the posthuman/ist time and space (present-future), thus implicitly subjugating the animal to a preposthuman time-space (past-present). Consequently, the word "posthumanism," as well as the utopic versions of the "posthuman," inadvertently do not account for what Derrida above rightly argues is the "multiplicity of organization" among and between both the animal and human realms and thus between the past-present and the present-future. This is because in the conventional discourse around animal-human relations, the human and the animal have been simplistically determined, defined, and opposed to each other: The human has been conceived as the potential acquisition of ipseity, and the animal as that which lacks ipseity. However, I would argue that if there are multiple organizations or realms that exist simultaneously—and which therefore would deconstruct the linear boundaries and progression of time as past, present, future—then there has to be various human realms (not all of which would be characterized as technologically advanced) as well as various animal realms (as we will examine shortly, with some of those animals moving in and out of the posthuman realms).

Given all of this, posthuman/ism, both as a referential word and in its utopic versions, makes the following assumptions. First, that all humans have access to all types of technologies that can take them beyond their organicism (through enhancement technologies, for instance). Yet what is overlooked is the fact that just as only some animals might benefit from the posthuman realms, not all humans (because of class, race, gender, and economic differences) have the opportunity to experience or benefit (or not) from the utopic vision of technology characteristic of a posthuman time and space. As Eugene Thacker (2005) points out, there is a conflation of "the stratifications between the discourse of science and political discourse. This is indicated by the common absence of the issues of race and ethnicity, gender and sexuality, public policy, governmentality, warfare, and global economics in most extropian [utopic] texts" (79). I would also add climate change to this list. This assumption is based on another, and it is an assumption that technology is outside, exterior to, the human body because it is a creation of "man," a mere "tool," and, further, that this technology is a result of our eschatoteleological evolutionary progression (as "gradual variation") discussed earlier. The result of this assumption is that it presupposes both "an ontological separation between human and machine" and between human and nonhuman animal. "It needs this segregation in order to guarantee the agency of human subjects in determining their own future and in using technologies to attain that future" (77).

Second, in the implicit assumption that technology is a human creation or tool used by and for humans (Hayles 1999, 3), technology is seen as that which is separate from the organic body. On this position, a prosthetics within the body, such as a steel plate or even a transplanted organ, can be clearly demarcated from the organic or the "original" body, respectively. Utopic versions of the posthuman also assume that only some forms of technology define the posthuman or a posthuman time-space (AI, robotics, cybernetics, enhancement technology that enables tissue engineering and regenerative medicine, etc.) and therefore does not acknowledge that the technological (in the form of sticks, stones, opposable thumbs, language, and so on) have defined the human from its very beginnings. What the utopic versions of the posthuman ignore, then, is, as David Wills (2008) argues, "the originary mechanics at work in the evolution of the species" (5), because

> there is technology as soon as there are limbs, as soon as there is bending of those limbs, as soon as there is any articulation at all. As soon as there is articulation, the human has rounded the technological bend, the technological turn has occurred, and there is no more simple human. Which, for all intents and purposes, means there never was any simple human.
>
> (3)

If we follow the logic Wills applies here to the technological turn (what he calls dorsality) and apply it to the term "posthumanism," we could argue that the human has always already been posthuman (in all its definitions), or other than human,[7] and that therefore the term posthumanism is redundant because we have never been *post*human (Hayles 1999, xiv, 279; Halberstam and Livingston 1995, 8). We could go so far as to say that it is technology (much like the animal, as we will see shortly) in all its forms throughout human "history" that has constructed and enabled "man" to define itself as human.

The two assumptions and the attendant consequences that the referentiality of the word posthumanism and utopic versions of the posthuman generate here only serve to perpetuate, if not highlight, this subjection of animal to human in and through the hierarchic and homogeneous opposition constructed between animal and human, where "the human" controls (or, more aptly, where the human perceives itself to control) animals and not vice versa. These assumptions, in other words, mean that animals are representationally and in practice relegated to preposthuman time-space precisely because if they do not have ipseity, then any idea that the animal can adapt to, or live in, or partake of human and posthuman time-space *without* human intervention is left unacknowledged, and thus the relation between human and animal could only be perceived (by humans) as relationally and ethically one-directional.

The eschatoteleological assumptions inherent in the word posthumanism and evidenced in the utopic versions of the posthuman therefore implicitly deny the complexity of the ethical and social relations between humans and animals. In other words, as alluded to at the start of this chapter, the consignment of the nonhuman animal (as a general category) to either a pre- or posthuman time and space is one where human ethics does not or cannot apply or is applied anthropomorphically, so that the coconstitution of the human and nonhuman animal (both morally and socially) is denied or is acknowledged only conditionally and within the bounds of humanistic discourse. In the following section, then, I will be using "dogs" as an example of how some animals can both undermine this pre- and posthuman time and space by existing in pre- and posthuman *realms*, thereby deconstructing our humanistic ethical and social assumptions of human-animal relations. For instance, are not dogs, and pets generally, an example of the heterogeneous multiplicity of organization taking place within the animal "realms" and between the various realms of the human and animal? Are not dogs, for example, those beings that put into question the constituted opposition between human and animal, preposthuman and posthuman time-space, which disrupt the humanist conception of this time and space? One way of answering these questions is to begin the next section with a recent account in evolutionary science of how our relations with dogs developed before turning to an exploration of how those relations expose the anthropocentrism of posthumanism. In turning to evolutionary science the aim is not uncritically to endorse behaviorist and thus biologically determinist accounts of the human-dog relation. Rather, evolutionary science in some ways (not all) helps us rethink the conventional and humanistic notions of the ethical formation of the human-animal relation precisely because the evolutionary account inadvertently undermines the opposition between animal and human and pre- and posthuman time-space, respectively.

The DNA of contemporary dogs stems back to the wolf, and wolves evolved into the dogs we know today *not* because of deliberate (artificial) domestication by humans[8] but initially (at the earliest stages of wolf and human interaction) because wolves "self-domesticated" (Dawkins 2009, 73). For Richard Dawkins,

> The idea is that the evolution of the dog was not just a matter of artificial selection. It was at least as much a case of wolves adapting to the ways of man by natural selection. Much of the initial domestication of dog was *self*-domestication, mediated by natural, not artificial, selection. Long before we got our hands on the chisel of the artificial selection toolbox, natural selection had already sculpted wolves into self-domesticated "village dogs" without any human intervention.
>
> (71)

While evolutionary evidence demonstrates that natural selection toward self-domestication in dogs is a result of the availability of food and shelter around human habitations, a question not often seriously posed is "who domesticated whom?" One of the reasons for not seriously posing this question is the common assumption among behavioral scientists and biologists, as well as in popular and lay conceptions of evolution, that humans did and do the domestication, always. If we take notice of this discourse, which is evident in Dawkins's statement above, we may also notice that it is the nonhuman animal that self-domesticates. It was wolves that were sculpted by natural selection, not humans. But perhaps if we want to challenge the human and animal opposition that is perpetuated in the posthuman discourses (and by the word "posthumanism" itself), we need to pose questions that do not privilege the human over the nonhuman animal. For instance, if dogs self-domesticated by scavenging food around human habitations, then how did dogs domesticate humans to respond to them further with empathy, sympathy, and morality? And *why* did humans respond to dogs in this way, or, to put it another way, why did humans self-domesticate to dogs? I will explore some of these questions further on, but for the moment I would propose that perhaps the self-constitution of the "human" as ipseity is arguably a result of the natural selection of dogs to tameness and hence "dog-ness." That is, maybe it is only because the dog actively sought cohabitation (rather than in humanistic terms it being deemed tamable) and relation with humans, in and through self-domestication, that the human was enabled to "name" or label, and thus define itself against and in distinction from, the dog—and the animal in general.[9]

In defining and distinguishing itself from the nonhuman, humans are then able to dissociate from their own animality. This dissociation is continued in the utopic versions' dream that humans will eventually transcend their animal bodies altogether, either by creating an immortal and autonomous body without death and disease through enhancement technologies or by dematerializing the body via uploading the mind into informational networks. Of course, this perpetuates the Cartesian mind-body split, but transcending the body altogether also allays the fear that, like the animal body, the human one is expendable and unstable and therefore subject to heteronomy and thus the loss of autonomy (Thacker 2005, 89).

Dissociating and detaching ourselves from the animal has further resulted in laying the foundation for an ethical relation between human and animal (in this case the dog) that is one-directional (that is, only humans can be ethical, not animals) and that is evident in the contemporary "rights" discourse surrounding animals. "Rights" discourse is founded on the assumption that nonhuman animals do not have an active will or intention (unlike humans, who shape their own thinking selves; see Hegel 1975). This assumption positions humans as having ipseity, therefore enabling them to be politically active, and considers animals to lack ipseity and thus politically inactive and unable to make ethical decisions about the environment in which they live (Anderson 2010b, 108). On this metaphysical position, animals do not have "rights" because in order to have them they have to have Reason[10] and to be able to act rationally, something they cannot do. The upshot of this is that humans have to confer rights on animals, and in the conferring humans reaffirm the hierarchical oppositions between human and animals (Anderson 2010b, 109). The utopic versions perpetuate this hierarchical opposition between human and animal, where the latter is relegated to preposthuman time-space because the animal is not believed to behave according to the principle of Reason. But as we have seen, this belief belies the ethical-social mutuality of self-domestication. For instance, a dog's response to being fed by humans, or at least being allowed to scavenge the scraps of human food, has resulted at some stage in our evolutionary cohistory in humans not being bitten or attacked but instead protected by dogs and therefore both humans and dogs developing trust and sympathy (see Darwin 1981, 77). And although in behaviorist terms dogs are simply protecting their food source, nonetheless this cosympathy and trust continues to this day to form moral communities between humans and dogs (Datson 2005, 48). However, according to the evolutionary behaviorist interpretation these traits are adaptive, a learned instinctive behavior, so that "pets are simply social parasites who have perfected the art of releasing and exploiting our innate parental instincts" (Serpell 2005, 124).

Whether we agree with this behaviorist account that pets are "social parasites," this cosympathy and trust has fostered a codomestication that rewards particular traits in dogs (and I would suggest, arguably and controversially, in humans too). Traits such as sympathy, trust, loyalty, and affection (even possibly our moral values and attitudes) may have coevolved between humans and dogs and have since been reinforced (along with particular physical traits) in dogs by humans through artificial selection. In the literature on human-animal relations the general argument is that artificial selection is an anthropomorphic response by humans (a response that again reinforces human superiority over animals). The common narrative in this literature is that, first, anthropomorphism is a human tendency that developed around forty thousand years ago in modern humans and that, second, this has resulted in the "incorporation of *some* animals into the social milieu, first as pets and ultimately as domestic dependents" (Serpell 2005, 123–124, italics mine). Generally the positive and negative positions on anthropomorphism go like this: On the positive side anthropomorphism is believed to be of benefit to dogs and other pets (Serpell 2005, 128) because it has enabled animals (*some*) to be reconceived as other than merely "Animal"; they are thought of as companions and friends ("man's best friend"). In behaviorist and evolutionary terms, the benefit to the dog is that it is provided with a constant source of food and shelter; the benefit to humans is that dogs in guarding their food source provide protection as well as social and emotional support. It is because of this mutual support (and Serpell goes so far as to claim that it is also because of "anthropological thinking") that there exist today the agencies that protect animals.

On the negative side, the philosophical objection is that anthropomorphism fetishizes the animal and thus reduces it "to the same" precisely because it fails to account for the differences between animals and between humans and animals (at least, in recent history, where pets are concerned). Thus, anthropomorphizing through artificial selection (that is, selecting for humanlike traits) is a form of dominating the animal in and through domestication (Derrida 2008, 16). Psychological and evolutionary behaviorists argue that artificially selecting animals for their "anthropomorphic appeal" means that those pets (dogs in this case) that do not exhibit anthropological behaviors are either abandoned, punished and abused, or killed simply because a dog's behavior has not been compatible with the anthropomorphic expectations of its human owner (Serpell 2005, 130–131). In other words, the criticism is that the psychological, social, and emotional behaviors we perceive in animals are simply projections of our own desires, conventions, expectations, and assumptions, rather than "universally objective descriptions of animals" (Knoll 1997, 21).[11]

I will return to the problems associated with both the positive and negative positions on anthropomorphism later, but in the meantime, I would suggest that the problematic implications around the common anthropomorphic narratives is, on the one hand, if anthropomorphism is a human tendency going back forty thousand years, then this is to suggest that anthropomorphism is biologically determined. But if, as the arguments against suggest, anthropomorphism is a projection of human desires and emotions, then where does Reason (a distinctly human trait, as the humanist story goes) fit with this tendency? Is Reason, too, an evolutionary construct and thus biologically determined?[12] An oblique answer suggests itself as we briefly explore, on the other hand, the argument that there are only *some* animals incorporated into the social milieu (Serpell 2005, 124, italics mine). I would add to this that only some of those, in turn, inhabit posthumanist realms, while others inhabit preposthuman realms. Those animals existing within the human and/or posthuman realms are generally there on condition that they meet humans' anthropological expectations (or can adapt to or fit into human spaces and activities), otherwise they are abandoned or punished and relegated to a preposthuman time-space, that of the merely "Animal": as we have seen, a label Derrida argues works to categorize all animals in opposition to the human.

While anthropomorphism is obviously humanistic, even the positions against anthropomorphism (either behaviorist or philosophical/ metaphysical) are not only humanist but, in turn, anthropomorphic.

This is because as humans attempt to apply objective descriptors to animal behaviors and as humans establish themselves as those beings possessing Reason, humans end up anthropomorphizing not only animals but themselves. For example, very rarely in the literature on anthropomorphism are the following questions raised or explored: Could dogs through natural selection and self-domestication have influenced the positive identification of human personalities in terms of animal characteristics?[13] Could dogs have made humans partly in their own image, what I coin "animorphism," perhaps? Thus not only posthuman/ism but also anthropomorphism is anthropocentric. Posthuman/ism is because it makes implicit reference to the animal and thereby privileges and hence only represents the human, which is characterized by ipseity and its eschatoteleological progress. Anthropomorphism is because the animal is propelled into the human and/or posthuman realms on condition that it complies with anthropocentric expectations, thus once again denying the absolute otherness of the animal by domesticating it to serve the purpose of the human species. Rather than enabling the possibility of an unanthropocentric relation with the other that is animal (not in the sense of objectifying animals and thus maintaining a distance but in treating animals as morally equal through respecting differences), animals instead are subjugated to the human/posthuman or relegated to preposthuman time and space.

⸺ ❧ ⸺

If both posthumanism and anthropomorphism are anthropocentric, then how is the nature of thought able to change, as Cary Wolfe calls us to do in relation to posthumanism? It is easy to "*realize* that the nature of thought itself must change if it is to be posthumanist" precisely by taking "yet another step, another post-" (Wolfe 2010, xvi, italics mine), but it is perhaps not so easy to know *how* to take this step, especially if even asking the question of how thought might change is a metaphysical and humanist gesture and thus an anthropocentricism, revealing that perhaps there is no way out of humanism. The critical versions of posthumanism (represented by Wolfe, Hayles, Badmington, Waldby, and so on), as we have seen, do attempt this further step by pointing out that humans are always open to the unknown, the unpredictable, and the uncontrollable precisely because humans are heteronymous and constituted by the other, such as the animal.

Yet even if posthumanism critiques and challenges the posthuman, both words are nevertheless still referential and representational of every humanism and modernism. That is, the referential discourse on the post-human and posthumanism continues to represent humanism in practice and, as we have seen throughout this chapter, one of the negative consequences is the formation of the human-animal hierarchical relation. And this is still the case even if we argue that posthumanism refers to what comes after humanism or refers to what is beyond humanistic thinking because (and there is no way out of it) the "post-" still references an escha-toteleological process or radical leap and thus creates a normative opposition. Thus any notion that there is a simple beyond humanism or a simple before posthumanism is to deny the past and the way in which the past is always tied up with the present and the future/post-, and vice versa. If there is a "post-"humanism or a "posthuman/ism," it is neither a one-directional leap into the future, nor is it a one-directional unfolding of time from past to future (and thus a perpetuation of an evolutionary continuum that "ends" in an ideal of human perfectibility); rather, as the word "realms" has attempted to convey, it is a move simultaneously forward and back to the future and the past. In this way the human species has never/has always been "post/humanimal." To acknowledge this—while not fetishizing our animality or denying our humanistic habits of thought—is to remind ourselves that humans are always already part of the biosphere, and this therefore perhaps allows us to learn to live with these nonhuman others rather than in opposition to, in domination of, or without them.

Yet, for all this, there is *still* a difference between a *realization* that humans are constituted by the unknown and the unpredictable and by the cohabitation with other animals and the *perpetuation*—through the continued use of the word—of the eschatoteleological meanings and representations inherent in humanism. Given this, should we continue to use the word posthumanism *without* neologizing the term to reflect the irony of humanism or to acknowledge our own particular animality? Or would it be more appropriate to abandon the word altogether? My notion of "pre- and posthuman realms" is one means by which to demonstrate the imbrication of the human and animal and the various and potential effects of these imbrications on both human and animal organizations and relations. The phrase doesn't deny the oppositions or abandon the words posthuman/ism; instead the phrase attempts to decenter the human *and* posthuman/ism with all the humanisms it entails. Therefore, without rejecting the word altogether and the way in which critical posthumanism has redescribed and reinscribed the meanings of the word, perhaps the best thing would be simply to put posthumanism under erasure (in the Derridean sense), not just metaphorically but literally: ~~posthumanism~~. By doing so not only are all the humanistic references retained along with the various utopic versions, but at the same time both will be exposed as inaccurate and inadequate. Thus, by erasing the word without totally erasing its various meanings and its past contexts, a new context (and the new meanings of the word created by various critical versions) is iterated and manifested while at the same time ensuring that the negative and positive ethicopolitical implications of the word's referentiality is not lost, veiled over, or subsumed by the dominant utopic discourses.

NOTES

1. There are various definitions and theories of posthumanism. But the posthuman and posthumanism are generally distinguished from each other. For example, transhumanism and posthumanism refer to and define the posthuman in differing ways. Throughout the chapter I will sometimes write "posthuman/ism" to convey the different definitions and interpretations of the posthuman and posthumanism. When I write it this way, however, despite the myriad definitions and interpretations, the referentiality of the word itself is what will be in question and deconstructed, and this referentiality is common to all the various definitions and political and ideological positions.

2. Ipseity is a word that encapsulates a variety of human/ist traits, such as autonomy, self-presence, self-reflection, reason and rationality, automobility, and so on.

3. 'Ultra' is defined by the *OED* as "beyond," "extremist," or "to an extreme degree; very." I use the word "ultra" here to convey what utopic theorists perpetuate in and through the appropriation of the "post" ("after"). The humanistic values that are taken up and perpetuated to an "extreme or heightened degree" in the idea of the posthuman include those that are currently privileged: ipseity. The posthuman is therefore the medium by which these ultrahumanistic values and beliefs are privileged to an extreme degree, thus projecting the human beyond how it is currently defined and experienced.

4. As I discuss later in the chapter, Derrida (2008, 16) argues that animals are homogenized under the label or unifying category of the "Animal" with a capital "A."

5. In evolutionary terms, gradual variation is defined by Darwin (2009, 123, 456) as "descent with modification," which proceeds gradually and continuously.

6. Many posthumanisms attempt to blur the distinction between organic and inorganic in order to intervene politically and disrupt traditional gender, identity, and racial categories. However, note that Derrida (1990) does not describe the difficulty of dissociating the organic and inorganic, life and/or death, under the umbrella term of posthumanism or the "posthuman."

7. Although the counterargument here is that if the human has never been "simply human" but always technological, this is what distinguishes, and always has, human from animal, human from ape. Humans in this sense have never been "ape." Again, this counterargument would play into the humanistic notions of creative evolution. It could also dangerously lend inadvertent support to Christian creationism.

8. Not only is artificial selection a technological tool that changes life forms and animal bodies to adapt to human traits (anthropomorphism), but dogs are also subject to bodily transformation using regenerative medicine. In the United States it is common to spend $6,500 per kidney for "renal transplant surgery" for people's pet dogs (Serpell 2005, 123). This is an example of a dog, for instance, living in the posthuman realms (and here we would define posthuman realms as ones that produce and are constituted by modern technological advances). Adapting to posthuman realms may perhaps be a result of anthropomorphism, as we will discuss further on.

9. In the past few years there has been biological scientific literature that acknowledges this claim. For example, see Hare and Woods (2013a, 2013b).

10. Reason, in the history of philosophy and metaphysics, is generally defined as "thought" independent of all empirical, aesthetic (moral) feeling, and religious influence. It is the "self-possessed [autonomous] legislation of the power of choice through reason" (Kant 2003, 659). To "think," then, is defined as the ability to draw on the logic of cause and effect, deduction and inference, and to move from one universal truth or true proposition to another. Thinking, then, is limited to deductive and logical reasoning associated with argumentative-theoretical and objective evidence.

11. This is a behaviorist perspective, one that comes to the fore in the twentieth century and ended the anthropomorphic methodology in both psychology and evolutionary-biological theory. Darwin and G. J. Romanes, for instance, anthropomorphized extensively throughout their writings, and for them, it was a "methodological necessary rule" (Knoll 1997, 17).

12. See Anderson (2010b) for a more in-depth discussion and problematization of Reason as biological.

13. The cartoon-character animals who take on human personality traits along with physical traits of "cuteness" or juvenilization are evidence of this. Steven Jay Gould (1980) discusses this phenomenon in relation to Mickey Mouse, where Mickey's change in personality (where he becomes over time the model citizen) evolves with a change in physical traits. In the early 1920s Mickey Mouse looks ratlike, with an elongated snout and small eyes. By the 1950s Mickey developed a round large head, large eyes, bulging cheeks, and more rounded lower ears. In other words, he took on the features of babyhood. Gould reminds us that this progressive evolutionary juvenilization is called neoteny, and the hypothesis for Mickey developing these traits is that according to Gould, "babyish features tend to elicit strong feelings of affection in adult humans, whether the

biological basis be direct programming or the capacity to learn and fix upon signals" (101). There is a direct correlation here between attributing cute, babylike features to animal cartoon characters and the artificial selection of similar features in dogs: large eyes, floppy ears, smaller snouts, etc.

WORKS CITED

Anderson, Nicole. 2010a. "Supplementing Claire Colebrook: A Response to 'Creative Evolution and the Creation of Man.'" *Southern Journal of Philosophy* 48 (1): 133–146.

——. 2010b. "(Auto)Immunity: The Deconstruction and Politics of 'Bio-Art' and Criticism." *Parallax* 16 (4): 101–116.

Badminton, Neil. 2003. "Theorizing Posthumanism." *Cultural Critique* 53 (Winter): 10–27.

Balibar, Etienne. 2009. "Eschatology Versus Teleology: The Suspended Dialogue Between Derrida and Althusser." In *Derrida and the Time of the Political*, ed. Pheng Cheah and Suzanne Guerlac, 57–73. Durham, N.C.: Duke University Press.

Bergson, Henri. 1928. *Creative Evolution*. Trans. Arthur Mitchell. London: Macmillan.

Bostrom, Nick. 2003. *The Transhumanist FAQ: A General Introduction, Version 2.1.* World Transhumanist Association. http://www.nickbostrom.com/.

——. 2005. "In Defense of Posthuman Dignity." *Bioethics* 19 (3): 202–214.

——. 2008. "Ethical Issues in Human Enhancement." In *New Waves in Applied Ethics*, ed. J. Ryberg, T. Petersen, and C. Wolf, 120–152. London: Palgrave Macmillan.

Bourdieu, Pierre. 1993. *The Field of Cultural Production.* New York: Columbia University Press.

Colebrook, Claire. 2010. "Creative Evolution and the Creation of Man." *Southern Journal of Philosophy* 48 (1): 109–132.

Darwin, Charles. 1981. *The Descent of Man, and Selection in Relation to Sex.* Princeton, N.J.: Princeton University Press.

——. 2009. *The Annotated Origin: A Facsimile of the First Edition of* On the Origin of Species. Annotated by James T. Costa. Cambridge, Mass.: Harvard University Press.

Datson, Lorraine. 2005. "Intelligences: Angelic, Animal, Human." In *Thinking with Animals: New Perspectives on Anthropomorphism*, ed. Lorraine Datson and Gregg Mitman. New York: Columbia University Press.

Dawkins, Richard. 2009. *The Greatest Show on Earth: The Evidence for Evolution.* New York: Bantam.

Derrida, Jacques. 1986. "The Ends of Man." In *Margins of Philosophy*, trans. Alan Bass. Chicago: University of Chicago Press.

——. 1990a. "Some Statements and Truisms About Neologisms, Newisms, Postisms, Parasitisms and Other Small Seismisms." In *The States of "Theory": History, Art, and Critical Discourse*, ed. David Carroll. New York: Columbia University Press.

——. 2003. "Autoimmunity: Real and Symbolic Suicides." In *Philosophy in the Time of Terror: Dialogues with Jürgen Habermas and Jacques Derrida*, ed. Giovanna Borradori. Chicago: University of Chicago Press.

——. 2005. *Rogues: Two Essays on Reason*. Trans. Pascale-Anne Brault and Michael Naas. Stanford, Calif.: Stanford University Press.

——. 2008. *The Animal That Therefore I Am*. Trans. David Wills. New York: Fordham University Press.

Fukuyama, Francis. 2002. *Our Posthuman Future: Consequences of the Biotechnology Revolution*. New York: Farrar, Straus and Giroux.

Gould, Steven Jay. 1980. "A Biological Homage to Mickey Mouse." In *The Panda's Thumb: More Reflections in Natural History*, 95–107. New York: Norton.

Halberstam, J., and I. Livingston, eds. 1995. *Posthuman Bodies*. Bloomington: Indiana University Press.

Haraway, Donna. 1985. "A Manifesto for Cyborgs: Science, Technology, and Socialist Feminism in the 1980s." In *Feminism/Postmodernism*, ed. Linda Nicholson, 190–233. New York: Routledge.

Hare, Brian, and Vanessa Woods. 2013a. *The Genius of Dogs: How Dogs Are Smarter Than You Think*. Dutton Adult.

——. 2013b. "We Didn't Domesticate Dogs. They Domesticated Us." *National Geographic* (March 3).

Hayles, N. Katherine. 1999. *How We Became Posthuman: Virtual Bodies in Cybernetics, Literature, and Informatics*. Chicago: University of Chicago Press.

——. 2003. "Afterword: The Human in the Posthuman." *Cultural Critique* 53 (Winter): 134–137.

Hauskeller, Michael. Forthcoming 2014. "Utopia in Trans- and Posthumanism." In *Posthumanism and Transhumanism*, ed. Stefan Sorgner and Robert Ranisch. Peter Lang.

Hegel, G. W. F. 1975. *Philosophy of Right*. Trans. T. M. Knox. Oxford: Oxford University Press.

Kant, Immanuel. 2003. "The Metaphysics of Morals." In *Moral Philosophy from Montaigne to Kant*, ed. J. B. Schneewind. Cambridge: Cambridge University Press.

Knoll, Elizabeth. 1997. "Dogs, Darwinism, and English Sensibilities." In *Anthropomorphism, Anecdotes, and Animals*, ed. R. Mitchell, N. Thompson, and H. Lyn Miles, 12–21. Albany: SUNY Press.

Kurzweil, Ray. 1999. *The Age of Spiritual Machine: When Computers Exceed Human Intelligence*. New York: Penguin.

——. 2005. *The Singularity Is Near: When Humans Transcend Biology*. New York: Viking.

Latour, Bruno. 1993. *We Have Never Been Modern*. Trans. Catherine Porter. Cambridge, Mass.: Harvard University Press.

Moravec, Hans. 1988. *Mind Children: The Future of Robot and Human Intelligence*. Cambridge, Mass.: Harvard University Press.

Serpell, James A. 2005. "People in Disguise: Anthropomorphism and the Human-Pet Relationship." In *Thinking with Animals: New Perspectives on Anthropomorphism*, ed. Lorraine Datson and Gregg Mitman. New York: Columbia University Press.

Taylor, Nikki. 2011. "Anthropomorphism and the Animal Subject." In *Anthropocentrism: Humans, Animals, Environments*, ed. R. Boddice. Boston: Brill.

——. 2012. "Animals, Method, Mess: Posthumanism, Sociology, and Animal Studies." In *Crossing Boundaries*, ed. L. Birke and J. Hockenhull. Boston: Brill.

Thacker, Eugene. 2005. "Data Made Flesh: Biotechnology and the Discourse of the Posthuman." *Cultural Critique* 53 (Winter): 72–97.

Waldby, Catherine. 2000. *The Visible Human Project: Informatic Bodies and Posthuman Medicine*. New York: Routledge.

Wills, David. 2008. *Dorsality: Thinking Back Through Technology and Politics*. Minneapolis: University of Minnesota Press.

Wolfe, Cary. 2010. *What Is Posthumanism?* Minneapolis: University of Minnesota Press.

Wesson, Robert. 1991. *Beyond Natural Selection*. Cambridge, Mass.: MIT Press.

2

POSTHUMANISM AND NARRATIVITY

Beginning Again with Arendt, Derrida, and Deleuze

FRIDA BECKMAN

ike the notion of the Anthropocene, which, coined at the dawn of the new millennium, was almost instantly absorbed into theoretical discourses, the notion of the posthuman is, arguably, a neologism that bears a potentially significant relationship to how we understand concepts from which it is derived, such as the human, humanity, and humanism.[1] Through both these notions (that is, the Anthropocene and the posthuman), age-old questions about life have been reactivated in a new ethical framework as they not only work to address the role of humans and their ethical responsibilities for their environments but also give new perspectives to the centrality of the human in humanist discourses. The Anthropocene, which can in one sense be seen as the ultimate implication of a worldview that consistently privileges the human, illuminates that life as we know it, or as we think we have known it, cannot be taken for granted. While the impact of human activity on the earth has certainly been a subject for philosophy before the notion of the Anthropocene emerged, not least by Michel Serres in *The Natural Contract* (1990), the essential claim of the Anthropocene—that the effects of human activity on the planet can now be measured—seems, as Claire Colebrook and Cary Wolfe (2013) put it, to be a "game changer." Because the Anthropocene introduces a difference not in degree but in kind, she argues, it has to change the way we do narrative. Its arrival, or maybe simply the recognition of it, challenges, if not breaks down, the parochialism of human narratives. Human history making is humbled in the light of the planet's own way of inscribing time, which will continue also after the end of the human. As Serres's work shows, we can reach the conclusion that we have arrived at a limit point in terms of history making also without the notion of the Anthropocene itself. "What has for several centuries been called history," he writes, is reaching a limit point where the destructiveness of the "subjective violence" of wars fought for dominance is outdone by the efficacy of the "objective violence" toward the Earth. At this frontier, he writes, "a certain history comes to an end" (Serres 1995, 12).

What these insights uncover, and what the Anthropocene exposes, Cary Wolfe suggests, is the radical disjunction between the evolutionary processes and dynamics that have generated the human as species on the one hand and the stories we tell ourselves about ourselves as human on the other. "To talk about the Anthropocene," he notes, "is not just to talk about the question of a radical finitude for the human, but it is also to talk about the extent to which the human can have a linear, intentional relationship of controlling and steering and mastering and healing that finitude or not" (Colebrook and Wolfe 2013). As such, he argues, the Anthropocene makes apparent what has always been true—that there is a nonlinear relation between the human's sense of its own intentionality and the processes of which it is part. The Anthropocene thus brings out a crisis for the human not only because it points to its possible future extinction but also because it challenges the very notion of the human as it has been constructed through, and by means of, its own narratives. When a certain history, or maybe a certain kind of history making, comes to an end, this forces us to rethink what it means to think through narrative. What is the human if it cannot fully grasp itself, and how can the human know itself and its own thinking if it is not in charge of its own narratives? "Are our thoughts," as Serres (1995) writes, "until recently rooted exclusively in their own history, rediscovering geography, essential and exquisite? Could philosophy, once alone in thinking globally, be dreaming no longer?" (6).

While questions of narrativizing and thinking the human have thus been intensified by discourses on the Anthropocene and the posthuman in recent decades, the question of our narratives about the human as well as our very capacity of knowing them is not necessarily related to the either of these current trends. As Gilles Deleuze suggests, this is, fundamentally, a question related to philosophy itself. We who call ourselves humans are repeatedly guilty not only of the stupidity (or *bêtise*) Jacques Derrida identifies as inherent in refusing to acknowledge the multiplicity of life forms other than our own but also, Deleuze poignantly argues, of the idiocy of taking for granted our capacity to think in the first place.[2] Deleuze, as I have noted elsewhere, argues that modern philosophy rests on an image of thought that fails to take into account its own starting point (Beckman 2009).[3] Introducing his first form of idiocy—a second form, as we will see, is developed with Guattari later on—he states in *Difference and Repetition* that philosophy tends to take for granted a capacity to think and an inclination toward good sense. Descartes's cogito, in returning to thought as a proof of itself, retains a blind spot precisely with this presumption. When Descartes (2003) states "*I think, therefore I am*" and that the self-evidence of this grounds the "first principle of philosophy" (23), the act of thinking remains a universal that he fails to question. The failure to recognize that his self-reflection is based on a presupposition regarding his own natural capacity for thought makes the philosopher an idiot. Appearing thus as an inquiry into theories of the modern subject more generally, questions of narrative and thinking can nonetheless be related to the more specific theoretical as well as scientific and geological developments outlined above. What these developments make clear is that while the scientific discoveries in the sixteenth century and the formalization of the scientific method from Francis Bacon onward prompted a reconceptualization of the human's place in the world, the climate in which the notion of the Anthropocene has emerged offers no consolation to the human, or the posthuman, in terms of the subject. Not only does it occur at a point in time when the Cartesian subject has long stopped being regarded as a stable point in an unsettled and unsettling universe; it also, importantly, exposes the remnants of our beliefs in such a human subject.

These remnants continue to leave their mark not only on questions of the subject but also on questions of its embodiment. A central concern for the field called the "posthumanities" is the extent to which studies within this field manage or fail to take us beyond a humanist framework as derived from a Cartesian dualism between mind and body. While the body has been brought back into the frame by twentieth-century philosophy, not the least through phenomenology, the concern for consciousness lingers in the phenomenological inheritance, as does a reliance on the ontological specificity of the human. The central preoccupation with the embodiment of the subject and of communication continues to be specifically human: It is seen to harbor a capacity for language that excludes nonhuman bodies as without language and without the ability to respond. As the later critiques, such as Derrida's, of such phenomenological presumptions suggest, the body as well as communication and consciousness do not hold as definitive marking points between human and nonhuman subjectivity. Indeed, this letting go of language and consciousness as two cornerstones of humanism from the disembodied Cartesian tradition to the embodiment of phenomenology constitutes a key to the posthumanist turn. Still, and from the very varied and frequently differing ways in which the notion of posthumanism and what it can do is understood, Wolfe (2010) notes that as it stands now, the human subject and its relation or nonrelation to the body continues to haunt and vex attempts to step into a truly posthumanist context (xv). Presumptions about embodiment and subjectivity risk making what is supposedly posthumanist into nothing but an intensification of humanism and the philosophical framework that underpins it. One rather dominant strand, he argues, does not succeed in taking us out of the humanist framework that the "post" seems to suggest. He notes an ambivalence in work such as that of N. Katherine Hayles, which on the one hand criticizes Hans Moravec for expanding rather than questioning the realm of the liberal autonomous subject and on the other itself expresses a "triumphant disembodiment" that continues to build on the humanist framework (xv). The tendency of opposing embodiment and posthumanism thus causes Wolfe to question the "post" of the term. In his own understanding, posthumanism is not actually posthuman at all, at least not as long as this term invokes a sense of coming after embodiment. To gain full force, posthumanism must take embodiment seriously as well as fundamentally reconfigure "the question of the knowing subject and the disciplinary paradigms and procedures that take for granted its form and reproduce it" (xxix). But even when researchers do take into account the embodiment of thought, he suggests, supposedly posthumanist thinking still tends to be based on humanist presumptions. We fail again and again to escape the Cartesian presumptions that we supposedly and repeatedly repudiate.

For example, Wolfe argues, even the materialist account that someone like Daniel Dennett offers of the workings of consciousness as a set of signals occurring between body and brain, professedly without a "Cartesian puppeteer" to govern the whole process, ultimately falls back on a representationalist conception of language that perpetuates a Cartesian disembodiment. As such, he suggests, "Dennett's apparent functionalism and materialism are unable to escape the spell of the very philosophical tradition—whose most extreme expression is Cartesian idealism—that he supposedly rejects" (129).

Readings of Dennett differ, as do readings of Hayles. Indeed, readings of Wolfe differ: Chris Peterson's (2011) response to what he sees as Wolfe's elitist and "ever-more-labyrinthine refinements of 'posthumanism'" suggests that this approach bears judgment on others' failures of escaping humanist frameworks while positioning itself in relation to a supposed blind spot that fails to recognize its own limitations (138). One of the central failings of Wolfe's project, according to Peterson, is not only that it schematizes the "good" and the "bad" versions of posthumanism—his "Humanist Humanism," "Humanist Posthumanism," "Posthumanist Humanism," and "Posthumanist Posthumanism"—but also that it projects a future of posthumanism as a goal "within a narrative of increasing perfection" (137). Wolfe (2011), unsurprisingly, rejects this reading, suggesting that Peterson attributes to his work positions that it argues explicitly against and that the perfectionism ascribed to it is not only irrelevant but "actively impossiblized" by his arguments (189–193, 190–191). Without entering into these debates in detail, what is important here is that they underline the challenge of escaping humanist presumptions about thinking in posthumanist discourse and that they do so explicitly in relation to ideas of narrative progression. The question of where and how to position posthumanism in relation to humanism is clearly fraught with difficulties as well as disagreements that also and inevitably refer back to the question of the human as that which knows itself and its own finitude. It is also closely linked, as I have begun to show, to the possibility of narrativizing history and mortality. In his introduction to *What Is Posthumanism?* (2010) Wolfe, via R. L. Rutsky, picks up on the importance of narratives to the possibility of a truly posthumanist discourse. The question of the importance of narratives is left lingering, however, and my ambition here is to try to pursue this thread and see where it takes us. For it seems clear that even as we try to break free from a humanist framework, the stories we tell ourselves as we "call ourselves human," as Derrida would have it, and the ways in which different forms of life are understood and valued depending on the role they are given in human narratives continue to shape posthuman thinking.

Crucially, we therefore need to differentiate between the posthuman as a discourse that comes after humanism, that is, an epistemic one, and the posthuman as a development in the nature of being itself, that is, an onto-logical one. Do we question conceptions of subjectivity as separable from the body, the human as separable from the animal, and so on because we believe they have always been wrong or because we think that they are no longer right? In the case of the former, the posthuman must be seen as a nonhistorical/atemporal mode if, indeed, it still makes sense to call it by that name at all. Thus, Wolfe, for example, insists that humanity "never is, never was, and never will be human" in the sense of having a proper, linear, intentional relationship to our own actions, to our own concepts, to our own thoughts simply because what brings the human into being, its very condition of possibility, is actually a technology—sometimes called symbolic behavior, language, communication, the semiotic in the broader sense (Colebrook and Wolfe 2013). The human is prosthetic—it comes into being through something that is not organic. In the case of the lat-ter, the posthuman is analyzed as a historical emergence of a form of life that is no longer fully human. Developments in technology, politics, and medicine constitute some of the reasons why the human is seen to be progressing into a state of posthuman-ness. The question in this instance thus becomes, as Hayles's famous book title puts it, *How We Became Post-human* (1999).

Depending on our starting point, then, our narratives and how we need to interrogate them differ in a fundamental way. We can either ana-lyze the narrative of how the "different versions of the posthuman *con-tinue to evolve* in conjunction with intelligent machines" and how a binary distinction between disembodied information and embodied human life is "*no longer* sufficient" (Hayles 2005, 2), in other words, narratives of human development, or, if we resist the idea that we have ever been human in the humanist sense, we could analyze what kind of thinking has shaped the concept of the human and, indeed, the posthuman, in the first place. This chapter is primarily interested in the latter. Taking off from a sense that both the human and the posthuman are concepts that cut out different aspects of life in Deleuze and Guattari's sense, it is intrigued less by how we became posthuman and more by how the stories we tell ourselves about our own humanity condition the possibility for posthumanist thought. If it is correct, as Eugene Thacker suggests, that classical frameworks such as Aristotle's continue to shape questions of life today; that posthumanist discourse continues to struggle with the ghost of the Cartesian subject even after decades of vehement interrogation, as Wolfe argues; and that this same discourse continues to have problems escaping "humanist narratives of historical change," as Rutsky points out (Thacker 2010, xiii), then it seems to me that we have to take the greatest care when identifying the possible humanist narratives that keep us tied to a particular idea of progression, however hidden such progressivist tendencies might be. We may also have to think again about the usefulness of a concept such as posthumanism. If the Anthropocene needs to change the way we do narrative because it exposes our inability to master our own narratives, as Colebrook suggests, it becomes clear that a narrative that takes us from the human to the posthuman perpetuates an idea about our own intentionality that is profoundly humanist.

This essay tries out a two-step model to see if there is a way in which thinking about life could take alternative routes to those commonly defined as posthumanist. The first step consists in claiming a particular aspect of Hannah Arendt's work, one that tends to be neglected in discourses on posthumanism. Arendt's work is frequently mentioned in studies of the posthuman, especially in relation to biopolitics. Her Aristotelian conception of bare life and political life and her work on totalitarianism have become influential not least after Giorgio Agamben's reading of Arendt and totalitarianism in *Homo Sacer* (1998). But Arendt also offers a conception of life that, while distinctly and even fervently human and thus not immediately of use to posthumanist thought, opens up possibilities of rethinking humanist narrative. There is a freedom to be excavated from Arendt, a freedom that for her is unambiguously human but for us may liberate us from the causalities that continue to shape how we think about the past, the present, and the future of what calls itself human. Arendt's thought is useful, but because it relies so heavily on the specifically human, we will also need a second step. To step beyond Arendt's Aristotelian reliance on *bios* and *zoe* and in order make Arendt's notion of action as relying on plurality useful beyond her humanist framework, we can put it in dialogue with theories of animality and idiocy as articulated by Derrida and Deleuze. This makes it possible to open up toward an understanding of life in terms of immanent singularities and thus to dimensions beyond humanist traces of reason, identity, and the subject. Having moved from Arendt to Derrida and Deleuze and from human life to life as immanence, I will, ultimately, revisit the usefulness of "posthumanism" as a concept.

LIFE/CONCEPT/NARRATIVE

Essentially, what I have suggested so far is that while the Anthropocene and the posthuman have emerged as concepts with the potential of radically changing the way we think and do narrative, the question is to what extent this potential is actualized. From a Deleuzo-Guattarian perspective, such a potential can only be actualized if we also allow the contours of these concepts to be open to respond to the problem beyond the way it has been narratively positioned. If we do not challenge the narratives of which they are part, these concepts do not have the capacity to tell us anything new at all. When Deleuze and Guattari (1994) maintain that philosophy is about the creation of concepts, this seemingly simple definition immediately throws us back into a larger one. In this light, the response to any philosophical problem "not only had to take note of the question, it had to determine its moment, its occasion and circumstances, its landscapes and personae, its conditions and unknowns" (2). When they ask us to take seriously the moment, the landscapes, and the conditions of a particular question, they are clearly not asking us to retrieve a narrative— this would make philosophy merely an account of a set of reflections and representations. Rather, they ask us to discover new ways of thinking— ways that are not determined by a representational history of thought. A philosophy as the creation of concepts must be perceptive to how concepts come to cut out and determine the contours of bodies and events. They are incorporeal but effectuated in bodies; they effectuate events but never essence (21). Always emerging as a response to a problem "without which they would have no meaning," concepts are not permanent but exist in a dynamic relation with other concepts (16). Thus, we can see, for example, how the concepts of "human" and "animal" emerge with a specific and antithetical relation through Descartes. The concept of the human has sought to determine the contours of one form of life while leaving all forms of life that fall outside them to become simply "animal." This is an epistemological issue with deep ethical implications. The rights that we who call ourselves humans have taken to decide on what forms of life deserve rights and what forms can be disposed of continue to be shaped in accordance with the exceptionality of the human. Deleuze and Guattari state that as the problem a concept responds to changes or merges with another, the contours of the concept may need to be recut (18). If a new or recut concept works "better" than an earlier version "it is because it makes us aware of new variations and unknown resonances, it carries out unforeseen cuttings-out" (18).

The concept of the posthuman can undoubtedly be used productively in a range of different disciplines as it comes to reflect a state of being that has been affected by developments in fields such as science, technology, and medicine. As such, the concept can have a direct and rather pragmatic purpose. From a Deleuzo-Guattarian perspective, however, its usefulness would also depend on its capacity to bring out new variations and resonances that the concept of the human did not. If we see the posthuman as a new stage in the development of life and technology, then this concept functions to draw our attention to how such developments have altered and continue to alter the human. Such an understanding, importantly, does not necessarily alter our conception of the human. The "post" in this sense indicates rather a new step in evolution. Like "past 'posts'" such as, for example, poststructuralism, postmodernism, and postfeminism, as Peterson (2011) puts it, posthumanism linguistically seems to promise a temporal shift, turn, or even a rupture in relation to humanism (127). Ironically, he continues, the rhetoric of posthumanism is suggestive of a progressive narrative that reflects the very Enlightenment principles of perfectibility that it centrally opposes (129). However, if we employ the concept of the posthuman to shed light on the ways in which the concept of the human has determined the contours of life, this puts the concept to work by making it clear that the human is but one, if an ever so powerful, concept and that other concepts can delineate/cut out the forms of life according to other patterns and logics. In this light, the concept of the posthuman reflects back on the concept of the human, emphasizing how this is a concept with specific characteristics responding to specific problems. Thus, for example, Descartes's concept of the cogito was a response to the challenges of uncertainty introduced by developments in science, Aristotle's distinction between *zoe* and *bios* a way of theorizing the difference between bare life and the political existence enabled by language, and Arendt's three concepts of labor, work, and action a means of determining the specificity of the human capacity for new beginnings. In each instance, conceptions of the human respond to specific problems, their moments, and their occasion, their circumstances, and conditions. By the same token, interrogations of the concept of the posthuman must try to determine the problems it responds to and, essentially, if these problems are themselves profoundly humanist. Depending on whether we want the concept of the posthuman to illuminate a new stage in the development of the human or to emphasize the ways in which incorporeal concepts shape life more generally, we must recognize how these different ways of using the concept rely on different kinds of narratives. Whereas the former rests more on identifying causal developments in a chain of events in a dialectical process whereby human becomes posthuman, the latter is interested in the history of the concepts themselves and the events they effect. The latter also carries a different relation to temporality. Concepts, Deleuze and Guattari (1994) argue, are not eternal, and they certainly appear in relation to specific historical problems, but neither are they temporal (27–28). It is not a question of replacing one with another so much as being aware that different concepts bring out different variations. Thus the concepts of the human and the posthuman are not necessarily to be understood as a chronological substitution of one for the other. What we need to understand, in that case, is not how we became posthuman, or how we became human, but how we become. At least from Deleuze and Guattari's perspective, philosophy should be a dice throw, an openness to the events that the concepts create (27–28).

Yet this openness poses particular problems when it comes to the faculty that is often regarded as that which most clearly makes the human as such, that is, not just the capacity for thought but the ability to use this capacity to reflect back on ourselves. Not only can the human think, but it can think about thinking. This is, after all, the very foundation for humanist thought—different as it may look. In the face of the void, the other, the animal, or the death of God, the human is established as a point of stability and reason in relation to whom we must seek our answers and our meanings. Already Linnaeus determined the human, not taxonomically like all other animals, but imperatively—*osce te ipsum* (know thyself). Heidegger's Dasein reflects on its own being—the human can tell itself it is human. The history of the heterogeneous borders of the abyss between "what calls *itself* man and what *he* calls the animal," as Derrida (2008) puts it, reflects, on the side of man, "an anthropocentric subjectivity that is recounted or allows a history to be recounted about it, autobiographically, the history of its life, and that it therefore calls *History*" and, on the other hand but not, importantly, antithetically, a "heterogeneous multiplicity of the living (30, 31). Again, the history of the human is intimately intertwined with the narratives it is capable of telling itself about itself. For a large part of our humanist past, "the Animal" has been positioned as other and without further differentiation. However, many narratives about the Animal have already been shown not to hold: The multiplicity of life has caused them to disintegrate. As Derrida has famously shown, the very notion of "the Animal" collapses as soon as this commonplace is challenged by "the infinite space that separates the lizard from the dog, the protozoon from the dolphin, the shark from the lamb, the parrot from the chimpanzee, the camel from the eagle, the squirrel from the tiger, the elephant from the cat, the ant from the silkworm, or the hedgehog from the echidna" (34). When it comes to the narratives of the human, on the other hand, it seems, as we have noted, that humanist narratives continue to determine its past and its future direction. As long as this is the case, we never seriously question the human or recognize the role of concepts. Caught up in such narratives, the "posthuman" as a concept runs the risk of being but a catchall phrase for our anxieties about the human. If we want to be truly open for thinking about life as being cut out in the shape of the human, or the posthuman, or in shapes that do not yet have a concept to determine them, we must find ways of understanding narrative itself differently. In Derridean terms, we could ask whom we are to follow. As long as we follow only ourselves, that is, as long as we return to ourselves as the point of knowing, it does not really matter what we call ourselves—we will continue to project our essential if contradictory humanism on the world. To achieve something different, or rather a difference, that is, one that is not predicated on being different from what precedes it, the concept of the posthuman must be a dice throw—it must not follow on the human; it must follow life in directions that we have not been able to foresee from what has come before.

IN SEARCH OF NEW BEGINNINGS

So how can we begin to vex the concept of the posthuman without losing sight of the questions that it, potentially, helps us ask? This question brings us to a first step, which, as I outlined above, entails claiming the work of Arendt as it brings together questions of life and narrativity in productive ways. If we configure Arendt's threefold conception of human life through posthuman discourse, her work and in particular her theory of action may be able to help us locate the potential for the new directions beyond the humanist paradigm. Arendt, as Julia Kristeva notes, consistently places life as a concept at the center of her work, and in the opening of *The Human Condition*—a book described as a "vehement defence of life" in the face of the nihilistic threats of totalitarianism—Arendt outlines her conception of human life as consisting of the three fundamental activities of labor, work, and action (Kristeva 2001, 3, 5). These are fundamental activities in that "each corresponds to one of the basic conditions under which life has been given to man" (7). The first, labor, is "life itself." This is life as it maintains itself as such—the basic functions of the body and its necessities. This aspect of life also has its own temporality: It is perishable at the same time as it is everlasting—this is the labor of perpetual satisfaction of basic needs. This aspect of man, man as *animal laborans*, is closest to the animal. The second notion is that of work and corresponds to the production of artifice, that is, things manmade. This, Arendt (1998) argues, is what humans construct—"The work of our hands, as distinguished from the labor of our bodies"—adjustments of nature that, unlike the permanent transience of labor, create some sort of stability (136). *Homo faber* produces objects, artworks, houses—things not eternal but that have enough durability to stabilize human life in a public manner. The third and final category of life, action, is the condition of political life in the sense that this is the only category that depends by definition on the plurality of men (7). Unlike the "mere bodily existence" of labor and the

production of objects and ways of being through work, action is an initiative that, by virtue of taking place among men, introduces a new beginning. To act is to begin, to set something in motion. Closely associated
with her concept of natality, Arendt's conception of action is suggestive
of the inevitability of new beginnings that characterizes the human condition. Not only does the event of birth constantly introduce newcomers
whose actions we cannot foresee, but there is also a "second birth" that
takes place through action. An initiative among men sets something in
motion, the consequences of which we are unable to predict fully. Because
the plurality of human coexistence entails that action cannot stand by
itself, we can never predict a final outcome of any one action. The potential for change is therefore inherent in the political life of humans. Here,
Arendt opens for an understanding of the new that differs radically from
any Hegelian or Marxian reading. There is neither a dialectical process to
be uncovered nor a causal narrative to be retrieved from history or to help
us predict the future.

For Arendt, action, or rather its possibility, is based on the two interrelated activities of forgiving and promising. Both these open for a particular relationship to temporality as released from causality. Forgiveness
releases the agent from a past, which, if not forgiven, will continue to
determine the direction of the future. Compared to its opposite in ven-
geance, which constitutes a reaction against an earlier deed that keeps
everybody "bound to the process," forgiveness opens up for unpredictability and the new (240–241). Unlike the "relentless automatism" of ven-
geance, forgiveness opens for true freedom—a space to act undetermined
by what came before. The promise—the rather astounding act of projecting yourself into the future you cannot foresee—entails a construction of
some continuity in an otherwise chaotic future (237). So, and seemingly
with opposite ends, forgiveness prevents the sedimentation of identity,
while the promise constructs new, more or less temporary, points of stability. Thus built on forgiveness and promises, action, rather than building
on a causal chain of action followed by reaction, entails a "constant establishment of new relationships within a web of relations" (240). Crucially,
both forgiveness and making promises rely on plurality—"men, not Man,
live on the earth and inhabit the world"—and both activities would be
superfluous if the similarities between men were similar enough to be
able to account for us all as "Man." Thus, and while Arendt sees men as

"all the same, that is, human," the human, for her, is such that "nobody is ever the same as anyone else who ever lived, lives, or will live" (7–8). We cannot forgive ourselves, and a promise made only to ourselves is not binding. Forgiving, for Arendt, is not necessarily individual or private but always personal because *what* was done is forgiven for the sake of *who* did it" (241). Importantly, we cannot forgive ourselves because we cannot fully perceive ourselves. "Closed within ourselves, we would never be able to forgive ourselves any failing or transgression because we would lack the experience of the person for the sake of whom one can forgive" (243). In other words, we cannot forgive, and thereby cannot take a new and unexpected route, without recognizing not only that we are dependent on the plurality of others but also that we are in ourselves incapable of tracing the circle of self-reflexivity back to ourselves. The thinking that inevitably takes us back to ourselves as human subjects prevents us from stepping outside the circle and "appear in a distinctness which we ourselves are unable to perceive" (243).

This reliance on plurality and the freedom it opens for us can be useful for a research field that is trying to find ways out of the humanist, and also occasionally posthumanist, narratives that keep it chained to a particular idea of human development and historical progression. Where Arendt speaks of plurality as involving more than one man, and for Arendt this is a distinctly human man, we may expand on this and think of it in terms of a plurality that is not limited to the conception of the human on which Arendt relies so heavily. For Arendt's understanding of action and of its reliance on plurality to be useful we must open it up toward a conception of life that is, first, not limited to the human and, second, does not limit multiplicity to interpersonal relations. Speech is a key point here because it constitutes a central aspect of Arendt's conception of action. Skirting the discourse about the role of logos and its relation to the human, we need to focus here on the particular relation between reaction and response. Because her conception of action relies exactly on the possibility of doing more than simply reacting, it would make sense that a creature incapable of responding would be excluded from this possibility. Forgiving, she maintains, "is the only reaction which does not merely re-act but acts anew and unexpectedly, unconditioned by the act which provoked it." But forgiving is also dependent on speech, of the disclosure of the "who" who, unlike the "what," may be forgiven (241). "In acting and speaking, men show who they are, reveal actively their unique personal identities and thus make their appearance in the human world, while their physical identities appear without any activity of their own in the unique shape of the body and sound of the voice" (179).

In a post-Derridean frame, however, Arendt's faith in speech as enabling a human communication that reveals the truth of the deed as well as its doer in some essential sense—and her dismissal of other forms of nonhuman response—appears highly problematic. Arendt's Aristotelian heritage shapes what she understands to be a major difference between human and animal: Unlike animals, humans have speech and thus the capacity not only to react but to respond. This distinction, which Lacan too inherits through Descartes and that, as Derrida shows, relies on a questionable and rather uncharacteristic from the Lacanian perspective presumption about "the purity, the rigor, and the indivisibility of the frontier that separates—already with respect to 'us humans'—reaction from response," fails in a fundamental way to take into account the infinitely differentiated difference that emerges within the "differentiated field of experience and of a world of life forms." Not only does it fail to account for the unconsciousness and the iterability that necessarily puts the supposedly human response as articulated by a pure and unadulterated subject position into question, but it also reduces the multiple differential field of the nonhuman animal and ties it to a position "subjected to the human subject" (Derrida 2008, 125–126). Communication as conceived of in terms of a language that conveys meaning between human subjects without contamination builds on what Ron Broglio (2013) describes as "a scaffolding for subject-object relations" that has worked to "immunize us from our own animality" (44). As he shows through Jean-Luc Nancy, already the voice as that which precedes language but that is capable of summoning the other without subject or signification opens up the possibility of a response to the world different from any distinctly human signification (44). Also in an Arendtian sense, the voice Broglio describes seems to point to the possibility of acting anew—"a utopic venture" as he calls it, an "attempt not to be alone but rather to imagine community through communication" (37). If Derrida works throughout his philosophy to elucidate the ways in which speech is always inscribed with a *différance* that makes it impossible to establish stable meaning, the objective throughout his essays collected under the common title *The Animal That*

Therefore I Am is to show also how it is necessary to reinscribe *différance* between reaction and response "within another thinking of life, of the living, within another relation of the living to their ipseity, to their *autos*, to their own autokinesis and reactional automaticity" (Derrida 2008, 126). The idea that the animal who calls itself human can have a stable relation to itself precisely because of this self-reflexivity quickly crumbles as the return to the subject is shown to be far from precise. By the same move, the relations of this animal to other human or nonhuman animals as well as to its own intentionality can never be as clear cut as they appear to be in Arendt. Derrida's deconstructive reading seeks not so much to restitute a capacity for a nonhuman animal to respond, although he is reluctant to exclude such a possibility, as it aims to question the extent to which the reflexive and reactive response ascribed to animals should not also be analyzed in a human context. In effect, this means forever disturbing the border that distinguishes human and animal on grounds of the capacity to respond or merely react (173).

Arendt (1998) maintains that speech is closely related to action because action must provide an answer to the question "Who are you?" (178). In the light of Derrida's work, we need to reconfigure this question. If released from the ties of speech and reason, the "who are you?" holds the potential of opening up an ethical relation that could be pursued in the Levinasian sense of the truly Other, but it could also be understood in the sense of a relation between living entities that is respectful exactly because it does not expect a stable response. In addition, this relation works only if the posers of the question themselves do not presume their own capacity to answer it. " 'But as for me, who am I?' And what would ever distinguish the response, in its total purity, the so-called free will and responsible response, from a reaction to a complex system of stimuli?" (Derrida 2008, 53–54). Arguably, our difficulties in escaping the narratives that shape our philosophical and ethical relationship to "the Animal," as well as to "the human" and "the posthuman," seem predicated at least as much on our conception of our own selfhood as humans as it is on our conceptions of nonhuman life. For Arendt's conception of action as opening toward the new to hold, not only in the case of wanting to claim it for the purpose of a posthumanist discourse but also more generally, must we not presume also that the "who" becomes new? To be born again, to open life truly toward new beginnings, as Arendt aims to do, must not the actor, which for Arendt remains crucial as somebody identifiable through the speech accompanying action, also begin again? As Arendt (1998) acknowledges: "It is not a beginning of something but of somebody, who is a beginner himself" (177). This could be an "I," as Derrida (2008) would put it, as "a sign of life, or life in presence, the manifestation of life in presence" whose self is never stable but whose traces we can follow (56).

It seems, then, that to make Arendt's conception of action into a useful tool for thinking about life in a posthumanist frame we need not only her notion of plurality—men rather than Man—but also a notion of singularity—"a life" rather than the generality "life." We are not truly plural, and we cannot truly begin again as long as we insist on a conception of the human as separable from the multiplicity of life. Bringing up Arendt in this framework also makes it clear that we cannot truly become posthuman—that is, we cannot open up toward a future not projected as a causal reaction to the humanist discourses of the past—as long as we do not allow for the singularities of life to replace debates about subject embodiment. Arendt's conception of natality, which relies exactly on the introduction of the unexpected and unforeseeable, gains considerable strength if it becomes truly open not only to the "who" but to what "who" means. Activating a Derridean and post-Derridean rethinking of speech enables us to claim Arendt's notions of new beginnings while also reintroducing the prepolitical "animal" dimensions of life. What helps us perform this move, however, is not so much related to conceptions of the animal but rather to an idiocy related to our understanding of the human. As soon as we accept the contamination of communication, Arendt's "who" opens toward a more radical conception of unexpectedness, which cracks open the narrative that, as I suggested in the opening of this chapter, contains discourses of posthumanism within discourses of the human.

FORGIVING THE HUMAN

Regardless of how we evaluate the notion of the Anthropocene, because the usefulness of this notion has been questioned, it seems fair to suggest, as I did earlier, that the concept itself puts the human into a crisis in a new way. The "violence toward others inherent in the formation of our social identities," as John Rajchman (2001, 15) puts it, is leaving marks on the planet, marks that, by the same move, expose the parochialism of human narratives and the delusion of a human subject in charge of its own narrative. On the one hand, the violence of the human to the planet arguably constitutes a major example of how centuries of putting the human at the center of existence are now leaving their marks, and as such, the implications of humanist frameworks on our ethical codes still need to be addressed. On the other hand, there is something essentially contradictory about trying to address the implications of this exposure *by means of* humanist frameworks because this keeps the human tied to a form of self-reflexive subjectivity that the arrival of the Anthropocene exposes as untenable. At the same time, as I have outlined here, research in the field called the "posthumanities" is fraught with ambivalence. Arguably, the understanding of the posthuman as occupying a place in a historical progression—that is, the ontological conception outlined above—delimits our possibilities of addressing the question even as it, even when it manages to let go of humanist presumptions about the subject, nonetheless continues to think the posthuman in relation to the human. The preoccupation with the connection between thinking and embodiment that takes up so much space in posthumanities discourse, while crucial in theorizing a number of important developments in how life is perceived today, continues, as Wolfe shows, to be predicated on humanist discourse. Rather than recognizing the delusions of a human subject in charge of its own narratives, such an approach continues the narrative by tracking the demise of such a subject. Revisiting this idea of narrative progression enables us to address the implications of a humanist approach to the world—what the concept of the human does to life—but also, crucially, how a concept of the posthuman may or may not be able to help us think differently. Our understanding of posthumanism and where it can take us depends on our starting point and the extent to which we are capable or willing to veer off from the narrative of human progression. As Wolfe (2010) shows, via Rutsky, for the posthuman to be "more than merely an extension of the human," we must find a way of recognizing the processes that go beyond the patterns, standards, codes, and information that we know (xviii).

To be able to go beyond the codes we know, we must recognize the idiocy that colors presumptions about our own ability to think. The Image of Thought that, Deleuze argues, presumes a subject capable of good sense and common sense, prevents thinking from truly occurring. Such idiocy arguably founds the basis of conceptions of the posthuman as a mode of being after the human because it continues to rely on our capacity to know the patterns, standards, and codes that Rutsky refers to. As I noted earlier, concepts such as idiocy, stupidity, and *bêtise* have been employed in slightly different ways by Deleuze and Derrida, and they have also been put to use differently by, for example, Avital Ronell and, in a more express-edly posthumanist context, by Broglio. The latter builds on this tradition to account for the "intensive rather than discursive mode of communi-cation of the body" in a posthumanist frame (Broglio 2012, 18). While Broglio's use of what I have accounted for here as Deleuze's earlier form of idiocy helps him toward "outwitting" reason by means of what it most clearly rejects (28), we get further still if we also include what I flagged earlier as another type of idiocy that Deleuze and Guattari develop later and that may also be useful to a posthumanist project beyond exposing its own blank spots. In *What Is Philosophy?* Deleuze and Guattari's (1994) conception of idiocy as the failure to recognize the presumptions of one's own capacity to think takes a new form as the conceptual persona of the idiot develops and gains new characteristics. The new philosophical idiot is a "Russian idiot" whose idiocy is characterized not by presumptions about his ability to think but, quite the opposite, by rejecting such pre-sumptions and affirming "the lost, the incomprehensible, and the absurd" (63). It is a matter of acknowledging the ruptures in the Image of thought, the falling apart, as Gregg Lambert (2002) puts it, at which point thought "does not accede to a form that belongs to a model of knowledge, or fall to the conditions of an action; rather, thought exposes its own image to an 'outside' that hollows it out and returns it to an element of 'formless-ness'" (127). What this Russian idiot shows, Rajchman (2000) notes, is "not only that philosophical thought is unlearned, but also that it is free in its creations not when everyone agrees or plays by the rules, but on the contrary, when what the rules and who the players are is not given in advance but instead emerge along with the new concepts created and the new problems posed" (38).

To read posthumanism first through Deleuze's earlier notion of idiocy and then through his and Guattari's developed version of it may provide a way of revisiting some of the presumptions that continue to haunt its practices. It may provide a way of accounting for and responding to the call of the nonhuman voice for a utopian future that "is in no particular temporal space of the past, present, nor future" (Broglio 2013, 38). Essentially, this is about letting go of the notion that we are the subject of our narratives about ourselves as human—be they on the level of biography or species. This is not to suggest an abandonment of ethics and responsibility but an acknowledgment of life as impersonal. It is only if we stop expecting a stable response, not only from the other but also from ourselves, that we can be respectful of life and relations that fall outside "fixed methods and prior forms." As Rajchman (2001) puts it, describing Deleuze's work: "To think is not to be certain nor yet to calculate probabilities. It is to say yes to what is singular yet impersonal in living" (18). For this to be possible we must believe not in God nor in the self but in the world, although it may be a most difficult task. Deleuze thus works to find a way out of nihilism—"a mode of existence to be discovered on our plane of immanence today" (18). Where Deleuze and Guattari's developed notion of idiocy points toward a thought that opens itself toward a world without presumptions, Arendt's notion of action can be claimed as a way of opening new directions in thinking about life that are not humanist and maybe not even posthumanist. This would be a direction that "forgives" the human as it has been constructed in a humanist sense and therefore does not follow in its, reverse, traces. If we still want to call such a mode of existence posthuman, this must be a "post" as in Lyotard's conception of postmodernism: that is, one that does not necessarily come after the human. "The posthuman," as Rutsky suggests,

> cannot simply be identified as a culture or age that comes "after" the human, for the very idea of such a passage, however measured or qualified it may be, continues to rely upon a humanist narrative of historical change. . . . If, however, the posthuman truly involves a fundamental change or mutation in the concept of the human, this would seem to imply that history and culture cannot continue to be figured in reference to this concept.
>
> (cited in Wolfe 2010, xvii)

In exploring the possibilities of making new routes that escape the circular footprints of humanism, I have tried to claim an aspect of Arendt that has not yet found its way into posthumanist projects to any great extent. Reading Arendt alongside Derrida and Deleuze and Guattari, it becomes clear that her work has something important to contribute to how we want to think about life. What this is more precisely, as it emerges together with Derrida's problematization of the relation between reaction and response as well as alongside Deleuze and Guattari's second form of idiocy, is a break from the narratives of progression that tie the posthuman to the human and the posthumanist to the humanist. This break would not arrive after humanism but emerge with the miraculous freedom of Arendt's action—a freedom from the processes of habit and causality, a freedom from the teleology that posits the posthuman in a determined relation to the human. The Arendtian promise—the daring gesture that makes a new beginning in the midst of a future, the nature of which we cannot foresee—could be not to be human. But maybe, this promise could also be not to be posthuman.

NOTES

1.   I would like to thank Charlie Blake and Ron Broglio for their generous feedback on drafts of this essay and for inspiring discussions on the topic more generally. I would also like to thank the anonymous reviewer(s) for their constructive comments.

2.   The relation between Deleuze's conception of idiocy and Derrida's understanding of stupidity is admittedly far more complex than this passing juxtaposition acknowledges and could, and has been, the subject for a separate essay. Bernard Stiegler (2013) has recently discussed this juxtaposition and what he recognizes as some problems of Derrida's reading of Deleuze in this respect. Indeed, he suggests that Derrida may even be *"playing the fool, fait la bête,* as one says in French" in his reading of Deleuze (163).

3.   In the present essay I revisit several points I made in this earlier article, albeit in a new context.

WORKS CITED

Agamben, Giorgio. 1998. *Homo Sacer: Sovereign Power and Bare Life.* Trans. Daniel Heller-Roazen. Stanford, Calif.: Stanford University Press.

Arendt, Hannah. 1998. *The Human Condition.* Chicago: University of Chicago Press.

Beckman, Frida. 2009. "The Idiocy of the Event: Between Antonin Artaud, Kathy Acker, and Gilles Deleuze." *Deleuze Studies* 3 (1): 54–72.

Broglio, Ron. 2013. "Abandonment: Giving Voice in the Desert." In *Glossator: Practice and Theory of the Commentary,* vol. 7, ed. Nicola Masciandaro and Eugene Thacker, 33–45. CreateSpace.

Broglio, Ron. 2012. "Incidents in the Animal Revolution." In *Beyond Human: From Animality to Transhumanism*, ed. C. Blake, C. Molloy, and S. Shakespeare, 13–30. London: Routledge.

Colebrook, Claire, and Cary Wolfe. 2013. "Is the Anthropocene . . . a Doomsday Device?" http://www.youtube.com/watch?v=YLTCzth8H1M.

Deleuze, Gilles, and Félix Guattari. 1994. *What Is Philosophy?* Trans. Hugh Tomlinson and Graeme Burchill. London: Verso.

Derrida, Jacques. 2008. *The Animal That Therefore I Am*. Ed. M. L. Mallet. Trans. David Wills. New York: Fordham University Press.

Descartes, René. 2003. *Discourse on Method* and *Meditations*. Trans. Elizabeth S. Haldane and G. R. T. Ross. New York: Dover.

Hayles, N. Katherine. 1999. *How We Became Posthuman: Virtual Bodies in Cybernetics, Literature, and Informatics*, Chicago: University of Chicago Press.

——. 2005. *My Mother Was a Computer: Digital Subjects and Literary Texts*. Chicago: University of Chicago Press.

Kristeva, Julia. 2001. *Hannah Arendt: Life Is a Narrative*. Trans. F. Collins. Toronto: University of Toronto Press.

Lambert, Gregg. 2002. *The Non-Philosophy of Gilles Deleuze*. New York: Continuum.

Peterson, Chris. 2011. "The Posthumanism to Come." *Angelaki: Journal of the Theoretical Humanities* 16 (2): 127–141.

Rajchman, John. 2000. *The Deleuze Connections*. Cambridge, Mass.: MIT Press.

——. 2001. "Introduction." In *Pure Immanence: Essays on a Life*, by Gilles Deleuze, trans. Anne Boyman, 7–23. New York: Zone.

Serres, Michel. 1995. *The Natural Contract*. Trans. E. MacArthur and W. Paulson. Ann Arbor: University of Michigan Press.

Stiegler, Bernard. 2013. "Doing and Saying Stupid Things in the Twentieth Centuries: Bêtise and Animality in Deleuze and Derrida." *Angelaki: Journal of the Theoretical Humanities* 18 (1): 159–174.

Thacker, Eugene. 2010. *After Life*. Chicago: University of Chicago Press.

Wolfe, Cary. 2008. "Thinking Other-Wise: Cognitive Science, Deconstruction, and the (Non) Speaking (Non)Human Animal Subject." In *Animal Subjects: An Ethical Reader in a Posthuman World*, ed. J. Castricano, 125–144. Ontario: Wilfrid Laurier University Press.

——. 2010. *What Is Posthumanism?* Minneapolis: University of Minnesota Press.

——. 2011. "Response to Chris Peterson, 'The Posthumanism to Come.'" *Angelaki: Journal of the Theoretical Humanities* 16 (2): 189–193.

3

SUBJECT MATTERS

SUSAN HEKMAN

As a result, gender is not to culture as sex is to nature; gender is also the discursive/cultural means by which "sexed nature" or a "natural sex" is produced and established as "prediscursive," prior to culture, a politically neutral surface on which culture acts.

—JUDITH BUTLER, *GENDER TROUBLE*

I am suggesting that the self is not only irretrievably "outside," constituted by social discourse, but that the ascription of interiority is itself a publicly regulated and sanctioned form of essence fabrication.

—JUDITH BUTLER, "PERFORMATIVE ACTS AND GENDER CONSTITUTION"

CONSTRUCTING SEX/GENDER

Judith Butler's claims about sex and gender in *Gender Trouble* had an explosive effect on the feminist community when the book was published in 1990. Her argument that there is no reality to gender, that it is irretrievably outside, a product of social discourse and that sex itself is a product of gender norms, a result of the functioning of publicly regulated gender norms, sparked intense debates in the 1990s. Most participants in these debates assumed, rightly or wrongly, that Butler was taking an extreme constructionist position that precluded the "reality" of the sexed subject and, worse, precluded the agency of the subject. Although Butler had many defenders, her detractors were vehement in their assertion that such extreme constructionism was dangerous for feminism. They claimed that it obviated the possibility of discussions of "woman," problematized the agency and autonomy of the subject, and called into question the articulation of a feminist politics.

Twenty years after these debates engulfed feminism we are still grappling with the question of the subject in feminist theory. But the debates we are engaged in today, far from resolving the question of the subject, have instead made the issue more difficult. The construction debates of the 1990s and beyond spawned reformulations and repositionings of the construction and constitution of subjectivity. Furthermore, in recent years an increasing number of feminists have approached the subject from the perspective of the "new materialism" that attempts to incorporate the material without abandoning social construction. But, if anything, this approach has made the issue of the subject even more problematic. Another dimension of recent discussions is the articulation of the posthuman. How discussions of the feminine subject can interface with the posthuman is very much open for debate. I think it is fair to say that at this point little or no consensus on how we should approach the subject has emerged from these debates.

The thesis I advance here will seem, on the face of it, counterintuitive: that our best guide in exploring the subject today lies in the work of Judith Butler. Although Butler's work has been identified with the extreme constructivist pole of the debate, I will argue that in her recent work she approaches the constitution of the subject from a perspective that is unique and uniquely helpful in trying to understand the complexity of subject constitution. I will not argue that Butler has repudiated her previous position but, rather, that she has complicated and enhanced it by looking at subjects from a broader perspective than the one she takes in *Gender Trouble*.

Butler's approach to the subject in *Gender Trouble* is not exceptional in the claim that gender is socially constructed and therefore not "real" but, rather, in the claim that sex cannot be distinguished from gender. The social construction of gender had been a staple of feminist literature by the time *Gender Trouble* was published. What sets Butler's theory apart is that she eliminates the ground on which most discussions of gender rested: the reality of sex. For Butler sex, like gender, is a social construction, and, even more provocatively, it is *produced* by gender.[1] To many of Butler's critics this seemed to go too far. If we remove the ground of biological sex, the result is that "there is no there there," no doer behind the deed. We are then left with a social dupe who cannot act and has no grounding in "reality."

I will not dispute whether this is a plausible reading of *Gender Trouble*. What I want to assert is that, even at this early stage, Butler introduces perspectives that challenge a reading of her subject as a social dupe. Early in the book she asks: If there is no recourse to a person that escapes the matrix of power, then where can we look for the possibility of the subversion of that power? Her answer: "If the regulatory fictions of sex and gender are themselves multiply contested sites of meaning, then the very multiplicity of their construction holds out the possibility of a disruption of their univocal posturing" (1990a, 32). Both sex and gender are socially produced, but that social production simultaneously creates the possibility of disruption. There is slippage in the construction process, and this slippage opens the door to other articulations of sex and gender. Expanding on this point at the end of the book, Butler argues that the coexistence and convergence of discursive injunctions produce the possibility of reconfigurations and redeployments. Thus, although there is no self prior to this convergence, the subject can take up the "tools" where they lie (1990a, 145). It follows that "construction is not opposed to agency; it is the necessary scene of agency, the very terms in which agency is articulated and becomes culturally intelligible" (1990a, 147).

Butler's goal in *Bodies That Matter* is, at least ostensibly, to answer some of the questions posed by the critics of *Gender Trouble*, particularly questions relating to the subject. In doing so she explores dimensions of the subject that were only latent in *Gender Trouble*. She begins her discussion with what sounds like a confession: "I began writing this book by trying to consider the materiality of the body only to find that the thought of materiality invariably moved me into other domains. I tried to discipline myself to stay on the subject, but found that I could not fix bodies as simple objects of thought" (1993a, ix). She then confronts the issue that was at the heart of the criticisms of *Gender Trouble*: How do we understand the constitutive and compelling status of gender norms without falling into the trap of cultural determinism (1993a, x)? The discussion that follows centers around two themes that structure not only this discussion but Butler's subsequent work as well: the "I" and the outside.

The "I," she declares, neither proceeds nor follows the process of gendering but emerges within the matrix of gender relations themselves (1993a, 7). This "I" (which is always in scare quotes) is not a simple entity.

In a discussion of Althusser Butler asserts that the "I" that is produced through the accumulation and convergence of calls of identity cannot extract itself from the historicity of these claims. The "I" is violating and enabling at the same time. The "I" who would oppose its construction is at the same time drawing from that construction to articulate its opposition (1993a, 122). This formulation also accounts for the issue of agency: The "I" draws its agency in part through being implicated in the very relations of power it seeks to oppose (1993a, 123).

The strategy that Butler employs in this analysis is articulated in a passage immediately preceding this discussion. What is required, she asserts, is to shift the terms of the debate from construction versus essentialism to the more complex question of how deep-seated constitutive constraints can be posed in terms of symbolic limits in their intractability and contestability (1993a, 94). It follows that "there is no subject prior to its construction, and neither is the subject determined by those constructions" (1993a, 124). In perhaps her most revealing comment Butler admits that in making these formulations she brackets the "I" but then comments, "I am still here" (1993a, 123).

The same deconstructive impulse informs Butler's discussion of the "outside." There is an outside to what is constructed by discourse, but it is not an absolute outside, an ontological thereness (1993a, 8). The task, she asserts, is to refigure this necessary outside as a future horizon, one in which the violence of exclusion is perpetually in the process of being overcome (1993a, 53). It follows that, as with the "I," the dichotomy is breached: Language and materiality are not opposed, "for language both is and refers to that which is material and what is material never fully escapes from the process by which it is signified" (1993a, 68).[2]

These two themes come together in Butler's discussion of an issue that will become a central theme of her later work: the abject. The constitution of the "I" always necessarily depends on exclusion and, specifically, the exclusion of the domain of the abject. Those who are not yet subjects form the constitutive outside of the domain of the subject (1993a, 3). Butler will have much to say about this exclusion in her subsequent work. What is important at this point, however, is that the domain of the "I" is bounded by the abject. Some are excluded from this domain, a domain is constituted and regulated by sexual norms. Those who are excluded lack subjectivity; they are not "I's" in the full sense.

It is undeniable that, for Butler, language is necessarily imbricated in the constitution of the subject. Her critics and defenders are undoubtedly correct on this point. But it should be clear from the above analysis that this does not lead to linguistic monism or determinism. It should also be clear that, from the outset, Butler is aware that the subject is more than language, that the "I" is not simply a product of culture but, rather, a complex mix of power, agency, and subjectification. By tracing the development of the subject in Butler's work, we can see that, after *Gender Trouble* and *Bodies That Matter*, she realizes the difficulty of the problem she has raised and in her subsequent work turns her attention to exploring that problem in more detail. This is what sets Butler's work apart from the social-constructionist position with which she is identified. Unlike other social constructionists, who make vague references to "reality" beyond discourse, Butler directly confronts the complexity of the problem and takes on the difficult task of defining a subject that is constituted conjointly by language, power, and materiality.

SUBJECTS AND SUBJECTIFICATION

Based primarily on a reading of *Gender Trouble* and *Bodies That Matter*, Butler's position on the subject has come to be emblematic of what Lois McNay (2000) calls the "negative paradigm" in feminist theory. My analysis of her early works reveals that Butler's understanding of the subject is much more complex and multifaceted than the "negative paradigm" allows. This complexity becomes more pronounced in her subsequent work. What emerges from this work is an understanding of the subject that, although not abandoning the constructed subject, presents the subject as a confluence of disparate forces that constitute a unique entity.

The central thesis that Butler develops in this work is what she calls the paradox of the subject. This theme will structure the various discussions of the subject that will occupy her in the books that follow *Gender Trouble* and *Bodies That Matter*. The paradox revolves around her claim that power is what we oppose *and* what forms us as subjects (1997, 1). The subject is "neither fully determined by power nor fully determining of power (but significantly and partially) both" (1997, 17). She continues to insist, as she did in *Gender Trouble*, that subjects are formed through subjection to power. What she emphasizes in her later work, however, is that this process is not complete, that the subject that is determined by power is not *completely* determined. Her task, then, is to explicate how that paradox operates, to explain how subjectification is not always subordinating (Allen 2008, 81).

In her attempt to explore the paradox of the subject, Butler finds an ally in Foucault. Both Butler and Foucault argue that there is no outside to power, that the self forms itself within the context of power:

> The self forms itself but it forms itself within a set of formative practices that are characterized as modes of subjectification. That the range of its passable forms is determined in advance by such modes of subjectification does not mean that the self fails to form itself, that the self is fully formed. On the contrary, it is compelled to form itself, but to form itself within forms that are already more or less in operation and underway.
>
> (Butler 2001b, 226)

The paradox of the subject is expressed here in terms of the character of power: It is both all encompassing and fragile at the same time. The "desubjugation" of the subject, Butler asserts, marks precisely the fragility and the transformability of power (2001b, 222). Along with Foucault Butler claims that we make ourselves, but it is a self-making that is never fully self-inaugurated. The self delimits itself, but it is always through norms that are already in place (2001b, 225).

In "Contingent Foundations: Feminism and the Question of 'Postmodernism,'" Butler takes on this paradox very directly: "But to claim that the subject is constituted is not to claim that it is determined; on the contrary, the constituted character of the subject is the very precondition of its agency" (1995a, 46). The relations of power that constitute the subject can be turned against itself, enabling a "purposive and significant reconfiguration of cultural and political relations" (1995a, 46). The paradox of power is that the power that constitutes us is at the same time the power that gives us the means to resist and the agency to employ those means. Actions can inaugurate effects that had not been foreseen. The question we should be asking ourselves, Butler argues, is: Where are the possibilities of reworking the matrix of power by which we are constituted?

Butler answers this question in various ways in her post–*Gender Trou-ble* writings. In "Competing Universalities" she articulates her thesis in terms of speech. Speech becomes something else by virtue of having been broken open by the unspeakable: "The unspeakable speaks, or the unspeakable speaks the unspeakable into silence" (2000b, 158). In her contribution to *Feminist Contentions* she explores how agency can be understood from this perspective. Agency, she asserts, is found at the juncture where discourse is renewed; the subject is open to formations that are not fully constrained in advance. Agency is the effect of discursive relations which, for that reason, do not control its use (1995b, 135–137).[3]

This, then, is the paradox of power: The subject is neither a ground nor a product but the position from which we weigh the possibilities of resignification (1995a, 47). Power provides its own possibility of being reworked. We are within power, and it is from this position that we resist power. Far from being unconcerned with agency and resistance, Butler is almost obsessed with the issue. Unlike her critics, however, Butler refuses to locate agency outside of power. Butler's position is that we find agency, along with everything else, inside of power. More specifically, agency appears at the juncture where discourse is renewed. Subjects are not con-stituted once and for all but again and again; the subject is open to formu-lations that are not fully constrained in advance (1995b, 135).

THE ONTOLOGY OF THE SUBJECT

Butler's exploration of norms, particularly as they relate to Foucault's work, continues to occupy her attention in *Undoing Gender* and *Giving an Account of Oneself*. On one hand, her discussion continues the theme of the paradox of power: Norms define us but also provide the possibil-ity of going beyond them. Norms, she asserts, are not static entities "but incorporated and interpreted features of existence that are sustained by the idealizations furnished by fantasy" (2000b, 152). The subject produced through operations of power is not "always already trapped"; resistance is not precluded (2000b, 151). In giving an account of myself I tell a story that is defined by its relations to norms, but this does not obviate moral agency (2005, 8).

But Butler's discussion of norms in these contexts also begins to explore a new aspect of the subject that she had not addressed specifically in her previous work: ontology. In *Undoing Gender* she asserts that norms have a double role: We need norms in order to live, and to live well, but norms also constrain us and do violence to us (2004c, 206). Norms tells us what kind of bodies and sexualities will be considered real and true; they give us an ontology. In *Giving an Account of Oneself* Butler argues that norms decide in advance who will and will not be a subject. Thus as I give an account of myself, I do so in terms of an ethical code already in place, an ethical code that defines who is a legitimate subject and who is not. It follows that to call into question a regime of truth is to call into question the basis of my own ontological status (2005, 22–23). I can give an account of myself, accept moral responsibility for my actions, only in terms of a moral code that precedes me and, most importantly, defines my existence. Norms are what we need to live, but challenging these norms, moving outside the ontological space they create, will efface us (2004c, 217).

The discussion of ontology marks a change in Butler's approach to the subject. But one must be very careful here in assessing this change. Butler is not positing an a priori realm of subjectivity, an essential "I." Rather, she is beginning to talk about a realm of the "I" that partially escapes subjectification. As she puts it in the introduction to the second edition of *Gender Trouble*: "I am not outside the language that structures me, but neither am I determined by the language that makes this 'I' possible" (1999, xxvi).

In her attempt to explore this ontological space, Butler turns to what seems to be an obvious topic: the psyche. But the psyche, like ontology itself, is dangerous territory. It is all too tempting to reify both ontological and psychic reality as a given. To offset this danger she makes it clear from the outset that it is a "significant theoretical mistake" to take the "internality" of the psychic world for granted (1999, xvi). The unconscious is not a psychic reality purified of social content; rather, like the conscious mind, the unconscious is an "ongoing psychic condition" in which norms are registered in both normalizing and non-normalizing ways. And here, also, norms can be undone: Subjects can pervert norms in identifications and disavowals that are not always consciously or deliberately performed (2000b, 153). The danger here, Butler argues, is to give psychic reality an independent ontological status. The ideals of personhood are socially produced; the emotions produced by the unconscious cannot be understood outside their social formation. "The specificity of the psyche does not imply its autonomy" (2000b, 154).

Butler's approach to the psyche, then, parallels her approach to the subject itself. Subjects exist; there is a reality to the "I." But we cannot discuss that reality apart from the discursive constructs that constitute it. What Butler seems to be saying about the psyche is that, as humans, we possess a conscious and unconscious mind; in that sense psychoanalysts are correct. But they are not correct in assuming that the unconscious is a realm of ontological reality prior to the social norms that are the ground of our identity. Her thesis is that both the conscious and the unconscious are constituted by those norms; neither is independent of the social. But, as with the subject, this is not all there is to it. Norms can fail, be undone, be perverted in both the conscious and the unconscious. Norms provide the possibility of their own transformation. Norms can be resignified and thus mobilized in unseen ways; power can be self-subverting (Thiem 2008, 81). Butler's perspective, although at odds with most psychoanalytic theories, is nevertheless a view consistent with her overall approach to subjectivity.

Butler's discussion of the psyche is closely related to an issue that occupies her attention in *Undoing Gender, Precarious Life,* and other works from this period: what I will call the necessity of identity. From the perspective of *Gender Trouble* and even *Bodies That Matter* this seems like a departure for Butler. The thesis of *Gender Trouble* is that we must eschew identity because it is the vehicle of women's oppression. Butler is famous for her anti-identity stance. It seems blatantly contradictory to attribute to her a theory that posits identity as necessary. But I think that Butler's concern with the necessity of identity is less a departure than a shift in focus. As Butler's concern with the subject deepens and takes on more complexity, she is forced to consider aspects of the subject that were not evident in her earlier investigations. And one of those aspects is the apparent fact that, for most of us, identity is a necessity for a livable life.

It is noteworthy that Butler's interest in this issue appears long before she becomes famous as the author of *Gender Trouble*. In an article originally published in 1987, Butler explores Foucault's approach to power and the body. She argues that Foucault redefines the body not as a substance, thing, or set of drives but as a site of transfer of power itself. It follows that for Foucault the subject is not merely the passive recipient of power but is also a site of agency and resistance. Agreeing with Foucault, Butler argues that although this relationship seems contradictory, it is better understood as complex. But there is another aspect to this relationship that Butler mentions, almost in passing, toward the end of the article: "Power attaches a subject to its own identity. Subjects appear to require this self-attachment, this process by which one becomes attached to one's subjecthood" (2004a, 190).

This is a very peculiar statement, particularly in light of Butler's other discussions of identity. She has little to say about this "requirement" until two books published in 2004: *Undoing Gender* and *Precarious Life*. She begins with the thesis that has informed her discussion of norms throughout her work: "I cannot be who I am without drawing on the sociality of norms that precede and exceed me" (2004c, 32). But she then proceeds to complicate this perspective seriously. Echoing her earlier thesis, Butler argues that identity, some form of stability, is a necessity for a "livable life." But what if none of the identities available in the norms of my society fit me? What if these identities fail to offer me a way of being? Sometimes, she asserts, "a normative conception of gender can undo one's personhood, undermining the capacity to persevere in a livable life" (2004c, 1). The consequence of this is that, in this case, I cannot "be." I am denied an ontology, a being, and, because I must have an identity to lead a livable life, I am denied that as well. It follows that "for those who are still looking to become possible, possibility is a necessity" (2004c, 31).

These passages add another dimension to Butler's previous discussion of norms. In those discussions she argues that, although norms define us, they do not wholly determine us: There is, in a sense, some wiggle room in the deployment of norms. What we get here is a darker picture. Norms allow us to "be," but for some subjects being is not a possibility. If we ask "What, given the contemporary order of being, can I be?" and that answer is negative, then I am denied what she calls "possibility." In *Undoing Gender* Butler discusses this in the context of transsexuals; in *Precarious Life* she refers to the prisoners at Guantanamo Bay. But the problem is the same. If, in both of these cases, these persons are not classified as human within existing norms, they are denied an ontology; they have no possibility.

Implicit in these discussions is an assumption that, although Butler never addresses it explicitly, is central to her argument. In *Undoing Gender* she states that if gender comes from elsewhere, then gender undoes the "I" who is supposed to bear it, and that undoing is part of the very meaning and comprehensibility of that "I" (2004c, 16). It is easy to misinterpret this passage. Butler is not positing an "I" prior to the assignment of gender; this would be antithetical to her whole corpus. But it is significant, and hugely so, that there are two entities here: the "I" that is defined by gender norms and the "I" that is the recipient of those norms. The two are not identical. Butler insists that the "undoing," that is, the resistance to gender, is implicit in the gender norms themselves, but it is still the case that there is an "I" that resists. And, furthermore, it is the case that the tension between the norms and the norms' recipient is what subjectivity is all about: "The 'I' is the moment of failure in every narrative effort to give an account of oneself. It remains the unaccounted for and, in a sense, constitutes the failure that the very project of self-narration requires. Every effort to give an account of oneself is bound to encounter this failure and to founder upon it" (2005, 79).

I think it is undeniable that Butler is offering what amounts to an ontology of the subject in these passages.[4] It is an ontology that departs radically from the ontology of the essentialist subject prior to discourse, but it is an ontology nonetheless. I would even go so far as to say that, for Butler, there *is* a there there, a subject who resists norms and "undoes" gender. But it is a subject who cannot be thought outside the norms that constitute it. The resistance itself is performed inside those norms and is made possible by them. And, most importantly, it is a subject whose ontology, whose being, is dependent on those norms.[5] Two important passages reveal the elements of Butler's position on ontology:

I see myself as working within discourses that operate through ontological claims—"there is no doer behind the deed"—and recirculating the "there is" in order to produce a counterimaginary to the dominant metaphysics. Indeed, I think it is crucial to recirculate and resignify the ontological operators, if only to produce ontology itself as a contested field.

(1998, 279)

We might reread "being" as precisely the potentiality that remains unexhausted by any particular interpellation. Such a failure of interpellation may well undermine the capacity of the subject to "be" in a self-identical sense, but it may also mark the path toward a more open, basic, even more ethical kind of being, one of or for the future.

(1997, 131)

The picture that emerges from this ontological view of the subject is more negative than that of Butler's earlier work. In her earlier view of the subject resistance was possible within the normative framework constituting gender. And, certainly, this option is still open to most subjects. Butler's ontological investigations, however, raise a problem that the earlier analyses had not addressed explicitly: What if norms make it impossible for the subject to *be* at all? What if norms preclude an identity for certain kinds of subjects? The only option for these nonsubjects is to challenge the norms that confer recognition, that grant being to some subjects but not others. What is required, Butler declares, is nothing less than to challenge what it means to be human.

In *Precarious Life* she takes on this challenge. She declares that it is "an ongoing task of human rights to reconceive the human when it finds that its putative universality does not have universal reach" (2004b, 91). We must, in the name of the human, allow the human to become something other than it is traditionally assumed to be. We must embrace the rearticulation of the human, not know in advance the form of our humanness. In politics, Butler argues, this demands a double path: simultaneously using the language of entitlement to assert the human while at the same time subjecting our categories to critical scrutiny (2003, 17–23). And the key to this is what Butler calls "cultural translation": "translation will compel each language to change in order to apprehend the other, and this apprehension, at the limit of what is familiar and parochial, will be the occasion of both an ethical and a social translation" (2003, 24).[6]

But how do we change our conception of the human? How do we extend the category of the human to include the excluded? Butler's answer to this question in her recent work revolves around a discussion of kinship. Discussions surrounding gay marriage and the question of how feminists, and particularly nonheterosexual feminists, should regard it stimulated Butler to look at issues of legitimacy, the state, and personhood. Legitimizing gay marriage would have the effect of expanding our norms of acceptable personhood, a goal apparently consistent with Butler's statements about extending the realm of the human. But only apparently. It also has the effect of granting the state the power to define what is legitimate and what is illegitimate in the realm of the sexual. Does the turn to the state, she asks, make it more difficult to argue in favor of the viability of alternative kinship arrangements? Does the turn to the state signal the end of a radical sexual culture (2002, 231)?

These are difficult questions for feminists and particularly for Butler. One aspect of the problem, however, is clear. Legitimizing gay marriage does extend the realm of acceptable personhood. The problem is that it has two unacceptable consequences. First, it acknowledges state power as the agent of legitimation. Second, it does nothing to change the delegitimization of other, more radical forms of sexual relations. Butler is not happy with the first consequence, but it is the second that she wants to pursue. There are realms outside the legitimate and the illegitimate in sexual relations that are not yet thought of as a domain. These "nonplaces" are foreclosed by the campaign for gay marriage. It is this domain that should command our attention. Butler asserts that "we should all be pursuing and celebrating sites of uncertain ontology and difficult nomination" (2002, 235). But how do we do this politically? How does one think politics from the site of unrepresentability (2002, 232–234)?

We need, Butler concludes, a critical challenge to the norms of recognition supplied by state legitimation. This conclusion causes Butler to turn to the issue that is at the root of these questions: kinship. Historically, marriage is inseparable from kinship, from the definition of who is legitimate and who is illegitimate, who belongs to whom. Every culture has kinship rules, and in most cases those rules are fundamental to the norms that define it. Thus challenging kinship entails challenging the basic structure of society. Butler cites recent work in anthropology that is doing precisely this. "Postkinship" studies in anthropology no longer situate kinship as the basis of culture but conceive it as one cultural phenomenon interlinked with other phenomena (2002, 250).

The turn to kinship provides the "critical challenge" that Butler is looking for. New kinship and sexual relations can compel a rethinking of culture itself. A more radical social transformation than that made possible by the legitimation of gay marriage is to refuse to allow kinship to be reduced to family or sexuality to marriage (2002, 254–255). Calling into question traditional forms of kinship displaces the central place of biological and sexual relations. New forms of association give sexuality a separate domain apart from kinship, allowing for durable ties outside the conjugal frame and opening sexuality to social articulations that do not always imply binding relations or conjugal ties. Sexuality outside the field of monogamy, she concludes, may open us to a different sense of community (2003, 206).

In *Undoing Gender* Butler goes into more detail on the radical potential of rethinking kinship. Once more citing "postkinship" studies in anthropology, she asserts that new kinship and sexual arrangements compel a rethinking of culture itself. When relations that bind are no longer traced to heterosexual procreation, then the homology between nature and culture is broken. This is the point of Butler's analysis in *Antigone's Claim* (2000a). Antigone is denied a livable life and, eventually, any life at all because her kinship ties are deemed illegitimate (Loizidou 2007). Kinship provides a unique link between nature and culture in human society. The biological relationships that define kinship are at the same time "natural" and at the root of cultural arrangements. Challenging kinship, thus, challenges the norms that define not only sexual relations but the realm of the human itself.

BUTLER, THE SUBJECT, THE MATERIAL, AND THE POSTHUMAN

That Judith Butler has been identified as the founding mother of linguistic constructionism in feminist theory, particularly as it relates to the subject, is established orthodoxy among feminists. She represents the "negative paradigm" in the subject debates that still occupy feminists. One of the purposes of the foregoing analysis has been to challenge this assessment. Particularly in her post–*Gender Trouble* writing Butler defines a subject who resists, who possesses agency, who participates in her subjectification. This subject is far from the social dupe that her critics have assigned to her.

A more important purpose of my analysis, however, is to explore the complexities of Butler's analysis and to suggest an affinity between her position and that of the "new materialists" as well as a significant connection to discussions of the posthuman. Like the new materialists, Butler integrates power and subjectivity, the material and the linguistic, into her theory. As Butler explores issues of resistance and agency, she does not abandon her constructivist position but, rather, integrates the discursive as an element, but not the only element, of her theory. This is particularly clear in *Frames of War* (2009), where she replaces "construction" with performativity to emphasize the material connection. The result is what she calls the "paradox of the subject." Subjects are both constituted and constituting, and the constituting is a result of the constitution. Subjects are acted upon and act; they embody elements of the discursive and the material.

The reading of Butler I have presented here has a notable resemblance to the "new materialism" of feminist theorists and certain philosophers of science. Feminist theorists such as Karen Barad and Nancy Tuana and the philosophers of science Bruno Latour and Andrew Pickering are articulating a theory that, while incorporating the insights of linguistic constructionism, also brings in the material (Alaimo and Hekman 2008, Hekman 2010). These theorists are doing the hard work of truly deconstructing the material/discursive dichotomy by refusing to choose one side or the other and, instead, integrating both elements. It is my contention that this is what Butler is doing with the subject. Her move to ontology is an attempt to articulate the "thereness" of the subject without relinquishing its discursive construction.

The result is a striking and strikingly complex understanding of the subject. Subjects are constituted by social norms, but this constitution is not complete. The performance of those norms opens up the possibility of resistance. Norms do not fully constrain subjects; there are openings—interstices—that can be exploited. Agency, likewise, is not antithetical to constitution. Agency is a product of constitution, not precluded by it. What this comes to is what Karen Barad calls an intra-action. The norms that constitute the subject give it its ontology, provide the possibility of a subject that constitutes itself. The material existence of the subject, a result of constitution, then becomes the ground of agency and resistance.

Although Butler has less to say about this material existence than the other authors who embrace this perspective, it is an important element of her analysis. Perhaps her most effective discussion of how the discursive, normative, and material intra-act in the constitution of the subject is her analysis of Venus Xtravaganza in *Bodies That Matter*. For Venus, gender is marked by race and class: "Gender is the vehicle for the phantasmatic transformation of that nexus of race and class, the site of its articulation" (1993a, 130). What *Paris Is Burning* suggests, Butler concludes, "is that the order of sexual differences is not prior to that of race or class in the constitution of the subject; indeed, the symbolic is also and at once a racializing set of norms, and that norms of realness by which the subject is produced are racially informed conceptions of 'sex'" (1993a, 130). All the elements in the mix that constitute Venus's subjectivity have both a material and a discursive component, but none can be isolated and analyzed apart from the others.[7]

Venus Xtravaganza's life—and death—also exemplifies Butler's understanding of resistance. Resistance only works for some of us, those who are recognized as subjects and who thus have an ontology, a being. For those of us, like Venus, who are denied personhood, denied a being by the norms that govern what is human and what is not, a livable life is, quite simply, impossible. It is my being as a subject that allows me to exploit the elements of my constitution to act as an agent and, most importantly, to resist. But if I am not constituted as a subject, this option is not open to me. My only hope is that the norms that exclude me from the human will change, will become more open and inclusive. Failing this, I cannot *be*.

Butler's discussion of kinship speaks to this possibility. Kinship is the central element defining who counts as a subject and who does not. It constitutes the material ground of subjectivity. If kinship can be destabilized, delinked from nature and biology, the definition of the human would correspondingly open up. But, as Butler freely admits, such a destabilization challenges the foundational norms of a society. As a consequence, the possibility of such change is very remote. What is important is that we recognize what is at stake here. Addressing the exclusion of some subjects from the realm of being must become our foremost political priority.

In *Precarious Life* and other works from this period Butler argues that we must challenge what it means to be human in order to allow the human to be something other than what is traditionally assumed. Although she does not cast this discussion in the idiom of the posthuman, there is a significant connection to this literature. Challenging the hegemony of the human strikes at the roots of the exclusions that constitute our cultural life. Butler characterizes this challenge as extending the concept of the human. It would be just as appropriate, however, to characterize it in terms of the erasure of the human/nonhuman dichotomy. If we challenge the hegemony of the human, the hierarchies that it creates would necessarily disappear. This applies not only to the hierarchy of the human and those excluded from being but also the privileging of the human over the nonhuman.

This connection between the new materialism and discussions of the posthuman is evident in the work of several writers who fall under the banner of the new materialists. Bruno Latour, one of the pioneers of the new materialism, argues that we should develop a politics that rejects the distinction between the human and the nonhuman. His "parliament of things" (1999) brings the human and the nonhuman together into one political unit. A similar impulse informs the work of writers such as Stacy Alaimo (2010) and Jane Bennett (2010). They call for an environmental ethics that does not privilege the human over the nonhuman. They argue that we must develop an ethics that encompasses all the elements of our world, not just the human. Butler's exhortation to extend the realm of the human, although cast in different terms, constitutes a closely parallel argument.

Where does this leave the subject? Challenging the privilege of the human allows us to bring in not only those excluded from being but also the entire realm of the nonhuman. As a consequence, our conception of the subject is radically altered. Subjects are constituted not only by discursive forces but by a whole range of other factors, among them the material. They are no longer clearly distinguished from the nonhuman world but are, rather, an intimate part of that world. In short, we have a radically new ontology of the subject that reflects a radically new world.[8]

NOTES

1. It is curious that Butler reverses her position on de Beauvoir's approach to sex/gender. In an article published in 1987 Butler claims that de Beauvoir, like herself and Wittig, takes the position that both sex and gender are fictions (1987, 134; 1986, 44–45). In *Gender Trouble*, however, Butler places Beauvoir, along with other social constructionists, in the camp of those adhering to the sex/gender distinction.

2. The strategy of rejecting the dichotomy also informs Butler's discussion of the materiality of the body. What is needed, she asserts, is to replace the "construction" with "deconstruction." It must be possible to claim that the body is not identifiable apart from the linguistic coordinates that establish it without claiming that the body is nothing other than the language by which it is known (2001a, 256).

3. For compatible interpretations of Butler, see Olsen and Worsham (2007) and Weeks (1998). For a conception of the subject similar to Butler's, see Smith (1988).

4. Chambers and Carver (2008, 93) claim that there is no ontological turn in Butler but that ontological questions always dominated her work.

5. Kathi Weeks (1998) defines her approach to the subject as pursuing a specific version of the ontological model of the subject located in a reading of Butler.
6. Lloyd (2008, 105) criticizes Butler's ontological assumptions, arguing that she does not sufficiently examine her position.
7. Butler's (1993b) discussion of Rodney King's beating is another example of the intra-action of the material and discursive in Butler's work.
8. For further discussion of these issues, see Hekman (2014).

WORKS CITED

Alaimo, Stacy. 2010. *Bodily Natures: Science, Environment, and the Material Self.* Bloomington: Indiana University Press.

Alaimo, Stacy, and Susan Hekman, eds. 2008. *Material Feminisms.* Bloomington: Indiana University Press.

Allen, Amy. 2008. *The Politics of Our Selves.* New York: Columbia University Press.

Bennett, Jane. 2010. *Vibrant Matter: A Political Ecology of Things.* Durham, N.C.: Duke University Press.

Butler, Judith. 1986. "Sex and Gender in Simone de Beauvoir's *Second Sex.*" *Yale French Studies* 72:35–49.

——. 1987. "Variations on Sex and Gender: Beauvoir, Wittig, Foucault." In *Feminism as Critique*, ed. Seyla Benhabib and Drucilla Cornell, 128–142. Cambridge: Polity.

——. 1990a. *Gender Trouble: Feminism and the Subversion of Identity.* New York: Routledge.

——. 1990b. "Performative Acts and Gender Constitution: An Essay in Phenomenology and Feminist Theory." In *Performing Feminisms: Feminist Critical Theory and Theatre*, ed. Sue-Ellen Case, 270–282. Baltimore, Md.: Johns Hopkins University Press.

——. 1993a. *Bodies That Matter: On the Discursive Limits of "Sex."* New York: Routledge.

——. 1993b. "Engendered/Endangering: Schematic Racism and White Paranoia." In *Reading Rodney King/Reading Urban Uprising*, ed. Robert Gooding-Williams, 15–22. New York: Routledge.

——. 1995a. "Contingent Foundations: Feminism and the Question of 'Postmodernism.'" In *Feminist Contentions: A Philosophical Exchange*, ed. Seyla Benhabib et al., 35–57. London: Routledge.

——. 1995b. "For a Careful Reading." In *Feminist Contentions: A Philosophical Exchange*, 127–143. London: Routledge.

——. 1997. *The Psychic Life of Power.* Stanford, Calif.: Stanford University Press.

——. 1998. "How Bodies Come to Matter." *Signs* 23:275–286.

——. 1999. *Gender Trouble.* 2nd ed. New York: Routledge.

——. 2000a. *Antigone's Claim: Kinship Between Life and Death.* New York: Columbia University Press.

——. 2000b. "Competing Universalities." In *Contingency, Hegemony, Universality*, by Judith Butler, Ernesto Laclau, and Slavoj Žižek, 136–181. London: Verso.

——. 2001a. "How Can I Deny These Hands and This Body Are Mine? In *Material Events: Paul de Man and the Afterlife of Theory*, ed. Tom Cohen et al., 254–273. Minneapolis: University of Minnesota Press.

——. 2001b. "What Is Critique? An Essay on Foucault's Virtue." In *The Political: Readings in Continental Philosophy*, ed. David Ingram. London: Basil Blackwell.

——. 2002. "Is Kinship Always Already Heterosexual?" In *Left Legalism/Left Critique*, ed. Wendy Brown and Janet Halley, 229–258. Durham, N.C.: Duke University Press.

——. 2003. "Global Violence, Sexual Politics." In *Queer Ideas*, ed. The Center for Gay and Lesbian Studies, 199–214. New York: Feminist Press.

——. 2004a. "Bodies and Power Revisited." In *Feminism and the Final Foucault*, ed. Dianna Taylor and Karen Vintges, 183–194. Urbana: University of Illinois Press.

——. 2004b. *Precarious Life: The Power of Mourning and Violence*. New York: Verso.

——. 2004c. *Undoing Gender*. New York: Routledge.

——. 2005. *Giving an Account of Oneself*. New York: Fordham University Press.

——. 2009. *Frames of War: When Is Life Grievable?* New York: Verso.

Chambers, Samuel, and Terrell Carver. 2008. *Judith Butler and Political Theory: Troubling Politics*. New York: Routledge.

Hekman, Susan. 2010. *The Material of Knowledge: Feminist Disclosures*. Bloomington: Indiana University Press.

——. 2014. *The Feminine Subject*. Cambridge: Polity.

Latour, Bruno. 1999. *Pandora's Hope*. Cambridge, Mass.: Harvard University Press.

Lloyd, Moya. 2008. "Towards a Cultural Politics of Vulnerability." In *Judith Butler's Precarious Politics*, ed. Terrell Carver and Samuel Chambers, 92–105. New York: Routledge.

Loizidou, Elena. 2007. *Judith Butler: Ethics, Law, Politics*. New York: Routledge-Cavendish.

McNay, Lois. 2000. *Gender and Agency: Reconfiguring the Subject in Feminist Social Theory*. Oxford: Polity.

Olson, Gary, and Lynn Worsham. 2007. "Changing the Subject: Judith Butler's Politics of Radical Resignification." In *The Politics of Possibility*, ed. Gary Olson and Lynn Worsham, 5–42. London: Paradigm.

Pickering, Andrew. 1995. *The Mangle of Practice: Time, Agency, and Science*. Chicago: University of Chicago Press.

Smith, Paul. 1988. *Discerning the Subject*. Minneapolis: University of Minnesota Press.

Thiem, Annika. 2008. *Unbecoming Subjects: Judith Butler, Moral Philosophy, and Critical Responsibility*. New York: Fordham University Press.

Weeks, Kathi. 1998. *Constituting Feminist Subjects*. Ithaca, N.Y.: Cornell University Press.

II

ORGANIC RITES

4

THEREFORE, THE ANIMAL
THAT SAW DERRIDA

AKIRA MIZUTA LIPPIT

Jacques Derrida's last interventions brought him to the threshold of the animal world, a horizon he made visible in his final seminar but that appears throughout his work from the beginning. This expanse is there already, everywhere, throughout Derrida's oeuvre. Derrida's address brought renewed energy to the nascent field of animal studies, a loose aggregate of disciplines organized around the status of animals in ethics and philosophy, law and literature, gender studies and visual culture, to name only a few fields. Derrida's contributions made possible the transition to new frontiers in the humanities, toward an extended critique of the humanities and a deconstruction of the figure of *man* at its core.

But beyond the attention that he brings to the question of animals and the rhetoric of animality and the many lines of inquiry his thought makes possible, Derrida's analyses are complex and extensive and cannot be reduced ultimately to a discourse on animals: The unfinished seminar remains to be thought through to an end not yet achieved. Derrida's seminar *The Beast and the Sovereign* (*La bête et le souverain*) intersects with numerous archives from classical philosophy to critical theory, fable to fiction, ethics to politics, psychoanalysis to gender studies, among others. In the end—if such an end exists or were to arrive—Derrida's work on animals and animality exceeds the very subject it refuses to exceed. This is its remarkable rhetoric and arithmetic: Without excess, without exceeding another, there is still always there, yet another. This other does not arrive in excess, nor is it essential, but rather it marks the possibility of a self realized in the eyes of another. Derrida discovers this supplemental logic of supplementarity in his seminar on animality. His discourse finds a beginning in the animal, in animals from which another subject or trace of a subject emerges, a figure that follows the animal and from which the animal becomes visible. Without exceeding, which is to say sacrificing, the original figure. (And such origins are not originary, simply before.) The second figure that Derrida imagines is, among many things, a figure for visibility as such; it enables a sustained reflection on the question of worlds and islands, being and solitude, and the ability to be alone with another.

Derrida's first articulation of a subject that nonetheless roams his work from the beginning took place in 1997. In July of that year, Cerisy-la-Salle in Normandy, France, hosted a *décade*, a ten-day conference around the work of Jacques Derrida. Derrida participated, providing the event's title, "The Autobiographical Animal," *L'animal autobiographique*. The combination of terms suggests a paradoxical erasure of each individual term in the phrase. What do animals have to do with autobiographies? What animals have written (could or would care to write) an "autobiography"? And what is an autobiography? Does it have any meaning outside of the ways in which human beings use it? Which is to say, can an animal or any other nonhuman being write or produce an autobiography? One would think the practice of autobiography, the very notion of autobiography, is reserved for human beings. Is there any other animal besides the human being that acts autobiographically? Autobiography is, one could say, the very essence of what makes one human, what makes an animal human—the capacity for autobiography. If so, if indeed the capacity not only to write but to write oneself—to write of and about, upon oneself— is reserved for human beings, then isn't the "autobiographical animal" merely another name for the human animal or human being? Isn't then the autobiographical animal a euphemism or pseudonym for human beings, like the expression "political animal," as Derrida later notes? Another name for human beings that have become human on the occasion of an autobiographical act, on the completion of a task one would call autobiographical? (In this sense, isn't all human life autobiographical in nature? Bare life becoming human in the reflective gesture of autobiography? Isn't this notion of human life a solipsism, the necessary solipsism by which humanity is established in the first instance?)

In his own contribution to the event, "The Animal That Therefore I Am (More to Follow)" (*L'animal que donc je suis* [*à suivre*]), Derrida does indeed speak of himself, but his presentation or seminar is not devoted to the metamorphosis of animal beings into human beings, animality into humanity, through self-reflection or self-inscription, through autobiography, but rather to a certain economy of animal being alongside or before human being. It is about animals to the extent that animals are about him, around him, before and after, here and there, beside him. He follows the animal, the question of the animal, and of animals—the question concerning the speech, lives, senses, beings, and rights of animals—to a place before him (Derrida), to the place before which Derrida finds himself naked, stripped bare before the animal (in this instance his cat), which looks at Derrida "since time," he says, "therefore" (Derrida 2008, 3). Since the beginning of time, before time, therefore the animal looks with the "gaze of seer, a visionary or extra-lucid blind one" (4).[1] It sees me, says Derrida, naked. This animal (Derrida insists on the singularity of the cat that sees him; neither figure nor metaphor, neither literary nor celebrity), this single, singular cat that sees Derrida knows nothing of its own nakedness.[2] "It is generally thought," says Derrida,

> Although none of the philosophers I am about to examine actually mentions it, that the property unique to animals, what in the last instance distinguishes them from man, is their being naked without knowing it. Not being naked therefore, not having knowledge of their nudity, in short, without consciousness of good and evil.
>
> (3–4)

"From that point on," Derrida concludes, "naked without knowing it, animals would not be, in truth, naked" (5). Nudity entails a self-awareness; to be naked one has to be aware of the distinction between being covered and uncovered, or discovered. It is, says Derrida, a matter of technics, of *technê*. At stake in nudity, in one's ability to be naked, but even more so to be aware of one's own nudity (the very precondition, according to Derrida, of nudity), is the consciousness of good and evil. Hanging in the balance of one's self-awareness, of the ability to see oneself as oneself, one's true self in the nude, is the consciousness of good and evil itself. A primal ethics rendered visible, a visual ethics that originates in one's naked self.

It fills Derrida with shame and provokes in him an epiphany: He sees himself in the eyes of his cat, this cat, therefore, before him, a sight that moves him to embarrassment.

But the visual economy that Derrida discovers in the eyes of his cat, the awareness of himself, and with it of shame and the "consciousness of good and evil," is not reflective in a superficial sense. This cat before him and before whom Derrida stands is not a mirror, a reflective surface against which Derrida finds himself naked. Or rather, Derrida introduces a species of mirror that precedes the cat and that stands before him and her, a "primary mirror" in which she too (this cat) is reflected beside him. The cat is not erased in reflection but rather is the very occasion for reflection, the medium and apparatus of reflexivity itself, *psyché*. Derrida's cat is a mirror that reflects itself, a narcissistic, autobiographical mirror:

> Henceforth I shall reflect (on) the same question by introducing a mirror. I import a full-length mirror [*une psyché*] into the scene. Wherever some autobiographical play is being enacted there has to be a *psyché*, a mirror that reflects me from head to toe. The same question then becomes whether I should show myself but in the process see myself naked (that is, reflect my image in a mirror) when, concerning me, looking at me, is this living creature, this cat that can find itself caught in the same mirror? Is there animal narcissism? But cannot this cat also be, deep within her eyes, my primary mirror?
>
> (50–51)

The cat is a primary mirror that reflects herself along with Derrida. This reflection that reflects the self along with another signals the critical economy of autobiography that Derrida seeks to elaborate. *I reflect you along with, beside me; I neither efface you nor myself in this reflection together. I am able to see myself, to show myself to you, who reflect me and you in a primary mirror.* Yet this reflection provokes shame in Derrida, a shame that comes from the simultaneity of showing himself and seeing himself, seeing himself show himself in the eyes of another, perhaps. In this scene, the cat remains. His cat is neither consumed nor negated (sublated) in the exchange that gives Derrida (over) to himself, in shame. One should pause for a moment on the feeling of shame that Derrida admits to before his cat; it is not necessarily the shame of a disgraceful or dishonorable act, from the Greek *aiskhyne*. The context suggests otherwise; his idiom points elsewhere. The Greeks distinguished this notion of shame from *aidos*, modesty or bashfulness. Following the latter conception of shame, one can imagine Derrida's modesty in discovering himself in the eyes of his cat, discovered by his cat, disrobed or uncovered by a cat that generates in Derrida, perhaps for the first time, an awareness of himself, naked, before the eyes of another. It is the self-realization as much as anything, the realization or actualization of oneself before, therefore, another that fills one with shame. The self-awareness brought about by the gaze of a cat brings with it a sense of embarrassment, shame. Is it that cat or the awareness that leads Derrida to blush?

Who is the subject in this encounter? Derrida's scene describes an experience of subjectivity, one's own subjectivity in the eyes of another, of the other, of an animal. The scene is familiar, and variations of this encounter have been described by Nietzsche (the happy, forgetful cows) and Adorno (the dying animal), among many others. In vastly different contexts, Jean-Paul Sartre (*Les mots*), Jacques Lacan (*The Four Fundamental Concepts of Psychoanalysis*), and John Berger ("Why Look at Animals?") have described the economy of self-discovery before the gaze of another: I see myself, I am *photographed*, says Lacan, in the gaze of another that inscribes me within a scene from which I am otherwise excluded. (One is also reminded of the epiphanies of self-discovery that James Joyce portrays in *Dubliners*.) *I am put into the picture of my own existence by another—into the topos or* locō *of a self that I do not see, that I did not see before. I see myself for the first time in being seen. I am seen, therefore, I see (myself). Before myself, I see myself only after.* (I am, in this scenario, therefore an afterimage; I follow myself, therefore I am.) One might call this logic *locomotive*, an emotive as well motional, emotional logic of space, or *locō*. A madness of space and movement, locomotion. Before the other that sees me and renders me naked, I find myself, ashamed. Still who is the subject of this discovery, the one that sees—animal, other—or the one that discovers oneself in the scene, the one that is discovered, uncovered in the picture, me?

The scenes that Sartre (the scene at the keyhole) and Lacan (the sardine can in the water and "Petit Jean") describe, like those imagined by Blanchot or Levinas ("*il y a*," "there is"), inscribe a structural, one might say architectural logic of the gaze. One is seen by another who is nonetheless absent. That is, the logic of the gaze, of an awareness or shame in the eyes of another does not require the presence of an actual other being there, before me, therefore. In the economy of the gaze, one is seen by the place of the other. Lacan makes this distinction between seeing and the gaze clear in his seminar on this subject. Only a being can see, but the gaze is a phenomenon that exceeds the dimension of individual beings: I feel myself to be seen, I am made aware of myself before another that may or may not be there. The feeling or emotion (self-awareness, shame) is entirely within me, even if the psychic economy that brings me to this realization circuits throughout the outside, projecting the figure or specter of another, there, therefore, before me. In the ordinary logic of the gaze, I am the subject of a projection; I project the figure of another there, who sees me here. I imagine the other that sees me, renders me, strips me of everything other than me. Because you are there, I am. You, therefore, I am. You are a projection there, a specter from which I am projected here.

In this passage of the gaze, I am photographed, inscribed in a scene that I do not see (at first); I become visible to myself by imagining my visibility to another, to you. The logic describes another solipsism, another closed economy of specters and subjects by which I enter into the space or spaces, the loco of my own being. A second person appears. *I am written through you, I write myself through you, before you, therefore.* The movement through another under erasure, the passage through the night, through obscurity (Levinas) on my way to a self-discovery and realization, engenders a dialectic of being and of becoming, the means by which I become who I am, by which I produce before the other, in the eyes of the other, my autobiography. *In the gaze, in the phantom fields of vision it extends, you are my projection. And in this projection, I see myself; I give myself over to myself in your eyes, in the eyes I imagine. You author my autobiography, authorize it.* And this second person is not necessarily human, nor is it not necessarily not me.

Before the animal, before the cat that sees me, I am also a second person seen, in this scene, there, therefore.[3] A cat who is there, now, and therefore not only "there is." It is a mobile cat, one that moves here and there, neither transfixed nor permanent, but a dynamic, mortal, vital cat that occupies space, that shares space with others here, there, all around. "The animal," says Derrida,

is there before me, there next to me, there in front of me—I who am (following) after it. And also, therefore, since it is before me, it is also behind
me. It surrounds me. And from the vantage of this being-there-before-me
it can allow itself to be looked at, no doubt, but also—something that philosophy perhaps forgets, perhaps being this calculated forgetting itself—it
can look at me. It has a point of view regarding me. The point of view of
the absolute other, and nothing will have ever given me food for thinking
through this absolute alterity of the neighbor or of the next (-door) than
these moments when I see myself naked under the gaze of a cat.

(11)

For Derrida, the animal, this cat, is not only there before him; it surrounds
him. It moves around him and in so doing establishes a point of view or,
rather, points of view. Surrounded by animals, Derrida takes place within
this zoography parenthetically. The perspective of this cat is not one but
many; it is, says Derrida, the perspective of absolute otherness. How does
one differentiate the architectural perspective of the gaze, the production of
phantom subjectivities and their perspectives from the living, mobile, and
actual perspectives of a cat or an animal, living perspectives that are, in the
end, equally distant, inaccessible and, in the logic of each formation, absolutely constitutive? What is the difference between the gaze as such and the
gaze of such and such an animal, between an abstraction and a living being
that discovers me? In whose discovery of me do I discover myself?

Derrida responds, as it were, to a series of questions he poses to himself, questions that assume a pose, naked before another to whom he feels
responsible: animal.[4] He says,

What is at stake in these questions? One doesn't need to be an expert to
foresee that they involve thinking about what is meant by living, speaking, dying, being, and world as in being-in-the-world or being-within-
the-world, or being-with, being-before, being-behind, being-after, being
and following, being followed or being following, there where *I am*, in
one way or another, but unimpeachably, *near* what they call animal. It is
too late to deny it, it will have been there before me who is (following)
after it. *After* and *near* what they call the animal and *with* it—whether we
want it or not, and whatever we do about this thing.

(11)

At stake in what "they call animal" is a spatial relation to being, not any being or being as such, but my own being. In some irreducible manner, whether following, near, behind, or after, I am always before the animal, therefore animal. I am myself with an animal, inscribed in the space not of solitude but of being with an animal. *Before and there before, therefore, autobiography.*

But further questions emerge, remain, come before the law of the second person that Derrida invokes. Why call this companionship, this being-before or -with, autobiography? Aren't autonomy, sovereignty, and even solitude preconditions for autobiography? Why is the animal, this or that animal before me, so critical to my being, to my ability to say, "I am"? Isn't the fantasy of autobiography related intrinsically to other fantasies of self-description and inscription, self-discovery and recovery? Of autosovereignty and self-determination? And of being naked before oneself, of being other to oneself, of seeing oneself from the vantage point of another, and of addressing oneself to oneself as if to another? And if so what is the animal, an animal, doing in this self-reflective space? Why has an animal been mobilized here, before me, coming between me and the specularity with which, through which, I recognize myself? Doesn't the animal in fact interfere with the alterity I invoke in the instance of my autobiography, with the ideal otherness through which I discover myself autobiographically? The logic is complex, demanding, and at times para-doxical. Why has Derrida placed an animal within the scene of autobiog-raphy, and which animal—the one that *he is* or the one that *he follows*—is the autobiographical animal? It is the animal that sees me naked and fills me with shame—and this, Derrida suggests, is the scene of writing, an obscene of writing, one is tempted to say, the moment therefore that I am able to see myself naked. It is not the apparatus of the gaze but an instance or event of otherness, of seeing, of being seen by an other that is there before me and therefore neither provocation nor projection but the alterity that is given to me, to see. When an animal appears before me, when I appear before it, I see with the absolute clarity of shame my own obscurity.

I see, like Blanchot's Orpheus, at the very limits of my vision, the very limits of visibility itself. At the end of vision, at its limit, endless vision appears. It is a bottomless vision, to use Derrida's idiom, which exposes the limits of visibility, of vision, and in this limit, humanity as such.

As with every bottomless gaze, as with the eyes of the other, the gaze called "animal" offers to my sight the abyssal limit of the human: the in-human or ahuman, the ends of man, that is to say, the bordercrossing from which vantage man dares to announce himself to himself, thereby calling himself by the name that he believes he gives himself. And in those moments of nakedness, as regards the animal, everything can happen to me, I am like a child waiting for the apocalypse, *I am (following) the apocalypse itself,* that is to say, the ultimate and first event of the end, the unveiling and the verdict.

(12)

Drawn into the bottomless gaze unleashed by an animal, the human being calls *himself by the name he believes he gives himself.* At the limits of the visible, at its threshold, the human being before the animal becomes autobiographical. An apocalypse opens from this bottomless gaze, an unveiling of human being that "dares to announce himself to himself," as Derrida says, an end of humanity at the very moment of its beginning. A revelation, uncovering that serves as both the beginning and end of humanity before the animal, before and after. Everything is given and destroyed in this apocalypse of the bottomless gaze at the limits of visibility. Everything is given and destroyed, clarified and obscured, inscribed and described in the obscure scene of writing that Derrida invokes before the animal. *I am revealed here, exposed, discovered, and uncovered before the animal there, therefore, I am.* I am not alone, apparently, in the scene of (my own) autoinscription. I neither write alone, nor live alone, ever.

The problem returns to the question of autobiography, this most human of activities. Is the very notion of autobiography a uniquely human conceit, a phantasm of humanity designed precisely to distinguish human being from other animals? Along with language and the capacity for death, the ability to say "as such" (Heidegger) and to die death rather than merely perish (Epicurus to Heidegger), is the ability to call oneself by the name that one believes proper to oneself—to give oneself in language life, one's own life—a feature that separates human from animal being? Following Descartes's skeptical solipsism, isn't the logic that leads from thinking to writing the foundation for saying "I am" the very precondition of being itself? It determines an act that must be undertaken alone, in the certitude of absolute solitude, yet is it, in the end, even possible?

The dilemma is this: How does one write alone, to and of oneself? Is such a writing, a self-inscription, a solitary writing—an autobiography—possible? Is a writing destined to oneself, a writing of and about oneself that renders a self, possible? Doesn't every writing invoke another, a specular, spectral addressee? Isn't every act of writing achieved in the second person? And might one consider the very notion of address, its very idiom in opposition to dress? Addressing as opposed to dressing; the addressee as the one who is undressed, *a-dressed*, as it were, naked, like Derrida, the recipient of a look that comes to one autobiographically and in shame? Isn't such an economy of writing already prescribed in its technics? In *Of Grammatology*, Derrida chides Saussure among others for having posited writing as the *clothing* of speech, as "matter external to the spirit, to breath, to speech, and to the logos": to living thought and the thought of life. "But has it ever been doubted," says Derrida, "that writing was the clothing of speech? For Saussure it is even a garment of perversion and debauchery, a dress of corruption and disguise" (1976, 35). How does one address, redress, undress the very debauched veiling at work in every act of writing? How to *address* oneself in writing, to remove the clothing and stand naked before oneself and before writing? Where does one undertake such writing, such address, such solitude? Where does it take place, and what sort of place does it take? Following Derrida's logic, a logic that is exemplarily deconstructive, autobiography is possible only before another, the other, but a different or other other that comes before, that moves before me into the space of my autobiography. *I am alone, able only ever to be alone, before the other that arrives before me, in front of me, on the occasion of my solitude.* For Derrida, this other is an absolute alterity, unlike any other in the singularity of its idiom, in the singularity of its proper name, designated in language by a cryptonym, a word that is not a word, a singular word given to designate a multiplicity that cannot be one, or said otherwise, therefore, "*l'animot.*"

L'animot, for Derrida, is the word for the word that human beings have given to animals, the collective singular name given to every form of animality other than human. "The animal," he says, "what a word!" It is, he continues, "an appellation that men have instituted, a name they have given themselves the right and the authority to give to the living other" (2008, 23). The authority to name the other, to give the other a designation or appellation, establishes at once the convergence of authority and authorship and in this condensation, the production, the projection of a singular and phantasmatic other at the scene of writing: a scene occupied by an author and an other. "Men would be first and foremost," says Derrida, "those living creatures who have given themselves the word that enables them to speak of the animal with a single voice and to designate it as the single being that remains without a response, without a word with which to respond" (32). In the linguistic economy that Derrida invokes, it is not only the condensation of many to one that is achieved by the name "animal," not only the reduction of multiplicity, diversity, and difference to sameness but also the organization of language, of the subject of language or of enunciation to a single authority. In saying "the animal," it is not only the object of this signifier that is reduced to one; it is the subject that speaks (it) as well. When I say "the animal," I give to myself a singularity of voice and subjectivity that is, to follow the linguistic idiom, performed on the occasion of my utterance. I become one, before the animal, when I say "the animal." The animal, therefore, I am.

In order to make visible the dilemma and violence inherent in the elimination of difference between subjects and objects (between but also within each), Derrida brings forth a neologism, transposed from French, "*l'animot.*" It is a homophonic word that sounds like the plural of animals in French, *les animaux,* but is spoken in the singular, "*le*" *animot*; it also fuses the word for "word," *le mot,* to animals, a word-animal, animal word, or the word for animal that engenders an animal-machine or *technê* that traverses life and language, bodies and words. Derrida explains his coinage:

> I would like to have the plural *animals* heard in the singular. There is no Animal in the general singular, separated from man by a single, indivisible limit. We have to envisage the existence of "living creatures," whose plurality cannot be assembled within the single figure of an Animality that is simply opposed to humanity.
>
> (47)

It is, like so many of the homophones that Derrida delivers, a visual economy of language that vanishes as it is spoken. A *différance* that can only be seen and that disappears on the occasion of its utterance. A trace that moves between the graphic and phonic, bound by no one economy between them. L'animot is linked in Derrida's thought intimately to the question of autobiography. To write of myself is to exist before an image of an other; an other that appears only in the form of an image, a grapheme. I am alone only when I am with such spectral companions. What unfolds throughout Derrida's account of the animal that he follows is the confusion between being, être, and following, suivre. Conjugated in the first-person singular, both words become suis. Being as following; to be is to be behind, in pursuit of another. Hunting, one is tempted to add. The logic is dynamic: I am or become what I am by following (another). The scene of writing, my scene of writing and the scene of my writing, its cogitatum, is marked by the presence of another who appears like a trace before me and in whose pursuit I become what I am. In following another, I discover myself after the other. And in following an other that precedes me to the scene of my own writing, I come into myself as the trace of this other. My autobiography is achieved in the shadow of another, and so, at the scene of my own writing, in the instance of my autobiography, I am obscure. A second person.

I find myself there, there before myself alone in the singular-plural orders of subjectivity and solitude, covered, covered over, concealed from view, obscure.[5] Among the various etymologies of the word obscure is the Greek skeue, "dress." What if the state of dress, of being dressed or concealed, covered over and obscured from view, was not only a matter of technics, as Derrida suggests, of a technicity proper to human beings, but was originary? What if human beings are not the only animal to dress itself but rather the only animal that appears in the first instance already dressed? One could say that human beings are born concealed, born obscure, and only achieve visibility later, after another. About clothing and its relation to human beings, Derrida says: "In principle, with the exception of man, no animal has ever thought to dress itself. Clothing would be proper to man, one of the 'properties' of man. 'Dressing oneself' would be inseparable from all the other figures of what is 'proper to man,' even if one talks about it less than speech or reason, the logos, history, laughing, mourning, burial, the gift, etc." (5). Perhaps the need to dress oneself, the act of concealing or covering oneself, is not only an effect of shame (before another human being or animal), of embarrassment at one's one nakedness, but the act of making visible an originary condition: that I am born obscure.

If a primary condition of humanity, of being human, was obscurity, then the moment of autobiography, of apocalyptic self-revelation, would always come in the form of undressing before another. The scene of writing, like Derrida's scene before his cat, is a primal scene of self-disclosure, of autobiography. I undress therefore I am; I undress before you, therefore I am. But this is not exactly right either. I am undressed by you, therefore I am (before you, naked, here, at last, myself). In this sense it is not only that a cat finds me naked, filling me with a sense of shame, self-awareness, and self-consciousness (isn't this already the essential first step of autobiography?); the cat undresses, discovers, and uncovers me in its gaze. It is in this power of the gaze (uninflected, one must insist, by violence, by sexual desire, for example) that I am able to step forth from my essential obscurity and see myself. My autobiography is made possible through this mobilization of the gaze, through the paradoxical plurality that engenders a self, myself.

Autobiography exists for Derrida, in movement, in motion or locomotion, in the form, he says, of "an immunizing movement." "Autobiography, the writing of the self as living, the trace of the living for itself, being for itself, the auto-affection or auto-infection as memory or archive of the living, would be an immunizing movement" (47).[6] Autobiography, in this view, is a toxin or virus (a "poison," Derrida says) introduced from the outside as a means of establishing a self, a body, this body before another antibody. An immunization against the other but also against oneself, myself, against the other that is already in some irreducible manner "I." In the scene of Derrida's body, a cat serves this immunizing role; the immunizing gaze of his cat, who sees Derrida naked and fills him with shame, serves as the precondition for Derrida's autobiography. Before the cat, therefore, I am. Before the cat I already am (follow). But in this locomotive scene of autobiography, the immunizing movement, who comes first, and who comes after whom? Do I stand before another that summons me, another whom I follow? (Levinas) Or do I name the other before me, bringing this other into my presence, sharing, as it were, the present with another, the gift of presence, the present of presence? When I see, for example, an animal, do I make myself known to it, or it to me? When we come before each other, which comes first before the other? Which, therefore other?

Everything is also at stake in this priority, in this primacy and temporality. Who enters the space of the present, its *locō* first? (The space of movement as well as of madness, if one hears both idioms of the word here.) Can one imagine a space and perhaps also a time of absolute simultaneity, like the middays that Nietzsche dreamed of, an ethics perhaps of time and space in which I am before an other who is also before me (at once)? Without priority, without precedence, another who is beside me and beside whom I am beside myself? What would such a place, such a *topos* or *locō* look like? Could one imagine such a place parenthetically, a *jouissance* beside the other?

The economy that forges seeing with writing passes, for Derrida, through the body—phenomenon or phantasm—of the animal. Through the bodies of animals en route to me. To write autobiographically, to inscribe oneself and consign a self to writing, a subject of writing, demands that one stand before an other, an animal other through which the forces of seeing and of visuality intersect with those of language and writing. Derrida raises this issue of autobiography, of the most human and humanizing of traits vis-à-vis the animal. "It happens," he says, "that there exist, between the word *I* and the word *animal*, all sorts of significant connections" (49). Distinguishing between the autobiographical capabilities, indeed necessities of human beings, and the more instinctive tracking and self-tracking that animals are thought to achieve, Derrida says:

> No one has ever denied the animal this capacity to track itself, to trace itself or retrace a path of itself. Indeed the most difficult problem lies in the fact that it has been refused the power to transform those traces into verbal language, to call to itself by means of discursive questions and responses, denied the power to efface its traces.
>
> (50)

For Derrida, there would be no self-inscription, no autobiography, no possibility even of autobiography without the presence (and absence) of an animal. Without the presence of the animal I am, that I was before and will be yet after, I inscribe myself in the place where I am, where I exist. Before the animal, therefore I am. Autobiography will have been the act of placing oneself, inserting oneself in the place where one ought to be, where one needs to be, there, beside and before another. It is the self-inscription of company or, to use Donna Haraway's idiom, of companionship. I write myself there, where I am, I inscribe a life, my life, and ascribe it to myself, here, before another, before the law of the other. To right oneself is also to write oneself, ethics and writing, positioning and inscribing brought together in the space of autobiography parenthetically. (Late Latin, *insertion of a letter or syllable in a word*, from Greek, from *parentithenai*, to insert: para-, *beside*; see para- + en-, *in*; see en in Indo-European roots + *tithenai*, to put; see *dhē-* in Indo-European roots.)

Before the animal, before and after it, therefore, I am with an animal, cotemporaneous, exposed to it as I am to myself—only exposed to myself, visible to myself because I am to it—parenthetically. To put in parenthesis is to put beside and inside, to be beside another inside. To be inside beside another, together, before and after oneself. The parenthesis establishes a unique spatiotemporal logic and economy: at once, which is to say simultaneously, and in sequence, one after the other. The law of parentheses is to be always at once and in sequence, before and after, and all around. For Derrida, the animal is there, animals are there but before them (I follow them), I write myself, I discover myself as the autobiographical animal. I am no less autobiographical before the other, before an animal, but I do not consume this animal in my autobiography; I am reflected in it. I write myself before the other, with the other, in the presence of the other. This is the only possibility of autobiography; I do not negate the other (in desire), nor do I deny the animal subjectivity (for Lacan, the animal cannot be the subject of the signifier). Yet I do not yield to the other either (Levinas); I do not imagine my existence as pure or only responsibility (to the call of the other). I am there, alongside, beside, before, surrounded by an other, animal.

The writing I do alone, the autobiographical writing, which is at once a solitary writing, a writing of solitude, and the writing of myself alone, is never, according to these insights, made possible by deconstruction, singular. Solitude, my solitude, is constituted always before another—not as an effacement, erasure, exclusion, consumption, or sacrifice—but in the presence of another who is there, beside me (around me) parenthetically. Autobiography is not a mode of writing that merely invokes an other, the other, that produces in the sense that Levinas, and to a certain extent Blanchot, imagined, a responsibility (to and from the other). It is not a genre that demands an audience, witness, reader; it is simply not possible without, in the first instance, another who is there, therefore, from the beginning with me. An other is there without being there in the form of a subject. It is there, he or she, with me and before me; *I am* to the extent that I follow this other, therefore me. It is a being that is not constituted or inscribed in writing: a trace, specter, shadow of another, a being that appears without entering into an economy of language—animal, *l'animot, animetaphor*. This scene of autobiography before the other animal must not be seen as a sacrifice, the other consumed (in and by difference) for the sake of a singular self. One must imagine, as Derrida suggests, the transvaluation not only of animal life, of animal being, not only of my relation to and understanding of animals, but in fact of autobiography. No longer an exemplarily solipsistic gesture, neither dialectical nor auto-motive, autobiography, the self-inscription of my life, perhaps of human animal life itself, begins already with a primal difference. I am in, in the first instance, other, obscure, concealed; only before the other, only by following the other, do I become who I am, giving myself over (to myself), exposed and uncovered. There alone, I am always at least two.

In virtually every contemporary invocation of the animal, figures and beings, animality comes to signify the site of an originary difference—the consciousness of good and evil, seeing and writing, life and technology (*physis* and *technê*), author and other, self and other. Each term of the dialectic mediated, as it were, by the figure of the animal. The very differ-ence of life itself appears to open up across the divide of human and ani-mal being. If the animal—signifier, figure, and living being—has served to establish the existential lines of a primary, primal difference, a primal scene of differentiation from which difference is instituted, then it is per-haps in the field of autobiography that this difference is deconstructed. Denied the capacity for transforming its tracks and traces into language, and in turn the capacity to efface these same tracks and traces, the ani-mal is incapable of transforming these into what is called conventionally autobiography. But what if, as Derrida suggests, there is no autobiography without the animal, no possibility of an autobiography that does not come before the animal, before an animal? Linked through some parentheti-cal frame, by an economy of radical antitheses (yet not a dialectic) that engenders a spectral, spatial, and causal reciprocity: the animal autobio-graphical, the autobiography animal? Animal therefore autobiography, autobiography therefore animal. Could one envision the profound het-erogeneity of an animal autobiography not as the eccentric result of a mad hybridity or experiment but as the very condition of autobiography as well as animal? The ability not only to write one's traces into language but also to efface them, as Derrida says; revealing a crucial condition of auto-biography as the ability also to erase oneself, to disappear before the other, to enter into a space of irreducible obscurity, fully naked and exposed? Would this not also be the task of autobiography revealed by animals to human beings who believe they call themselves by the names they have given themselves, the act of losing oneself before the other, forgetting one-self, one's own name, the name proper to oneself, myself, before the other? Isn't therefore the autobiography destined, in its completion to produce before the other, therefore the other, my own obscurity? Wouldn't my autobiography always take place before the other, animal other, obscure in its nakedness, therefore animal?

(And wouldn't then the autobiography always be destined to another, *addressed* to an other, and designed to name in an absolutely singular idiom the other's name? A proper name other than my own uncovered in my own. A second name or name before the name, according to Derrida's logic of autobiography, is therefore a name that cannot be one, that can never adhere to a single individual. Instead, a singular plural [or plural singularity] that follows. I become a second person, therefore, you.)

NOTES

1. "Since so long ago, can we say the animal has been looking at us?" (Derrida 2008, 3).

2. "I must immediately make it clear, the cat I am talking about is a real cat, truly, believe me, *a little cat*. It isn't the *figure* of a cat. It doesn't silently enter the bedroom as an alle-gory for all cats on the earth, the felines that traverse our myths and religions, literature and fables" (Derrida 2008, 6).

3. One should write a history of cats in cinema, as Jean-Claude Lebensztejn has written on cat music, *Miaulique* (2002): the ambiguous cat among the adoring Nazis that looks down on Hitler's motorcade as he makes his way toward the 1934 Nazi Party Congress in Nuremberg in Leni Riefenstahl's *Triumph of the Will* (1935), or the obscure cat Kitch that watches and blocks the view as Jamey Tenney and Carolee Schneemann make love in *Fuses* (1967), or the many cats of Chris Marker's worlds, to name only a few.

4. Elsewhere, Derrida (2008, 23) makes explicit the multiple registers of posing, from hypotheses to nudity, postures to positions. "I'll attempt the operation of disarmament that consists in *posing* what one could call some hypotheses in view of theses; posing

them simply, naked, frontally, as directly as possible, *pose* them as I just said, by no means posing in any way one indulgently poses by looking at oneself in front of a spectator, a portraitist, or a camera, but 'pose' in the sense of situating a series of 'positions.'"

5. *OED*: c.1400, from O.Fr. obscur "dark, dim, not clear," from L. obscurus "covered over, dark, obscure, indistinct," from ob "over" + -scurus "covered," from PIE *(s)keu- "to cover, conceal;" source of O.N. sky, O.E. sceo "cloud," and L. scutum "shield" and Gk. skeue "dress" (see sky). The verb is first recorded 1432. Obscurity is attested from 1481 in sense of "absence of light;" 1619 with meaning "condition of being unknown." Obscurantism (1834) is from Ger. obscurantismus (18c.).

6. He continues the thought: "But an immunizing movement that is always threatened with becoming auto-immunizing. . . . Nothing risks becoming more poisonous than an autobiography, poisonous for oneself in the first place, auto-infectious for the presumed signatory who is auto-affected" (Derrida 2008, 47).

WORKS CITED

Derrida, Jacques. 1976. *Of Grammatology*. Trans. Gayatri Chakravorty Spivak. Baltimore, Md.: Johns Hopkins University Press.

——. 2008. *The Animal That Therefore I Am*. Ed. Marie-Louise Mallet. Trans. David Wills. New York: Fordham University Press.

5

THE PLANT AND THE SOVEREIGN

Plant and Animal Life in Derrida

JEFFREY T. NEALON

It's a little bit odd that Jacques Derrida has become such a linchpin figure in the field of animal studies. Indeed, it's ironic (though not I think inaccurate) to suggest that if animal studies publications continue coming out at their present pace, Derrida will be remembered a decade from now primarily as a thinker of animals rather than as a theorist of *écriture* or a deconstructor of Western metaphysics. I suppose there's nothing particularly strange in the fact that his late essays, collected in *The Animal That Therefore I Am* (2008 [French 2006]), became a jumping-off text for Cary Wolfe's foundational academic work on animality and that Derrida remains a focal point for countless other theorists of the animal who've followed in Wolfe's tracks. The oddity, perhaps, is that Derrida posthumously finds himself lauded in the North American *popular* press when the discussion turns to animals—in, for example, Jonathan Safran Foer's bestseller *Eating Animals* (2010), where Derrida is approvingly quoted a few times, and even in a place where he never got any love when he was alive, the *New York Times*. While making it clear that they are sticking to their script (grumbling that "his writing is almost impossible to capture in a quotation"—referencing their own infamous 2004 obituary headline for Derrida, "Abstruse Theorist Dies at 74"), by 2012 the *Times* informs us, however begrudgingly, that "Jacques Derrida has had an equally strong influence" as Peter Singer when it comes to "the way we think about animals."[1]

Throughout *The Animal That Therefore I Am*, Derrida goes to great lengths to point out that he had been interested in the question of the animal for a long time. Over several pages, he summarizes the "horde of animals," the "innumerable critters" that run through his texts (2008, 37–41). Though for someone seeking a review of all the forms of life treated in Derrida's work (including, say, plant life, which we're interested in here), his own summary in *The Animal* text remains a bit strange. For example, in beginning to review his own work on "the question of the living and of the living animal" (34), Derrida goes first and directly to "White Mythology," an essay that I had always taken to be largely about the heliotropic—which is to say, vegetal—nature of metaphor in the text of Western thinking (metaphorical tropes as those flowers of rhetoric that turn toward the sun of truth). In "White Mythology," Derrida points out that Western thinkers since Aristotle have "been carried along by the movement which brings the sun to turn in metaphor; or have been attracted by that which turned the philosophical metaphor towards the sun. Is not this flower of rhetoric (like) a sunflower? That is—but this is not an exact synonym—analogous to the heliotrope?" (1984, 250 [French 1971]). Derrida goes on to note that the plant life of metaphor is both inside and outside the logos that defines anthropos for Aristotle. Metaphoricity (the movement from the sensible to the intelligible) makes understanding possible but simultaneously makes it impossible for intelligibility fully to escape its dependence on these flowers of rhetoric, which are always already overgrown: "Each time polysemia is irreducible, when no unity of meaning is even promised to it, one is outside language. . . . At the limit of this 'meaning-nothing,' one is hardly an animal, but rather a plant" (248).[2]

In *The Animal* book, however, Derrida suggests that "White Mythology" is not primarily about language's plant figures but rather about what he later names "the animality of writing" (2008, 52): "As I see it, one of the most visible metamorphoses of the figural, and precisely of the animal figure, would perhaps be found, in my case, in 'White Mythology'" (35).[3] Indeed, even when Derrida recalls the founding biblical story of murderous intrahuman antagonism (Cain and Abel) and notes its originary connection to animal sacrifice, he seems oddly uninterested in the (even more abject) status of plant life within this foundational story: "Cain, the older brother, the agricultural worker, therefore the sedentary one, submits to having his offering of the fruits of the earth refused by God who prefers, as an oblation, the first-born cattle of Abel, the rancher" (2008, 42). While certainly Cain and Abel is a story about the fundamental role of animal sacrifice at the basis of Judeo-Christian relations to the human other as well as to the divine—it demonstrates, in Derrida's concise phrase, that "politics presupposes livestock" (96)—it's also a story about the abjection of plant life in forming that crucial relation of value and sacrifice (and thereby the plant's elision within the whole apparatus of "life-as-sacrifice" that, as Agamben has argued, is crucial to Western thinking and politics). In short, it is Yahweh herself who opens this foundational abyss between a privileged, value-laden animal life (whose sacrifice *means* something—is maybe even the condition of possibility for religious and cultural meaning) and a necessary but abjected other of sessile plant life (a form of life that is essentially meaningless). As Derrida reminds us, God "refused Cain's vegetable offering, preferring Abel's animal offering" (43), and thereby we see inaugurated a linchpin of Judeo-Christian culture: Animal sacrifice is the only sacrifice worth the name. As *Glas* notes, from the beginning it may be that "the flower is in the place of zero signification" (1985, 31b [French 1974]).

In the course of his *Beast and the Sovereign* lectures (2011 [French 2008, 2010]), Derrida wonders out loud several times about whether what he has to say about animal life might be extended to plant life as well. For example, he opens the second volume of lectures with a kind of summary of the first and its questioning of any category of "the animal in general," but he just as quickly insists that there is common ground among animal and human lives:

> Once we have given up on saying anything sensible or acceptable under the general singular concept of "the" beast or "the" animal, one can still assert at least that so-called human living beings and so-called animal living beings, men and beasts, have in common is the fact of being living beings (whatever the word "life," bios or zoe, might mean, and supposing one has the right to exclude from it vegetables, plants and flowers).
>
> (10)

Here Derrida clearly marks the exclusion of vegetal life from the discussion of "living beings" but leaves it unexamined. Derrida even more keenly notes several times the suspicious eliding of plant life in Heidegger's famous 1929–1930 lecture series on *World, Finitude, Solitude* (where Heidegger infamously holds that the stone is worldless, the animal poor in world, and man is world forming). At one point, wondering aloud about "the ambiguity of vegetables and plants" in Heidegger's schema, Derrida asks flat out: "Would Heidegger have said that the plant is *weltlos* like the stone, or *weltarm* like the living animal?" (2011, 6).

For an answer to this excellent question, we look to Derrida's next sentence: "Let's leave it here for now: the question will catch up with us later" (6). When it does reappear later in the lecture series, Derrida continues:

> Heidegger wonders more than once how life is accessible to us, be it the animality of the animal or the vegetable essence of the plant; and twice— this is highly interesting in my view—Heidegger classifies the plant, the plant-being of plants, the vegetable, as they say, among the phenomena of life, like the animality of the animal; but he will never grant to the living being that the plant is the same attention he will grant the living being that the animal is.
>
> (113–114)

OK, you might think, now we're on to something. But Derrida's next sentence, following the protocol already established, immediately closes this parenthetical observation and leaves the question of vegetal life completely, and oddly, in abeyance. It asks, "So how is life accessible to us, given that the animal, Heidegger notes, cannot observe itself?" (113–114).

Similarly, while discussing iterability, repetition, and originality within the commentary on Heidegger, Derrida offers a brief aside on cloning:

> *Klon* is, moreover, in Greek, like *clonos* in Latin, a phenomenon of *physis* like that young sprout or that (primarily vegetable) growth, that partheno-genetic emergence we talked about when we were marking the fact that, before allowing itself to be opposed as nature or natural or biological life to its others, the extension of *physis* included all its others. There again it seems symptomatic that Heidegger does not speak of the plant, not directly, not actively: for it seems to me that although he mentions it of course, he does not take it as seriously, qua life, as he does animality.
>
> (2:75–76)

Interesting. In the next sentence, Derrida muses over the ranunculus, an aquatic flower whose name means little frog. Then right back to *Robinson Crusoe*. Nothing more on the exclusion of the plant in Heidegger.

Turning back to *The Animal That Therefore I Am*, the single reference to plants arises when Derrida quotes a passage from Heidegger's 1929–1930 lecture course, wherein Heidegger muses, "We do not say the stone is asleep or awake. Yet what about the plant? Here we are already uncertain. It is highly questionable whether the plant sleeps, precisely because it is questionable whether it is awake" (148). At this juncture, Derrida tantalizingly comments, "one should spend a long time on this" (148) question of sleep and vegetable life raised by Heidegger. But rather than do so, Derrida again abandons the question immediately, and without further comment, to stay on the trail of the animal. It seems that when focusing on animal life, the Derridean stage direction for discourse surrounding plant life remains akin to Shakespeare's famous directive in *The Winter's Tale*: "*Exit, pursued by a bear*" (3.3.1550).

At one level, it's completely understandable to background the question of plant life within an inquiry dedicated to animal life (to follow a singular trail, to see where it leads specifically), but there are other places where Derrida seems not so much to wonder about but actually to *further* this exclusion of plants from the larger discussion about life. For all of his relentless interrogation of Heidegger's views on animality and its discontents, Derrida never ceases to follow and endorse Heidegger's insistence that the question of life is entangled at all levels with the question of world—which is to say, a singular being's relation to possibility, futurity, and to itself (as an other). Derrida will consistently interrogate Heidegger's characterization of animals as "poor in world," and Derrida will spend a lot of pages and seminar time deconstructing the "as-structure" that Heidegger wants to reserve solely for human worlds (recall that humans are the only ones, Heidegger will insist, who relate to the world *as* world, to a field of possibilities as such; while animals in Heidegger have some limited access to world and to agency, they are not world building or forming. They merely behave within their environment; they can't transform it, primarily because they have no access to the logos).[4] Countering this sense, Derrida will insist throughout his reading of Heidegger that being poor in world is still having some access to world and thereby to the privileges of the human realm.

However, it's crucial to recall that Derrida's project in deconstructing the human/animal binary will *not* find its warrant in "granting" human privilege to animals—as if that were even possible—but will take the form of wondering whether humans are also "*weltarm*," creatures poor in world who merely behave within an environment rather than exercising sovereign agency over it. In the end, the haunting Derridean question will be whether both "animals" and "humans" share the fate constructed for animals in Western thinking: Having no access to life or death "itself," we—like they—don't die but merely cease to exist. As Derrida (2008) writes about his "governing strategy" in discussing Heidegger and animality, "it would not simply consist in unfolding, multiplying, leafing through the structure of the 'as such,' or the opposition between 'as such' and 'not as such,' no more than it would consist in giving back to the animal what Heidegger says it is deprived of; it would obey the necessity of asking whether man, the human itself, has the 'as such'" (159–160).

But even within this thoroughgoing critique of Heidegger, Derrida continues to accept (and in fact builds on) Heidegger's comparative method, defining "life" in terms of "world." To put it another way, the most obvious point of agreement between Heidegger and Derrida is their shared rejection of a biological or metaphysical thematization of life. Recall Heidegger (2001) from *World, Finitude, Solitude*: "The main thrust of our considerations does not rest upon a thematic metaphysics of life (of plants and animals)" (193); that "main thrust" will instead fall on the ontological question of world. Which is at least partially to say, life is defined by Heidegger as a specific set of emergent *relations* or *abilities* rather than as a kind of originary substrate animating all living things. Indeed, Heidegger's "comparative method" sets out first and foremost to disrupt this naïve "metaphysics of life." Heidegger's project hopes "to free us completely from the naïve view from which we originally started, namely that the beings in question—stone, man, animal, and indeed plants—are all given on the same level in exactly the same way" (207).

Derrida wants first and foremost to question Heidegger's confidence concerning humans' privileged and exclusive access to this thing called world (and to wonder whether there's ever any access to anything "as such"). But throughout Derrida's late work, he continues to follow Heidegger in defining life (and death) through this overarching thematic of world. For example, when it comes time to publish the French edition of his many memorial-funeral pieces (originally published in English as *The Work of Mourning*), Derrida renames the collection *Each Time Unique, the End of the World* [*Chaque fois unique, la fin du monde*]. In the "avant-propos" of *Chaque fois unique*, Derrida (2003) lays out what he calls the "thesis" of the collection:

> The death of the other, not only but especially if one loves that other, does not announce an absence, a disappearance, the end of *this or that* life, that is to say, of the possibility of a world (always unique) to appear to a *given* living being. Death declares each time the end of the world in totality, the end of every possible world, and *each time the end of the world as unique totality, therefore irreplaceable and therefore infinite*. As if the *repetition* of the end of an infinite whole were once more possible: the end of the world *itself*, of the only world that exists, each time. Singularly. Irreversibly. For the other and in a strange way for the provisional survivor who endures this impossible experience. It is this that I would like to call "the world."
>
> (9)

In a characteristically Levinasian challenge to Heidegger's sense of world as Dasein's relation to the abstract possibilities of Being, the Derridean world is by contrast opened by the radical singularity of the other(s), a finite world of future possibility that is each time unique. Understood as opening to the necessary finitude or alterity that marks "our" lives, the world is haunted always by the end, insofar as the finite temporality of mortality is what makes "life" possible in the first place.

As Martin Hägglund (2008) explains in *Radical Atheism: Derrida and the Time of Life*, it is the Derridean "trace structure of time that is the condition for life in general. Whatever we do, we have always already said yes to the coming of the future, since without it nothing could happen" (96). If life is defined by a singular being's relations to future possibilities (a being that is alive is inexorably open to the future and is likewise necessarily open to mortality), then death comprises the radical end of the world. At the end of a living thing's ability to endure through time, when it is no longer alive, that being does not escape time somehow but is left by death to live on (or not) through that very same trace structure (through, among other traces, the memories of the "provisional survivor"). Hägglund glosses this Derridean sense of "world": "The other is infinitely other—its alterity cannot be overcome or recuperated by anyone else—because the other is finite. . . . When someone dies it is not simply the end of someone who lives in the world; it is rather the end of the world as such, since each one is a singular and irretrievable origin of the world" (111). Hägglund argues at great and convincing length that the Derridean world is a site where the only definition of life is the life of finite beings (there is no God—no principle of untouched purity or everlastingness— or even the desirability thereof, hence the "radical atheism" of Hägglund's title). And when one of those singular finite beings is no more, the unique "world" that person had opened and inhabited ends as well.

Quite simply, if there is no other world, then death is each time the end of the whole world. Derrida (2001) eloquently sums this up in his eulogy for Louis Althusser:

> What is coming to an end, what Louis is taking away with him, is not only something or other that we would have shared at some point or another, in one place or another, but the world itself, a certain origin of the world—his origin, no doubt, but also that of the world in which I lived, in which we lived a unique story. It is a world that is for us the whole world, the only world, and it sinks into an abyss from which no memory—even if we keep the memory, and we will keep it—can save it.
>
> (2001, 115)[5]

Which likewise explains Derrida's obsessive interest in the line from Paul Celan's "Vast, Glowing Vault": "The world is gone, I must carry you."[6] After death, we "live on" (or not) only as a series of traces in a series of other archives, other memories, other acts, other databases, other citations.

Importantly, Derrida's crucial work on animality extends these concepts of life, death, and world beyond the exclusively human realm (thereby posing his most serious question to Heidegger's analysis of death and world as exclusively the provenance of Dasein). For example, in the second year of the seminar on the death penalty (the session of 10 January 2001), Derrida (2014) insists that

> the death one makes or lets come . . . is not the end of this or that, this or that individual, the end of a who or a what *in the world*. Each time something dies [*ça meurt*], it's the end of the world. Not the end of a world,

but of the world, of the whole of the world, of the infinite opening of the
world. And this is the case for no matter what living being, from the tree
to the protozoa, from the mosquito to the human, death is infinite, it is
the end of the infinite. The finitude of the infinite [*le fini de l'infini*].

(118–119)

There is, of course, a lot that one could say about this astonishing Derridean opening of "world" and finitude beyond the human, and we will
return to the question of Heidegger and the end of the world shortly.

But here I'd like to zero in on one very specific point: Derrida thema-
tizes world as a question of life and death, a unique relation wherein a
"living being" or a singular entity relates somehow to an "infinite opening" (world as a drama wherein "no matter what living being," human
or otherwise, is dealing in some way with the "finitude of the infinite").
However, given the definition of life that also gets configured here (as a
singular entity's relation to the finitude of the world), it's not at all clear
that most plants are technically among "the living" in Derrida's scheme
of things. (This despite Derrida's inclusion here of the tree—as Deleuze
and Guattari insist, that most arborescent and individuated plant form.)
As Rodolphe Gasché has written about Hägglund's work, it's becoming
clearer that "Derrida is essentially a philosopher of life, but of the only
life there is—the life of finite beings."[7] Here, I don't want particularly
to put pressure on the Derridean sense that temporality is the decisive
defining factor for "finite beings" (the sense that living is essentially
surviving—"living on" in relation to an indefinite, but mortal, future).
Rather I'd like to emphasize the Derridean requirement that life and
world accrue only to singular, each time unique "beings." In short, if a
kind of openness to the finite future within an individuated organism (a
relation to world) is the ante, it's not clear many plants—especially those
that are rhizomatic collectivities—can be said to be playing the game
of life. In any case, it seems clear that the question of what's alive and
what's not in Derrida remains tied to a question of world, to futurity, and
at least some inkling of an individual (though not necessarily human)
being's relation to mortality and futurity "as such."

Hägglund is the recent commentator who has done the most extensive work on Derrida and this question of life, though I would note for
the interested reader that Richard Doyle's *On Beyond Living: Rhetorical*

Transformations of the Life Sciences (1996) inaugurated this debate about life within Derrida scholarship twenty years ago. In a postpublication exchange, Hägglund was asked about the ambiguity surrounding "the question of the scope of 'life'" in *Radical Atheism*—about the distinction between *différance* as a condition of (im)possibility for anything at all and as the condition narrowly of living things and their logic of survival. Hägglund (2009) responds:

> My answer is that everything in time is surviving, but not everything is alive. . . . The isotope that has a rate of radioactive decay across several billion years is in fact surviving, since it remains and disintegrates over time, but it is indifferent to its survival, since it is not alive. The two midges [that Derrida recalls from a fossil dated fifty million years ago], on the other hand, have a project, need, and desire. Like any other living being, they cannot be indifferent to their own survival. This distinction is decisive for the definition of life in *Radical Atheism*. The reason I focus on life is because only with the advent of life is there desire in the universe. Survival is an unconditional condition for everything that is temporal, but only for a living being is the affirmation of survival unconditional, since only a living being cares about maintaining itself across an interval of time.
>
> (245–246)

One could say a lot about this quotation, but in general this seems quite persuasive to me as a reading of Derrida. "Survival [living on] is the unconditional condition for everything" in Derrida's work, but not everything is "a living being": Isotopes or volcanoes exist in and change over time (they survive as entities), but they are not "alive" insofar as they have nothing analogous to interests or desires (they have no phantasmatic world)—"since only a living being cares about maintaining itself across an interval of time."[8]

So the pressure I'm interested in applying to the question of plant life in Derrida has less to do with plants' existing in time or with the struggle for continued survival as a condition for a definition of life.[9] Surely this drive to live on is evidenced in plants, from the forest fight for sunlight to the roots' competition for water or soil nutrients, or plants' agency in repelling predators or attracting pollinators. However, I'm wondering whether many species of plants (a wall of ivy, a meadow of switchgrass, a pondful of lily pads—not to mention plants' ancestors, the algae) can be thematized as a unique singular "being" or an "itself"? A field of Kentucky bluegrass is more a collectivity (a smear of life across series of emergent sites) than it is a singular, bounded living "being." To put the question quite pointedly: On the grasslands that still cover more than 30 percent of the earth's surface, are the individual blades of grass or daisy stems "alive" in Derrida's sense? Do all plants have access to a phantasm of futurity (or the lack thereof)? Does every single blade of grass have a "world," each time unique? Perhaps when Sean Gaston (2013) writes in *The Concept of World from Kant to Derrida*, "this fiction of the world is an inescapable part of life and death, of animals and humans living and dying together. There is always the possibility that there is no world, but the world remains a necessary fiction" (133), he puts the question most succinctly: Does this "fiction of the world" shared by "animals and humans" extend to vegetable life?

My sense is that, in terms of this Derridean discussion of life and world, a wide swath of vegetable life remains *weltlos*—and thereby not technically "alive"—insofar as any given plant is not necessarily a singular entity and thereby cannot have a unique relation to futurity or to the "each time unique" phantasm about (in)finitude that is the "world." If, as Timothy Clark (2013) argues in "What on World Is the Earth?" the Derridean world is "something delusory in which one cannot not believe, simply by being alive" (18), then it would seem that plants aren't alive, or at least aren't alive in any similar way to humans or animals, because plants (as far as we know) do not or cannot project the fictional phantasm of a world. To put the question a different way: When Hägglund argues that living beings in Derrida "cannot be indifferent to their own survival" (to their own future), that's largely because they are *beings*, individual organisms whose survival or future can be thematized (however naïvely or phantasmatically) as "their own." Similarly, when Clark argues that world is something that "one cannot not believe," I might stress less the Derridean question of "believing" (or not) in the world and put more pressure on the question of the "one" who's doing the believing (or not). Surely, the desiring "one" remains as thoroughly phantasmatic as the possible "world" for Derrida, but the necessity for something like consciousness-projection or desire remains on both ends of this definition of "life." In short, if something is not such an individuated being (if it is not a phantasmatic site for living out a singular world of its ownmost desiring possibilities), then that thing is not, I don't think, technically "alive" on Derrida's terms.

Of course there are several caveats that immediately attach themselves to this seemingly perverse claim that plants are not alive for Derrida: First, plants are obviously "alive" in the everyday biological sense, for Derrida and for everyone else who treats these difficult questions, and I'd also have to admit that things are complicated considerably by Derrida's specific inclusion of the tree in his list of singular living beings with access to world. Things here get very murky indeed, as a series of large oak trees, each one rising from an individual acorn, I think can easily be considered unique living beings, with a singular world of phantasmatic openness to other living things (this bird, squirrel, or set of humans) and marked from the beginning by the eventual finitude of those possibilities, but a stand of aspens, on the other hand, comprises a rhizomatic collectivity (that's unique no doubt, but each individual tree is not really an "each one," genetically speaking). Maybe plants that reproduce sexually (through pollination and seed distribution) have "worlds" insofar as their offspring are "each time unique" beings (akin to a human or animal newborn), while plants that reproduce rhizomatically (underground through extending stolons) are *weltlos*. Here we might note that Derrida does at points offer childbirth as a privileged example of the event itself (absolute arrival that can't finally be calculated in a solely rational way), though his comments on cloning and "originary technicity" (the sense that we are always already technological copies rather than organic originals)[10] would, I think, complicate any wholesale privileging of sexual over asexual reproduction right out of the gate—not to mention the suspicious anthropomorphism inherent in granting a world only to things that reproduce in the manner that humans do.

We would also have to wonder at this point about Derrida's mention of the "world" of the protozoa, an example that I presume he chooses as much for its Greek etymology (*protos—zoon*, first life or first animal) as anything else. Of course, consideration of the single-cell kingdoms opens up an even larger and more difficult series of questions for this discourse on life and world, and even more so for any discourse on the privileges of the unique human as the benchmark for life itself, insofar as there are ten "other" cells (literally trillions of microorganisms) for every "human" cell in each of our individual bodies.[11] In short, at the micro level, the "each time unique" human or animal entity begins to look like it's made up of a smear of other forms of life, most of which reproduce asexually. Looked at from the world of the protozoa, living bodies (such as human bodies) begin to seem less like individuated beings and more like a kind of prairie: a swarm of tangled and connected life forms, with much of the life activity taking place at a level unseen by the naked eye. In any case, there are a lot of questions here, and I'll admit right now that I'm not sure I have any definitive answers concerning the liminal status of vegetable life in Derrida.

However, as I'll try to suggest in the rest of this essay, this Derridean swerve around plants in relation to life is not merely evidence that Derrida callously abjects or ignores the vegetable kingdom. Rather, the excluded question of "plants, flowers, and vegetables" in Derrida becomes transversally attached less to the discourse of animal life than to the thematics of *physis* itself (often translated as "nature" but tending to have the sense in Greek of that "power of growth" linked to vegetable life, the condition of possibility for all emergence). Like *différance*, *physis* comprises a power of emergence that remains indifferent to this or that form of "life," to this or that individual being or its world: a necessary condition of (im)possibility that (in the end, or from the beginning) is wholly indifferent to any individual stone, animal, plant, or human.

To be clear, then: I'm not interested here in accusing Derrida of zoocentrism in his work on animal life. Though at the same time, one would be remiss in not wondering about the uncomfortable silence on the question of vegetative life both within Derrida's recounting of philosophy's shabby treatment of animals and even more so in the wider flood of recent work on biopower and animality—where Derrida's work has found a new and amenable home. I'm *following* Derrida rather than *critiquing* him here—tracking him like the animal that therefore he is. But in doing so, I'm also following him like a plant—tracing the innumerable seeds he's disseminated from the flower of deconstruction.

To lay my cards fully on the table, I want to suggest that abundant resources for thinking about plant life in Derrida's work are to be found in his (now, it seems, all but forgotten) 1974 masterwork *Glas*. Among the many, many other debates that *Glas* engages (Hegel versus Genet, the law

versus the outlaw, philosophy versus literature, normativity versus queer-
ness, proper versus improper), I want to linger here with *Glas*'s staging
of the animal versus the plant—life understood as animality versus life
understood as vegetality, animal reproduction as appropriation versus the
flower's expropriative deluge of pollen or nectar, the animal religions of
ennobling sacrifice (from Judaism's myriad animal sacrifices all the way
to Christianity's Lamb of God) versus the flower religions of excessive
expenditure (from the murderous Cain, whose fruits of the soil displeased
God, to the ancient flower religions of India and the cults of the Bacchan
grape). I even want to go as far as suggesting that Derrida's final words
in his final seminar (volume 2 of *The Beast and the Sovereign*) turn not
on the animal and the question of "higher" life forms but turn instead
back toward the vegetable "world" of uncontrolled, cancerous growth and
indifference, not to the individuated animality of *bios* or *zoe* but to vegetable *physis* as the inhuman power of emergence or event. Perhaps, in
the end, the plant (as the privileged figure for *pheuin*, growth or emergence itself) doesn't have access to world because it embodies and deploys
something prior to (and indeed superior to) the problem of world: the
Plant and (or maybe even *as*) the Sovereign?

If nothing else, I want to demonstrate that the notable absence of
deconstructive thinking about vegetal life in Derrida's late work is more
than made up for in *Glas*, which performs what one might call a classic deconstruction of the animal/plant opposition. *Glas*'s thoroughgoing
investigation of what Derrida (1985) names "the question of the plant, of
phuein, of nature" will lead him to posit straightforwardly in *Glas*: "Practical deconstruction of the transcendental effect is at work in the structure
of the flower" (15b). I'm likewise following out what Michael Naas (2009)
in *Derrida from Now on* has wryly called

> an almost failsafe hermeneutic principle when trying to check any
> hypothesis or thesis regarding the work of Jacques Derrida. The princi-
> ple runs something like this: make your case by ranging widely through
> Derrida's corpus . . . but *then* turn at the end of the day to *Glas* to see
> whether the whole thing was not already laid out for you, from start to
> finish, in 1974.

(208)

GLAS

As John Leavey, one of the English translators of *Glas,* ironically noted in his 2012 Modern Language Association conference presentation, Derrida's experimental 1974 text is a consistent winner in the academic "best book you've never read" game that sometimes breaks out after several glasses of wine at college-town dinner parties. Thereby if nothing else, *Glas*'s obscurity constitutes an ironic bookend to Derrida's immensely popular work on animality. And one should be quick to point out that *Glas* is a justifiably infamous monster—consisting of two columns on each (extra-wide) page of the text, one concerning Hegel, the other Jean Genet. And of course to say "it's written in two columns" seriously underestimates the difficulties that *Glas* presents, as Derrida complicates his own binary schema by consistently inserting other texts within the columns ("tattoos" of different-fonted text in boxes, along with citations, dictionary definitions, and other breaks where myriad voices or texts intervene). Whatever else it is, a text like *Glas* is a tour de force comedy (and the genre is literally comic, a circle—the beginning is the end, *Finnegans Wake* style). And the joke is aimed squarely at that most serious and tragic of philosophical projects, Hegel's. The last word of the Hegel column (such as it has a "last word") is literally "comedy."[12] In a world that has posthumously decided Derrida is either a thinker of animals or a mournful thinker of "prayers and tears," *Glas* intensely recalls for us another Derrida, the philosopher of malicious performative joy and relentlessly avant-garde artistic production. And the Derrida who is a thinker of vegetable life.

Readers well versed in *Glas* will forgive me if I lay out a brief introductory roadmap for those unfamiliar with the text. First, and most obviously, the text's two columns are set up to test the Hegelian laws of dialectical sublation—Hegel's famous *Aufhebung,* the engine of dialectical progress: affirmation, negation, synthesis. The most obvious question posed by the text (as well as performed in any attempt to "read" it) concerns the first column, which takes up the project of dialectical progress posed by Hegel—"a higher calculus without remains, what consciousness wants to be" (60a). The question that inaugurates the book and "remains" throughout it is whether this consciousness can, as Hegel would lead us to believe, "sublate" column B (which in this case happens to concern the proper name for impropriety itself, the literary project of transgression and masochistic queer desire performed in Jean Genet's literary writings). As a voice (maybe Derrida's?) puts it some sixty-five pages into the text:

> Why make a knife pass between two texts? Why, at least, write two texts at once? What scene is being played? What is desired? In other words, what is there to be afraid of? Who is afraid? Of whom? There is a wish to make writing ungraspable, of course. When your head is full of the matters here, you are reminded that the law of the text is in the other, and so on endlessly. . . . You are no longer let know where the head of the discourse is, or the body.
>
> (64–65b)

While reading *Glas* seems far from the ordinary experience of responding to texts, Derrida's gambit will be to show that such multiplying intertextuality is the most basic practice of all reading and writing: Even "if I line myself up and believe—silliness—that I write only one text at a time, that comes back to the same thing, and the cost of the margin must still be reckoned with" (66b), and the text demonstrates this productive undecidability in a wide variety of discursive regimes. Indeed, *Glas* moves like a plant or a vine, by "agglutinating rather than demonstrating" (75b), by cross-pollinating, becoming invasive, coiling (104–105b), and, above all, by what Derrida calls (vegetable) grafting or cutting of one text and hybridizing it with another (108). Perhaps the only thing *Glas* doesn't allow is dialectical sublation—and Derrida will go on to suggest that this is in fact the one and only law of the dialectic: its condition of possibility is simultaneously its condition of impossibility.

Of course, for Hegel a certain kind of undecidability ("tarrying with the negative") is absolutely necessary for the progress of spirit—which must constantly risk knowledge and life by confronting (and overcoming) ignorance and death. This is the dialectic in a nutshell, negating difference and raising it (however momentarily) to a higher level of synthesis through the movement of desire. The dialectic teaches, if nothing else, how you make the negative productive—with death, for example: You need to take death within the itinerary of your life, tarry with it and raise it up, make it into an engine for progress and discovery, rather than treating death's negativity as a sinkhole of despair. Otherwise, death simply defeats you. The otherness confronted by the Hegelian negative is then not paralyzing but is far rather what consciousness requires to become stronger, more confident, to progress: Consciousness must feed on the remains each time an opposition is adjudicated, reappropriating these remains, raising them up. If Hegel is often smeared as a totalizing thinker, it's important to note that his vision of the absolute is not a static state but the endless movement of desire: Absolute spirit is perhaps nothing other than the command that you constantly have to lose yourself in order to find yourself—endlessly risk death in order to overcome it.

As such, Derrida demonstrates in *Glas* that Hegel is a thinker of animal desire as the engine of knowledge and progress, but this animal desire must be negated and raised—made culturally respectable by spirit. As Derrida writes, in Hegel "Human feeling is still animal. The animal limitation, I feel it as spirit, like a negative constraint from which I try to free myself, a lack I try to fill up. . . . Man passes from feeling to conceiving only by suppressing the pressure, what the animal, according to Hegel, could not do" (25a). And of course this Hegelian tension between "the animal moment and the spiritual moment of life" (25a) begins a biopolitical project that becomes very familiar to us throughout the nineteenth century, to our own day: the project of sorting out man's animal desire and refining it (or not) through education, repression, or normalization by culture. The theme of man's subtending animality runs straight through all the thinkers who usher in what we like to think of as "our" biopolitical era (Nietzsche, Freud, Darwin, Schumpeter), and Derrida's reading of Hegel comes to a conclusion similar to Foucault's in *The Order of Things*: Modernity is not, as some thinkers in animal studies would have us believe, born by jettisoning or abjecting animality but rather by fully incorporating animal desire into our bedrock definition of the human.[13]

And for his part, Derrida doesn't jettison this "animal" Hegelian project at all but in fact intensifies it, puts it to its own test: Does or will this animal appetite for appropriating otherness succeed in the end? Can the negative be so easily economized, made to pay off as knowledge at a higher level? As Derrida sums up the obsessions of *Glas*, "What remains irresolvable, impracticable, nonnormal, or nonnormalizable is what interests and constrains us here" (5a): The text lingers over all those remains that refuse to be lifted up, normalized, incorporated, known.

But *Glas* does not merely stage for us a Manichean struggle of norms versus excess, with some flavor of excess as the inevitable winner. This is a popular rendering of the project of deconstruction, but nowhere is it clearer than in *Glas* that Derrida does not simply celebrate a negative undecidability—the flower, literature, queer desire, the mother, the sister, etc. It's not that these "undecidables" are simply the other face of Hegel's project of accounting for everything, but they comprise their own (for Derrida, oddly "superior") form of totalization, a form that can in fact account for its own necessary failure, for its own remains, in a positive rather than negative way. In short, the drama that *Glas* stages for us is not merely Hegel versus Genet (totalization versus undecidability, philosophy versus literature, animal versus plant), but the book performs two differing modes of totalization, what Derrida calls two different ways of "saying everything."[14]

In any case, Derrida incessantly tests Hegel's logic of raising the remains by contrasting Hegel with Genet on a number of topics—from death and burial (literally, normativizing dangerous remains—making mourning out of melancholia) through the question of the family (which is at base the normalization of an otherwise dangerous sexual desire, raising it to the level of propriety—though there is much discussion of Hegel's bastard son and his odd attraction to his crazy sister). But I will argue that the binary clash that is privileged within *Glas* is less philosophy versus literature than it is animal versus plant.

PLANTS AND ANIMALS IN *GLAS*

At the beginning of the Hegel column, Derrida very economically focuses the Hegelian corpus on two moments that he wants to interrogate: "two very determined, partial, and particular passages, two examples. But perhaps the example trifles with the essence": "First passage: the religion of flowers. . . . Second passage: the phallic column of India" (2). Why, one wonders, these two strange moments in Hegel's texts, the first from the *Phenomenology* and the second from the *Aesthetics*? How can the religion of flowers and the "Dionysiac celebration" of fertility cults (practices recalled from ancient India and Africa) be privileged moments to begin thinking about modern Europe's most ambitious Enlightenment thinker? Initially, one might begin to answer the question "why would Derrida begin here?" by venturing that these are both originary or maybe even preoriginary moments in Hegel. Everything in Hegel moves in threes, of course (affirmation, negation, synthesis), and his sweeping eagle-eye views of history and religion are no exceptions: Like the opening of the *Phenomenology* itself, which moves from sense-certainty, through unhappy consciousness, to the first inklings of something like knowledge, history for Hegel moves from the ancient world (and its naïve "sense-certainty" religions of flower innocence and fertile desire) through the Judeo–Early Christian world of guilt and culpability, toward their dialectical sublation in the Christian Enlightenment underway in Hegel's own day.

So the religion of flowers or the fertility cults of the ancient world are less originary moments in Hegel than they are preoriginary moments, naïve nonstarters. Like sense-certainty (the necessary but by no means sufficient condition for perception), the religion of flowers comprises a historical phase that exists in Hegel's system only to be overcome. As Derrida writes:

> Flower religion [in Hegel] is not even a moment or station. It all but exhausts itself in passage, a disappearing movement, the effluvium floating above a procession, the march from innocence to guilt. Flower religion would be innocent, animal religion culpable. Flower religion . . . no longer, or hardly, remains; it proceeds to its own placement in culpability, its very own animalization, to innocence becoming culpable and thus serious.
>
> (2a)

Derrida then quotes Hegel from the *Phenomenology*: "The innocence of the flower-religion, which is merely self-less representation of the self, passes into the seriousness of warring life, into the guilt of animal religions; the quiet and impotence of contemplative individuality pass into destructive being-for-self" (2a) and thereby into the dramas of progress, history, law, and politics. Like the places from whence they came (Africa and India), flower religions are literally prehistorical for Hegel; they are merely a "quiet and impotent" state of nature that has yet to be negated, yet to pass over into the dialectical realm which is culture. As Derrida quotes

Hegel later on: "The purpose of nature is to kill itself and break through its shell of the immediate, of the sensible, to consume itself like a Phoenix, in order to upsurge, rejuvenated, from this exteriority, as spirit" (117a). Nature must be raised and transformed; nature must become culture.

And let's recall that a certain resurgence of innocent flower religion (going by the modern name of "romanticism") is one of Hegel's primary targets in the *Phenomenology*—most specifically, he takes aim at Goethe's and Schelling's nature-obsessed romanticism, "a view which is in our time as prevalent as it is pretentious" (Hegel 1976, 4), Hegel snickers. Functioning as a naïve, "bad infinity"—nature as preoriginary in-difference, unable to account for or direct progress—this newfangled flower religion is famously mocked by Hegel as "the night in which all cows are black" (Hegel 1976, 9).

And yet, Derrida asks, and yet . . . Just as sense-certainty will be difficult to leave behind on consciousness's ascent toward knowledge, one might summarize the whole of *Glas* as an extended question posed to Hegel's discourse: If Hegel's thinking of culture and law is a graduated process of extracting cultural sacrifice, progress, and knowledge from an indifferent, asignifying nature—animalizing the plant, so to speak, and then humanizing the animal—will this project really ever be able to gain and maintain escape velocity from its primary other and condition of possibility, a kind of preoriginary *physis*? The question remains: Can the dialectic "escape" or sublate the otherness upon which it depends for its fuel: nature, physis, the plant, growth, and emergence—*this preoriginary indifference that (secretly) calls for and drives the law?* Can you "train" the event of emergence with desire-as-lack, shame, or propriety? Or is the event of emergence indifferent to culture—is it just not the kind of thing that will tamely undergo such higher-purpose "progress"? And doesn't such a system of cultural differentiation require (endless?) production of ever more indifferent remains for the whole system to be able to move to a higher level? More remains, more desire, higher plateaus for spirit: "By removing itself from nature, by denying nature within itself, by relieving [sublating], sublimating, idealizing itself, desire becomes more and more desiring. Thus human desire is more desiring than animal desire; masculine desire is more desiring than feminine desire, which remains closer to nature. More desiring, it is then more unsatisfied and more insatiable" (Derrida 1985, 169a). As nature must be cultured, so the plant must be animalized: "The more one is raised in(to) the differentiating hierarchy of animalness . . . the more the organism is capable of assimilating foreign bodies or differentiated organic totalities" (114a).

To fast-forward through all the avatars of plant religion in *Glas*—literature, the queer, the convict, the erection, the crazy sister, the illegitimate child, all sorts of nonsignifying desire—the question for all of them will be similar: Can these "preoriginary," non-normative states finally be normativized, animalized, economized, made to pay, made to progress toward higher knowledge? Can the inert or indifferent condition of possibility for dialectical sublation merely be taken up into the movement of knowledge, negated and lifted to the absolute by the engine of spirit?

At the macro level, *Glas* stages for us a faceoff between the animality of appropriating, civilizing movement of spirit and the "law" of plant-nature—which Hegel will grant is growth yes, emergence yes, but a kind of undifferentiated and uncontrolled growth that is finally anathema to the progress of spirit or law—and not because nature or the plant (or literature, for that matter) is merely inert but because they're dangerous. The "bad-infinite" plant must be schooled by the animal in the "good infinite" of culpability, desire for objects that can be worked on by spirit. Animal desire is "superior" for Hegel because that desire can be repressed, sublated, trained. Plant desire merely grows without telos, without a proper end (or a proper death)—and the logic here is classic: In Aristotle, the vegetable *psukhe* or soul deploys a single power, the ability to grow. But that power is without entelechy, as plants have no ideal form toward which they grow—they just grow uncontrollably until they grow no more.

So in this vein, for example, Derrida will recall for us Hegel's account of the feminine and its philosophical prejudices concerning the woman and the plant:

This discourse on sexual difference belongs to the philosophy of nature. It concerns the natural life of differentiated animals. Silent about the lower animals and about the limit that determines them, this discourse also excludes plants. There would be no sexual difference in plants. . . . In this sense, the human female, who has not yet developed the difference or opposition, holds herself nearer the plant.

(Derrida 1985, 114a)

Or as Hegel puts it, "The difference between man and woman is that of animal and plant" (qtd. in Derrida 1985, 191a).

So for Hegel the plant must be "animalized" in the same way that the woman must be manned, the family must be nationed, and mere matter must be lifted up by spirit: Without that sublating moment, there's nothing but cancerous "natural" growth, without regard to betterment or higher ends: nutritive life, without a "world."

Again it is the plant, which Derrida consistently links to the question of *physis*, that remains the linchpin site of inquiry in *Glas*. Not the animal, which Derrida demonstrates time and again is already incorporated into the movement of spirit by Hegel (animality may in fact be said to be the ground of that movement of desire). Hegel's animal desire is the engine of the dialectical discourse, whereas the "plant desire" endlessly on display in the Genet column doesn't "lead" anywhere, or anywhere proper: masturbation, queer desire, prison, the thief's underworld, cathexis onto the mother—can any of these be raised? Like a flower, and connected centrally to the erections that populate Genet's text like flowers, this "other" desire emerges intensely and lingers for a moment before it releases its seed to the wind and it's gone. It falls, as remains, downward, as it were, back to the bad-infinite ground—as opposed to those remains constituting an engine for higher knowledge. The Genet column of *Glas* resolutely resists that raising, remaining with nonreproductive desire, with sense-certainty, with the plant: *physis*. In a crowning irony, on Hegel's account, Genet's queer desire is shown finally to be "natural desire"—the plant, the flower, the queer outlaw, all remaining indifferent to culture, to raising. And Derrida clearly demonstrates this in the Genet column through the figure of the erection, that privileged sign for male sovereignty and reproductive power, which is of course "raised" by means that are largely out of any subject's rational control. In the Genet column, erections grow like flowers and spill their seed all over the place, but their activity doesn't lead to anything that Hegel would recognize as a "higher" state.

And of course this non-normative activity is all the more necessary because cultural progress depends on a remainder that it finally cannot control: Any progressive dialectic must consistently produce ever more new, indifferent, asignifying remainders—emergent undocumented material that needs to be rendered meaningful, negated, and raised.

As Derrida summarizes *Glas*'s reading of the Hegelian dialectic in concise fashion,

> There is no choosing here: each time a discourse contra the transcendental is held, a matrix—the (con)striction itself—constrains the discourse to place nontranscendental, the outside of the transcendental field, the excluded, in the structuring position. The matrix in question constitutes the excluded as transcendental of the transcendental, as imitation transcendental, transcendental contraband. The contraband is not-yet dialectical contradiction. To be sure, the contraband necessarily becomes that, but its not-yet is not-yet the teleological anticipation, which results in its never becoming dialectical contradiction. The contraband remains something other than what, necessarily, it is to become. Such would be the (nondialectical) law of the (dialectical) stricture.
>
> (244a)

Within this general law of the dialectic, the emergence of difference can only look like indifference, a remainder that is not-yet subject to a law, made whole. But of course the emergence of this not-yet, this indifferent difference, is at the end of the day the structural principle of the entire discourse. The dialectic is as stale as last night's beer without the emergence of the event, the remainder, the contraband. The "excluded" (the bad infinity of supposedly in-different nature) finds itself in fact "in the structuring position," "the (nondialectical) law of the (dialectical) structure." The structure whereby this remainder is ineluctably produced by dialectical sublation is most concisely worked out in *Glas* according to the logic of the plant, though again it's worth remembering that the plant is linked by Derrida directly to the *phuein* of *physis*—the power of growth and emergence that is in turn linked with "nature" in most Western thinking: "The flower appears in its disappearance, vacillates like all the representative mediations, but also excludes itself from oppositional structure" (246a).

When summing up the stakes and movements of *Glas* late in the text, Derrida asks straightforwardly, "How far have we got?" In answering that question, Derrida returns to his opening focus on plant and animal in Hegel:

Why, plant and animal, plant then animal? . . . . No opposition can form itself without beginning to interiorize itself. This organicity already binds itself again to itself in the plant, but life represents itself therein only by anticipation. The actual war, as opposition internal to the living, is not yet unchained. The plant, as such, lives in peace: substance, to be sure, and there was not yet any substantiality in the light, but peaceful substance, without this inner war that characterizes animality. Already life and self, but not yet the war of desire. Life without desire—the plant is a kind of sister.

(245a)

In the end, Derrida suggests that Hegel elides plant life because "the subjectivity of the plant is not yet for itself. The criterion for this is classic: the plant does not give itself its own place" (245a).

FROM *GLAS* TO THE FINAL WORD

To return from *Glas* to the logic of Derrida's later work on animality, it's hard not to conclude that Derrida's analysis of the plant in Hegel recoils immediately on Hegel's central theme, the animal privileges of human consciousness. Recall again that Derrida's work on animality is not offered in the name of lifting up the animal (suggesting the animal is similar to humans in sensation, suffering, death, relation to world, etc.), but rather it wonders whether all the things that Western thinking has said about the animal (it behaves rather than acts, it has no access to entities "as such," it has no knowledge of finitude) are finally all that can be said about humans as well.

In *Glas*, this seems clearly the gambit and logic of Derrida's obsession with the plant as well: Are humans ever lifted to the level of being "for themselves," giving themselves their own place? Or are humans somehow akin to plants—at the end of the day subject wholly to their environments, living, growing without entelechy, and dying indifferently; responding as they can to the events that happen around them, events or emergences within a physical world largely indifferent to any particular life form "as such"? In short, *Glas* asks us: Can what Hegel calls the "innocent indifference of plant life" (246a) finally be incorporated, mastered, and left behind by the differentiating work of spirit, "this determinateness and this negativity" (246a)? Does human-animal desire triumph over vegetable *physis* in the end? Or is it far rather the other way around? Suggesting that *physis* is less "innocent" than it is "sovereign."

In fact, one is left to wonder whether Derrida's work on plant life in *Glas* finally ups the ante around the question of life, even more so than the work on animality—insofar as *Glas* suggests that a kind of plant logic is "superior" to the animal logic of Hegel's humanism. (Hence the upped ante—Derrida's work on animals demonstrates that the logic of animality may in fact be identical to the logic of humanism, but to my knowledge he never suggests that the logic of animality is somehow superior to the logic of humanism).

I think the privileged place to go from here in exploring the upshot of this question is right to the very end of Derrida's itinerary, to his final lecture course before his untimely death from pancreatic cancer: the second series of *Beast and the Sovereign* lectures, where he both reopens and is tragically forced to end his long engagement with Heidegger and the question of animality. This last lecture course is unique, as Michael Naas has definitively shown in *The End of the World and Other Teachable Moments*, insofar as it's marked by a discontinuous but obsessive investigation of a single word that emerges as key in Heidegger's texts of the 1930s, the German word *Walten* (as a kind of sovereign, originary violence). As Derrida (2011) puts it in the final lecture of what turned out to be his final seminar, "late in my life of reading Heidegger, I have just discovered a word that seems to oblige me to put everything in a new perspective" (2:279). As he summarizes what he takes from Heidegger's work on animality in the second volume of *The Beast and the Sovereign*,

I hang onto this curious non-sequitur that consists in defining animality by life, life by the possibility of death, and yet, and yet, in denying dying properly speaking to the animal. . . . What is lacking is not supposedly access to the entity, but access to the entity as such, i.e. that slight difference between being and beings that, as we shall see, springs from what can only be called a certain *Walten*.

(2:116)

In his last lecture course, Derrida returns one final time to Heidegger's work on animality and world, recalling for us Heidegger's original gambit: to think life not as an essence, property, or thing but to think various forms of life always within a series of relations to their origins and possibilities, to *physis* and world. Following a fundamental ontology, Heidegger insists that we must return to *physis* in order to rethink what human life is. As Derrida reminds us:

> Remember that for Heidegger—because we'll need to think seriously about this—*physis* is not yet objective nature but the whole of the originary world in its appearing and its originary growing. It is toward this originary "world," this *physis* older than the objective nature of the natural sciences . . . that we must turn our thought in order to speak anew and differently about the being-in-the-world of man or of Dasein and animals, of their differential relation to this world that is supposed to be both common and not common to them.
>
> (2:12–13)

Throughout this final lecture course, Derrida tracks Heidegger's incessant use of the word *Walten* and its variants within the conversation about the questions of animality, world, and death, suggesting that for Heidegger there is a kind of originary granting even before the "world" of Dasein's possibility—or at least suggesting that what is gifted in Heidegger's notion of *physis* is tinged less with Hegel's preoriginary "innocence" than with a certain kind of originary violence. As Derrida writes, "*walten*: always the force of this same word that bespeaks a force, a power, a dominance, even a sovereignty unlike any other ['more sovereign than all sovereignty,' he says in a footnote]—whence the difficulty that we have in thinking it, determining it, and of course translating it" (2:123). Derrida finally ventures this translation of *Walten* in Heidegger: "a sovereignty of the last instance . . . a superpower that decides everything in the first and last instance" (2:278). And Derrida consistently notes the linkage of *Walten* to the question of *physis* in Heidegger: "*Physis* means this whole *Walten* that prevails through man himself and over which he has no power, of which he is not the master. . . . In other words, this all-powerful sovereignty of *Walten* is neither solely political nor solely theological. It therefore exceeds and precedes the theologico-political" (2:41). Derrida goes on to defend "my insistence on *Walten* here, as a figure of almost absolute power, of sovereignty before even its political determination," through its originary links to *physis*:

> Heidegger explains to us that if we translate more intelligently and clearly, if we (that is, he) translate *physis* not so much by growth (*Wachstum*) as by *Walten* ["the self-constituting, self-formed, sovereign predominance of beings in their totality"], if, then, we translate *physis* by *Walten* rather than *Wachstum* (as sovereign power rather than growth), that is, as Heidegger expressly says, because it is clearer and closer to the originary sense, the intentional sense, the meaning of the originary sense or the originary meaning of *physis*.
>
> (2:40)

In circling back to *Walten*, which concerns the sovereign, originary reign of *physis*—Derrida seems to hint at thinking life and death not in terms of an animal or human access to world but in terms of a "vegetable" power of growth or emergence that's indifferent to man or animal and thereby not so much animating or reassuring but crushing in its indifference. At the very end of the day, one might say that *physis*-as-*walten* may in fact comprise the ironic realization of Hegel's all-encompassing dream for philosophy: an originary force that in fact constitutes and lords over everything, all the rest. But that force turns out to be *physis*, not *Bildung*; vegetable, not animal; nature, not culture. The condition of possibility for a life, but not alive.

In the very end, there undoubtedly will be a kind of totalization waiting for the human—as there has been for the 99.9 percent of all living species which have already gone extinct over the history this planet. But that sovereign totalization happens *to* us, not *for* us. And Derrida wraps up his analysis of Heidegger on precisely this point: "the as, the as-structure [of access to world] that distinguishes man from the animal is thus indeed what the violence of *Walten* makes possible" (2:288). The human-animal "as-structure" of world, in other words, depends wholly upon a prior field of indifferent forces, "a superpower that decides everything in the first and last instance." Just as humans violently displaced *physis* to make way for the Anthropocene, that very same *walten*-as-*physis* will—in the "last instance"—violently displace and overtake each and every one of us.

Which returns us again to *Glas* and the serene indifference of the plant as the most intense figure for the power of *physis*—all the more a superior logic for its originary, sovereign force of emergence and growth, coupled with its utter indifference to this or that living thing. Like the deep time of *différance*, the Derridean plant doesn't *have access* to a past or a future world, but maybe that's only because the originary *walten* of vegetal *physis* comprises both the absolute past and the posthuman future. An intense irruption of *physis*, the plant as such is not technically alive in Derrida (it has no phantasmatic world or singularity), but it is the condition of (im) possibility for "life." To circle back to *Glas*, it shows us in the end that the supposed privileges of human-animal desiring life arise by necessity out of a nonliving principle—*physis*, *différance*, vegetable life, call it what you will—that necessarily afflicts and marks the supposed living presence of *Geist* or world. Emergence happens in time, but neither emergence nor time is to be confused with the correlationist phantasms of human desire. The end of the world indeed, but maybe the beginning of something like "critical life studies."

NOTES

This essay is reprinted, with permission, from Jeffrey T. Nealon, *Plant Theory: Biopower and Vegetable Life* (Standford, Calif.: Stanford University Press, 2016).

1. See "Animal Studies Cross Campus to the Lecture Hall." January 2, 2012, http://www .nytimes.com/2012/01/03/science/animal-studies-move-from-the-lab-to-the-lecture -hall.html.

2. Derrida here cites Aristotle's *Metaphysics* (1006a12–15) on the irreducibility of noncontradiction: "It is absurd to reason with one who will not reason about anything, insofar as he refuses to reason. For such a man, as such, is seen already to be no better than a mere vegetable."

3. While I don't want to deny that "White Mythology" takes up the question of animality—it most certainly does—I'm just marking here that the essay is also seriously engaged with vegetal life and the status of metaphor. The essay begins, "From philosophy, rhetoric. That is, here, to make from a volume, approximately, more or less, a flower, to extract a flower, mount it, or rather to have it mount itself, bring itself to light—and turning away, as if from itself, come round again, such a flower engraves" (Derrida 1984, 209).

4. Recall Heidegger: "the main points of our approach are encapsulated in three theses: [1.] The stone is worldless; [2.] The animal is poor in world; [3.] Man is world-forming" (2001, 184). Later, Heidegger goes on to state quite straightforwardly that "What we call here world-formation is also the ground of the very inner possibility of the logos" (335).

5. See Derrida's (2001) thoughts in his letter to Max Loreau's widow: "I lack the strength to speak publicly and to recall each time another end of the world, the same end, another, and each time is nothing less than an origin of the world, each time the sole world, the unique world, which, in its end, appears to us as it was at the origin—sole and unique— and shows us what it owes to the origin, that is to say, what will have been, beyond every future anterior" (95). See also Derrida's recollection for his friend Jean-Marie Benoist: "death takes from us not only some particular life within the world, some moment that belongs to us, but each time, without limit, someone through whom the world, and first of all our own world, will have opened up in a both finite and infinite—mortally in-finite—way" (107).

6. See Derrida's 2005 (especially 141ff.) lectures, as well as numerous places throughout the *Beast and the Sovereign* (2011), for his extensive commentary on this line from Celan.

7. This quotation is from Gasche's endorsement of Hagglund's *Radical Atheism* (2008), on the back cover.

8. The Derridean world in this sense is indeed a "fiction" or a "phantasm." As Derrida (2011) writes, "It seems to be as if we were behaving as if we were inhabiting the same world and speaking the same language, when in fact we well know—at the point where the phantasm precisely comes up against its limit—that this is not true at all" (2:268). The world, in other words, is a kind of enabling fiction or projection, one that keeps at bay what Derrida calls the "infantile but infinite anxiety that *there is not the world*" (2:83). As Michael Naas (2009) explains in his authoritative work on the subject, "The [Derridean] phantasm can thus be described as a projection on the part of the subject that is then taken to be something external to the subject, a projection that then has real effects . . . and effects that then reinforce the phantasm" (255). As such, then, the phantasm is at the origin of the notion of world, as a kind of necessary fiction that, because of its Necessity to the social world, can never be debunked. As Naas points out, for Derrida "the phan-tasm is not an error to be measured in relation to truth; it is not some imitation, image, or representation to be measured against the real but is akin to what Freud, in *The Future of an Illusion* and elsewhere, terms an 'illusion.' Not a representation or misrepresentation of the way things are but a projection on the part of a subject or nation-state of the way one would wish them to be—and thus, in some sense, the way they become, with all their real, attendant effects" (207).

9. This sense of life as living-on or survival remains crucial to Derrida right up to the end. See his "Last Interview": "I have always been interested in this theme of survival, the meaning of which is *not to be added on* to the living and dying. It is originary: life *is* living on, life *is* survival [la vie *est* survie]. To survive in the usual sense of the term means to continue to live, but also to live *after* death. . . . All the concepts that have helped me in my work, and notably that of the trace or of the spectral, were related to this 'surviving' as a structural and rigorously originary dimension" (2007, 26).

10. On childbirth as absolute arrival, see Derrida's "Artifactualities" (2002, 20); on originary technicity, see Bradley (2011).

11. See the NIH Human Microbe Project, http://www.nih.gov/news/health/jun2012/nhgri -13.htm.

12. The penultimate paragraph in the Hegel column is the "final" one insofar as the very last section is in fact the beginning. Page 1 is a continuation from the final page, and the column begins in midsentence, "what, after all, of the remains, today, for us, here, now, of a Hegel?" (Derrida 1985, 1a). The penultimate (though secretly final) paragraph of the Hegel column concludes: "The syllogism of spiritual art (epic, tragedy, comedy) leads esthetic religion to revealed religion. Through comedy then" (262a).

13. See my "The Archaeology of Biopower" (forthcoming) for a fuller elaboration of this argument.

14. When asked by Derek Attridge what he means by literature, or what it meant to him to be invested in literature as a young man, Derrida answers: "literature seemed to me, in a confused way, to be the institution which allows one to say everything, in every way. . . . The institution of literature in the West, in its relatively modern form, is linked to an authorization to say everything, and doubtless too to the coming about of the modern idea of democracy" (Derrida 1991, 37).

WORKS CITED

Bradley, Arthur. 2011. *Originary Technicity: The Theory of Technology from Marx to Derrida*. New York: Palgrave.

Clark, Timothy. 2013. "What on World Is the Earth?" *Oxford Literary Review* 35 (1): 5–24.

Derrida, Jacques. 1984. "White Mythology." In *Margins of Philosophy*, trans. Alan Bass. Chicago: University of Chicago Press.

——. 1985. *Glas*. Trans. John P. Leavey. University of Nebraska Press.

——. 1991. *Acts of Literature*. Ed. Derek Attridge. London: Routledge.

——. 2001. *The Work of Mourning*. Ed. and trans. Pascale-Anne Brault and Michael Naas. Chicago: University of Chicago Press.

——. 2002. "Artifactualities." *In Echographies of Television*. London: Wiley.

——. 2003. *Chaque fois unique, la fin du monde*. Ed. Pascale-Anne Brault and Michael Naas. Paris: Éditions Galilée.

——. 2005. *Sovereignties in Question: The Poetics of Paul Celan*. Trans. Thomas Dutoit. New York: Fordham University Press.

——. 2007. *Learning to Live Finally: The Last Interview*. Trans. Pascale-Anne Brault and Michael Naas. Hoboken, N.J.: Melville House.

——. 2008. *The Animal That Therefore I Am*. Trans. David Wills. New York: Fordham University Press, 2008.

——. 2011. *The Beast and the Sovereign*. Vols. 1–2. Trans. Geoffrey Bennington. Chicago: University of Chicago Press.

Derrida, Jacques. 2014. *The Death Penalty*. Vol. 1. Trans. Peggy Kamuf. Ed. Geoffrey Bennington, Marc Crépon and Thomas Dutoit. Chicago: University of Chicago Press. [*Séminaire: Peine de mort, volume II (2000–2001)*. Ed. Geoffrey Bennington, Marc Crépon, and Thomas Dutoit, Paris: Éditions Galilée.]

Doyle, Richard. 1996. *On Beyond Living: Rhetorical Transformations of the Life Sciences.* Palo Alto, Calif.: Stanford University Press.

Foer, Jonathan Safran. 2010. *Eating Animals.* Boston: Back Bay Books.

Gaston, Sean. 2013. *The Concept of World from Kant to Derrida.* Lanham, Md.: Rowman & Littlefield.

Hägglund, Martin. 2008. *Radical Atheism: Derrida and the Time of Life.* Stanford, Calif.: Stanford University Press.

——. 2009. "The Challenge of Radical Atheism: A Response." *New Centennial Review* 9 (1): 227–252.

Heidegger, Martin. 2001. *The Fundamental Concepts of Metaphysics: World, Finitude, Solitude.* Trans. William McNeil and Nicholas Walker. Bloomington: Indiana University Press.

Hegel, G. W. F. 1976. *Phenomenology of Spirit.* Trans. A. V. Miller. London: Oxford University Press.

Naas, Michael. 2009. *Derrida, from Now on.* New York: Fordham University Press.

——. Forthcoming. *The End of the World, and Other Teachable Moments from Jacques Derrida's Final Seminar.* New York: Fordham University Press.

Nealon, Jeffrey T. Forthcoming. "The Archaeology of Biopower: From Plants to Animals in Foucault's *The Order of Things* and *History of Madness*." In *Biopower: Foucault and Beyond,* ed. Nicolae Morar et al. Chicago: University of Chicago Press.

6

OF ECOLOGY, IMMUNITY, AND ISLANDS

The Lost Maples of Big Bend

CARY WOLFE

My title is taken from the name of Lost Maples State Natural Area, in the Hill Country of Texas, west of San Antonio—a place named for some of the most southerly occurring maple trees in the United States. But my focus will be on the lost maples (and other "relict" species, such as oaks) that live in isolated sections of Big Bend National Park, on the U.S./Mexico border, especially in areas of the park such as Boot Canyon and Pine Canyon. Big Bend, which takes its name from the "big bend" that the Rio Grande river takes southward toward the bordering Mexican states of Chihuahua and Coahuila, was established as a national park in 1944. It is a UNESCO biosphere reserve the size of the state of Rhode Island, a dramatic landscape in the Chihuahuan Desert made up of five distinct ecological zones bordered by the river for 118 miles, "an arcing linear oasis," as the National Park Service brochure you receive upon entering puts it, "a ribbon of green that cuts across the dry desert and carves deep canyons." Because of its geological and climatological complexity, Big Bend is a site of remarkable biodiversity, home to 1,200 different species, including 430 species of birds and over fifty endangered, threatened, or otherwise listed species, some of which, such as the Carmen Mountains white-tailed deer, exist nowhere else in the United States (Wauer and Fleming 2002, 21–22). As one of the most isolated, least-visited—and darkest—places in the lower forty-eight states, Big Bend is in many ways an island of biodiversity in the vast, arid wasteland that is western Texas.[1]

But in another sense, the Big Bend area is not isolated at all because it sits on the U.S./Mexico border in an era of considerable national debate about immigration across the border from points south—a fact reflected in the curious recent history of the border crossing at the east end of the park, at Boquillas del Carmen, which was closed in 2002 in the wake of the 9/11 attacks because of anxiety over "homeland security," reopening on April 10, 2013 (Grebowicz 2015, 3). Yet, as Margret Grebowicz notes in her wonderful little book *The National Park to Come*, Boquillas del Carmen is itself a kind of island; "the village itself would not exist were it not for the park, because it is completely isolated from the rest of Mexico by the protected land which surrounds it, three distinct Mexican wilderness areas under protection, together comprised in the 'sister park' to Big Bend" (44). Two islands, then, linked by the ebb and flow of world-historical events.

Like Grebowicz, I want to explore these two registers together—the ecological and the political—and how the conceptual apparatus of the island functions at their point of articulation in the place that is Big Bend. We tend to think of islands in horizontal and nautical terms, but with Big Bend we find another kind of island: the island of the Chisos Mountains—the southernmost mountains in the United States—mountains that sit in the center of the park, rising to a height of 7,800 feet, "surrounded by an ocean of Chihuahuan desert," as one observer puts it, dropping to 1,800 feet above sea level at its lowest point, where temperatures from late spring into early fall routinely reach well above 100 degrees (Wauer and Fleming 2002, 6). As the park's entrance brochure describes it, "if the Rio Grande is Big Bend country's linear oasis, then the Chisos Mountains are its green island in a sea of desert," a horizontal oasis in a scorching, inhospitable landscape, creating a distinct microclimate in terms of both temperature and moisture. These "sky islands" inhabit a geological terrain, as John McPhee (1981) writes in *Basin and Range*, that is not "corrugated, like the folded Appalachians, like a tubal air mattress, like a rippled potato chip. This is not—in the compressive manner—a ridge-and-valley situation. Each range here is like a warship standing on its own . . . as if they were members of a fleet without precedent" (47). Because the basins separating such sky islands are typically quite vast and dry, McPhee continues, "animals tend to be content with their home ranges and not to venture out across the big dry valleys." As one of his interlocutors explains, "the faunas in the high ranges here are quite distinct from one another. Animals are isolated like Darwin's finches in the Galápagos" (48).

And the same is true of the flora in these sky islands, nowhere more so than in places like Boot Canyon in Big Bend, which, because of its particular geology and aspect, is an island within an island, part of the Moist Chisos Woodland Formation, which consists of only about eight hundred acres along the northeastern slope of the Chisos—in a park comprising over eight hundred thousand acres. As one naturalist's guidebook notes, radiocarbon dating in Big Bend shows that fifteen to twenty thousand years ago the vegetation in Big Bend's desert valleys were much like what we find in the Chisos today, suggesting that conditions in the area during the Pleistocene were much moister than they are now. About ten thousand years ago, the climate in Big Bend began to change from relatively high levels of precipitation to scant rainfall, becoming dryer over time (Wauer and Fleming 2002, 6–7). One result of this is the presence of so-called relict species that can live nowhere else in the park, such as big tooth maple, Arizona cypress, quaking aspen, and several species of oaks, which were stranded in Boot Canyon and Pine Canyon with the retreat of the last ice, far from their normal alpine habitats farther north in the Rockies and Sierras. What we find here, then, is an island that is both spatial and temporal; the trees are an index of geological time and climate change against which the vagaries of human time and human civilization—such as the closing and opening of the Boquillas Crossing—may be contextualized. But they are also traces that register the presence of an absence—not just the absence of the ecosystems in which the maples, oaks, and aspens are typically found, much farther north, but a much more profound absence that challenges the commonplace notion in ecological thought that "everything is connected," an absence that challenges, that is, the notion of "world," in which islands would just be nodes, points of interconnection in a larger, encompassing fabric of life.

To get at what I mean, I want to linger over Jacques Derrida's counterintuitive contention, in his remarkable reading of *Robinson Crusoe* in the second volume of *The Beast and the Sovereign* seminars, that "there is no world, there are only islands" (2011, 9). Taking up, as he has before, his engagement of Heidegger on the question "what do beasts and men have in common?" (8), Derrida offers in reply this quite remarkable passage made up of a movement through three possible theses:

1. Incontestably, animals and humans inhabit the same world, the same objective world even if they do not have the same experience of the objectivity of the object. 2. Incontestably, animals and humans do not inhabit the same world, for the human world will never be purely and simply identical to the world of animals. 3. In spite of this identity and this difference, neither animals of different species, nor humans of different cultures, nor any animal or human individual inhabit the same world as another, however close and similar these living individuals may be (be they humans or animals), and the difference between one world and another will remain always unbridgeable, because the community of the world is always constructed, simulated by a set of stabilizing apparatuses, more or less stable, then, and never natural, language in the broad sense, codes of traces being designed, among all living beings, to construct a unity of the world that is always deconstructible, nowhere and never given in nature. Between my world . . . and any other world there is first the space and time of an infinite difference, an interruption that is incommensurable with all attempts to make a passage, a bridge, an isthmus, all attempts at communication, translation, trope, and transfer that the desire for a world . . . will try to pose, impose, propose, stabilize. There is no world, there are only islands.

(8–9)

Now it is the first thesis, I would argue, that dominates ecological thought and environmental ethics—we share the same world, we just experience it differently—but what I want to argue here is that it is actually the *third* thesis that is the most radically ecological. To show why, I would like to make brief recourse to systems theory, but before I do, we should remember that, for Derrida, this is not an affair of humans and animals only. Or more precisely, we can never really be certain, once and for all, what forms of life fall under this description. What we *do* know, he asserts, is that death is

not the end of this or that, this or that individual, the end of a who or a what *in the world*. Each time something dies, it's the end of the world. Not the end of a world, but of the world, of the whole world, of the infinite opening of the world. And this is the case for no matter what living being, from the tree to the protozoa, from the mosquito to the human, death is infinite, it is the end of the infinite. The finitude of the infinite.[2]

(Naas 2014, 188n14)

There is no world, then; there are only islands, and this is true, at least potentially, for all forms of life, "from the tree to the protozoa."

Derrida's assertion might seem counterintuitive, but it will seem less so if we remember that in the terms of biological systems theory, "there is no world" precisely for the reasons we may trace back to Jakob von Uexküll's work on human and animal *umwelten* (2010) and forward to those who work on the biology of consciousness and cognition, such as Humberto Maturana and Francisco Varela, whose theory of "autopoiesis" demonstrates that what counts as "world" is always a product of the contingent and selective practices deployed in the embodied enaction of a particular autopoietic living system, which is always *closed* and self-referential on the level of its particular mode of "organization" but *open* to its environment and its perturbations on the level of "structure" (Maturana and Varela 1998).[3] "If we deny the objectivity of a knowable world," they ask, "are we not in the chaos of total arbitrariness because everything is possible?" The way "to cut this apparent Gordian knot," they write, is "to bypass entirely this logical geography of inner versus outer" by noting, for example, that

> the operational closure of the nervous system tells us that it does not operate according to either of the two extremes: it is neither representational nor solipsistic.
>
> It is not solipsistic because as part of the nervous system's organism, it participates in the interactions of the nervous system with its environment. The interactions continuously trigger in it the structural changes that modulate its dynamics of states. . . .
>
> Nor is it representational, for in each interaction it is the nervous system's structural state that specifies what perturbations are possible and what changes trigger them.
>
> (135, 169)

Crucial here—and easy to miss—is that *time* is constitutive in this ongoing navigation of the system/environment relationship, as the phrase "dynamics of states" suggests. And this is true not only ontogenetically but phylogenetically, in evolutionary terms. There is no "world," in other words, not just because what counts as "world" depends upon any particular life form's autopoietic self-reference, its own internal mode of organization, but also because the system/environment relationship is above all a phenomenon of *temporalized complexity*. Indeed, as the theoretical biologist and Macarthur fellow Stuart Kauffman argues, the world is "enchanted" precisely *because* there are no "entailing laws" that govern, in Newtonian fashion, the evolution of the biosphere and its various forms of life. As Kauffman puts it, even before we reach the level of what he calls "Kantian wholes" such as megafauna and -flora, we have to ask:

> Has the universe in 13.7 billion years of existence created all the possible fundamental particles and stable atoms? Yes. Now consider proteins. These are linear sequences of twenty kinds of amino acids that typically fold into some shape and catalyze a reaction or perform some structural or other function. A biological protein can range from perhaps 50 amino acids long to several thousands. A typical length is 300 amino acids long. Then let's consider all possible proteins [of a] length [of] 200 amino acids. How many are possible? Each position in the 200 has 20 possible choices of amino acids, so there are 20×20×20 200 times or 20 to the 200th power which is roughly 10 to the 260th power possible proteins of length 200. Now let's ask if the universe can have created all these proteins since its inception 13.7 billion years ago. There are roughly 10 to the 80th particles in the known universe. If they were doing nothing, ignoring space-like separation, but making proteins on the shortest time scale in the universe, the Planck time scale of 10 raised to the −43 seconds, it would take 10 raised to the 39th power times the lifetime of our universe to make all possible proteins of length 200 just *once*. In short, in the lifetime of our universe, only a vastly tiny fraction of all possible proteins can have been created. This means profound things. First, the universe is vastly non-ergodic. It is not like a gas at equilibrium in statistical mechanics. With this vast non-ergocity, when the possibilities are vastly larger than what can actually happen, history enters.

(Kauffman n.p.)

Of course, Kauffman argues, this principle obtains even more radically at the level of "Kantian wholes," and from this vantage, what we confront in the lost maples of Boot Canyon is, precisely, a materialized "trace," as Derrida would put it, whose inscrutability haunts the present with retentions from an evolutionary past and protentions of an evolutionary future whose radical alterity resides in the fact that they are constituted by a complexity of recursive system/environment relations that are in principle, as Kauffman argues, nontransparent to those who would control, direct, or predict them, try as they might. In other words, there is no "there" there, there is no "world"—not in the sense of the world not being "real" but in the sense of a philosophically *realist* account of the world not being real. (And it is the utter confusion of these two claims, an ontological claim and an epistemological claim, the question of materialism and the question of realism, that has caused so much confusion in the contemporary theoretical landscape, as if disagreeing with the position called "philosophical realism" automatically means believing that the material world isn't real.)[4] No one dispatches this problem more elegantly and lucidly than the philosopher Richard Rorty in *Objectivity, Relativism, and Truth*, where he points out that philosophical "idealism" and philosophical "realism," which seem opposites, are in fact two sides of the same coin called philosophical *"representationalism."* As Rorty (1991) puts it:

> For representationalists, "making true" and "representing" are reciprocal relations: the nonlinguistic item which makes S true is the one represented by S. But antirepresentationalists see both notions as equally unfortunate and dispensable. . . . More precisely, it is no truer that "atoms are what they are because we use 'atom' as we do" than that "we use 'atom' as we do because atoms are as they are." *Both* of these claims, the antirepresentationalist says, are entirely empty. Both are pseudo-explanations.
>
> (4)

Nevertheless—and I have explored this question in some detail elsewhere[5]—in the face of this complexity and infinitude, we are forced to make decisions all the time about what Derrida calls the "killing" and the "letting die" of various forms of life, human and nonhuman—including, for example, the "letting die" of the big tooth maple and the Carmen Mountains white-tailed deer through the ongoing global warming that will, in time, force areas such as Boot Canyon beyond a fatal threshold for these species in terms of both temperature and aridity. As Derrida puts it in *Philosophy in a Time of Terror*:

Can't one terrorize without killing? Does killing necessarily mean putting to death? Isn't it also "letting die"? Can't "letting die," "not wanting to know that one is letting others die"—hundreds of millions of human beings, from hunger, AIDS, lack of medical treatment, and so on—also be part of a "more or less" conscious and deliberate terrorist strategy? We are perhaps wrong to assume so quickly that all terrorism is voluntary, conscious, organized, deliberate, intentionally calculated.

(2003, 108)

And this leads in turn (in both Derrida's discussion and in my own) to the question of *community*—whom do we "put to death"? whom do we "let die"? who is protected?—and to the *immunity* and *autoimmunity* with which it is, for Derrida and others such as Roberto Esposito and Donna Haraway, ineluctably bound up. These are precisely the sorts of questions that Grebowicz is interested in when she argues that the idea of the national park produces a kind of phantasmatic form of democratic community, which then gets naturalized by particular concepts of "nature" and "wilderness." As she puts it,

nature not only becomes a useful locus for the democratic ideal; it allows us to imagine that democracy is not a form of modern politics, but some original human state. The idea that this kind of original experience, this opportunity to be "found," should be available to everyone and not just the elite creates the effect of a people who can become truly, once and for all, themselves, speaking as themselves in this, the final form of social organization, final because it is original.

(2015, 25)

It is worth exploring for a moment, I think, the extent to which this characterization might apply to a figure who is certainly one of the most important and penetrating philosophers to probe the relationship between community and *im*munity, especially in the context of attempting to think a form of "affirmative" biopolitics that would take seriously the claims of nonhuman life on our forms and norms of community, namely, the Italian political philosopher Roberto Esposito. Like Derrida (and this is not surprising, given that Esposito freely acknowledges the indebtedness of his concept of immunity to Derrida's notion of the *pharmakon*), Esposito

(2013) realizes that "the idea of immunity, which is needed for protecting our life, if carried past a certain threshold, winds up negating life," so that "not only is our freedom but also the very meaning of our individual and collective existence lost: that flow of meaning, that encounter with existence outside of itself that I define with the term *communitas*" (62). And like Derrida, Esposito sees the semantics of the "person" and the "individual" as inadequate to the task of thinking questions of community, democracy, and so on: "One cannot base a philosophy of community on a metaphysics of the individual," he writes (16).[6]

For Esposito, we must not think of community as "an external injunction that addresses us from elsewhere but something more inherent. We need community because it is the very locus or, better, the transcendental condition of our existence, given that we have always existed in common" (15). We will want to question this last assumption in a moment, but for now, we should note that, for Esposito, community is at the same time "both necessary and impossible" because it is "something that determines us at a distance and in difference from our very selves, in the rupture of our subjectivity, in an infinite lack, in an unpayable debt, an irremediable fault. . . . We are lacking that which constitutes us a community, so much so that we must conclude that what we have in common is precisely this lack of community" (15). If the Derridean resonances of this formulation are clear enough, they become even clearer when Esposito asks:

> What else is community if not the lack of "one's own"? . . . This is the meaning that is etymologically inscribed with the very *munus* from which *communitas* is derived and that it carries within itself as its own nonbelonging to itself, as a not belonging, or an impropriety, of all the members that make up community through a reciprocal distortion, which is the distortion of community itself. . . . If community is nothing but the relation—the "with" or the "between"—that joins multiple subjects, this means that it cannot be a subject, individual, or collective.
>
> (29)[7]

But if community is impossible—if indeed it is nothing but the conjunctive and, as it were, formal semiotic relation of difference of the "with" or the "between"—then it is not at all clear how the question of community countenances the entire semantic nexus of "lack," "guilt," "fault," "debt," "perversion," and so on that eventuates in Esposito's key claim that "melancholy is not something that community contains along with other attitudes, postures, or possibilities but something by which community itself is contained and determined," resembling "a fault or wound that community experiences not as a temporary or partial condition but as community's only way of being" (28). The problem is not just that such a formulation would seem to thrust us back into the domain of "the person" and the "the individual" (isn't it only persons that experience melancholy?) that Esposito has already declared beside the point for thinking questions of biopolitics—an especially acute problem for a thinker who wants to argue that, in this day and age, "the entire world seems increasingly to be a body united by a single global threat that holds it together and at the same time risks smashing it to pieces," that "a single destiny binds the world, the whole world, and its life. Either the world will find a way to survive together, or it will perish as one" (76).

Rather, the primary problem can be brought into sharper focus by remembering the distinction Derrida makes in many places in his early work between the *lack* and the *absence* of Being, center, presence, *arché*, *telos*, and so on—including the being in common that is community—an absence generated by the force of iterability, the trace, *différance*, and so on: an absence that must be thought "without *nostalgia*," as he puts it in *"Différance."* Indeed, he writes, "we must *affirm* this, in the sense in which Nietzsche puts affirmation into play, in a certain laughter and a certain step of the dance" (1982, 27). Rather than launching into a long discussion of Derrida's early work at this juncture, I will rely here on Matthias Fritsch's able summary in a piece focused very much on politics and political theory, specifically on Chantal Mouffe's attempt to ground a theory of constitutive social antagonism and its centrality to democracy in the Derridean economy of *différance* and iterability. As Fritsch (2008) writes,

> *Différance* names the empty gap, the differential relation, between elements without which they cannot function. Secondly, however, the deferral aspect of *différance* also signals that the differentiation process never comes to a close, but is begun anew with each new instance of an ele-

ment's use or occurrence. If the set of elements were in principle finite, as in Saussure's theory of *langue*, the distinguishing relation would always pass through the same elements for all elements, thus providing them with a stability of meaning of which users of the structure of *la langue* can avail themselves in their parole, or actual speech.

If, however, one follows Derrida in denying the boundary between the structure and its use, between *la langue* and *la parole*, a necessary infinity of distinguishing references enters the system, which is not a quasi-system in the sense that its structurality consists in nothing other than its use or its event. Every use of an element must seek to return to its identity "after" its necessary detour through all other signs, but this detour turns out to be infinite. An element must refer to the future of its identity, but this same future is simultaneously deferred as a "to come" that never arrives as such. The future to come thus implies that identity is infinitely divided against itself, and hence, in open-ended and unforeseeable ways, infiltrated with the otherness needed to establish it.

(179–180)

And what this means—and it's why Derridean *différance* cannot be assimilated to or used to ground Schmitt's friend/enemy distinction as Mouffe wishes—is that "the relation between identity and its other is not exclusionary of a clearly demarcated 'outside', as for Derrida identity is not so much marked by excluding defined others, but by the infinite porosity of a supposed inside and outside, and hence its constant re-negotiation" (181).

A few crucial points follow from this. First, *time* is thus constitutive of the problem of the system/environment relationship for Derrida as it is for systems theory—a point that Martin Hägglund has made well in his book-length study (2008)—which is why Derrida writes that the play of *différance* designates "the unity of chance and necessity in calculations without end" (1982, 7)—precisely as described earlier by Kauffman. Or to combine the terms of systems theory and deconstruction, the formal element of a system can never be characterized by any kind of basal simplicity because its meaning depends upon the real-time dynamic state of the system in which the element functions—what Derrida would call the moment of the element's "performativity" or "iteration," in which case the same formal elemental features may have divergent, even opposite, functions and meanings at different points in time. For Derrida, this constitutive role of what he calls "the becoming-space of time or becoming-time of space" mitigates against all forms of sovereignty, conceived as what he calls "*ipseity*" or the "self-same," which are fatefully imbricated in this dynamic once they performatively attempt to enact or declare themselves as such. As he writes in *Rogues* (2005), sovereignty "always contracts duration into the timeless instant. . . . Sovereignty neither gives nor gives itself the time, it does not take time" (46, 109).

Second, a related point that derives from this first: Esposito may be right that our "mortal finitude" assumes the form of "reciprocal 'care'" and that "care, rather than interest, lies at the basis of community. Community is determined by care, and care by community" (2013, 25–26). But what Esposito seems both to invite and ignore—invite by his seemingly Derridean insistence that community is nothing but the spacing of the "with" and "between" of individual subjects and ignore by his reinscription of what Derrida would call a phantasmatic "being-able" that sets up such finitude as a "fault," "wound," or "perversion"—is a more profound understanding of what I have elsewhere called "double finitude": our finitude not just as embodied and vulnerable beings who need care but also as ones who, to enter into communicative relations and social bonds with others at all, are by necessity subjected to the "not me" and "not ours" of semiotic systems characterized by *différance* and the trace that, as Derrida puts it, must "be extended to the entire field of the living, or rather to the life/death relation, beyond the anthropological limits of 'spoken' language" (Derrida and Roudinesco 2004, 63).[8]

Third—and crucially—this has profound implications for Esposito's rendering of the immunitary paradigm of biopolitics, which in this light seems to reify the inside/outside relation in arguing that "whereas *communitas* opens, exposes, and turns individuals inside out, freeing them to their exteriority, *immunitas* returns individuals to themselves, encloses them once again in their own skin. *Immunitas* brings the outside inside, eliminating whatever part of the individual that lies outside" (2013, 49). What such a characterization misses is not just the "infinite porosity" and "constant re-negotiation" of the inside/outside relation noted by Fritsch but also what I have called the "second-order" turn of systems theory, which holds that—contrary to the understanding of autopoietic systems as solipsistic—the operational closure of systems and the self-reference based upon it arise as a practical and adaptive necessity precisely because systems are *not* closed, that is, precisely because they find themselves in an environment of overwhelmingly and exponentially greater complexity than is possible for any single system. To put it another way, systems have to operate selectively and "blindly" (as Luhmann puts it) not because they are closed but precisely because they *are not*, and the asymmetrical distribution of complexity across the system/environment difference is in fact what *forces* the strategy of self-referential closure and autopoiesis.[9] Indeed, the "second-order" turn, as I have argued elsewhere,[10] is to realize that the more systems build up their own internal complexity through recursive self-reference and closure, the *more* linked they are to changes in their environments to which they become more and more sensitive. They can buy more time in relation to environmental change, as it were, but there is also more to respond to.

From this vantage, we can now see more clearly the cascade of problems that eventuates from Esposito's seemingly innocent and commonsensical assertion that "we have always existed in common"—an assertion that not only seems counterfactual in light of the question of "world" as rearticulated through work in biological systems theory reaching back to Uexküll but that also raises the question of "from what Archimedean vantage point is such an assertion to be made, surmounting as it appears to do, the constraints and finitude of self-reference?" But what if we begin instead not with commonality but with difference and alterity, the finitude and situatedness from which we all blindly set out, and, with Donna Haraway (1991), see that "immune system discourse is about constraint and possibility for engaging in a world full of 'difference,' replete with non-self" (214)? In which case, as she writes, "immunity can also be conceived in terms of shared specificities; of the semi-permeable self able to engage with others (human and non-human, inner and outer), but always with finite consequences; of situated possibilities and impossibilities of individuation and identification; and of partial fusions and dangers" (225).

In this light, we can appreciate the more-than-metaphorical resonance of Derrida's assertion that "not only is there no kingdom of *différance*, but *différance* instigates the subversion of every kingdom. Which makes it obviously threatening and infallibly dreaded by everything in us that desires a kingdom, the past or future presence of a kingdom" (1982, 22). And this is precisely why Derrida suggests that "the thinking of the political has always been a thinking of *différance* and the thinking of *différance* always a thinking *of* the political, of the contour and limits of the political" (2005, 39). We are thus back to the question of sovereignty as what Derrida calls the "self-same" and "*ipseity*," "the power that *gives itself* its own law . . . the sovereign and reappropriating gathering of self in the simultaneity of an assemblage or assembly, being together or 'living together,' as we say" (11). Back, in fact, to the questions Grebowicz is interested in in *The National Park to Come*: questions of nation and national boundaries and how those map and are mapped onto a certain notion of "nature" and "wilderness" as "places where the political and the natural collapse into each other in a way that makes both democracy and wilderness policy appear transcendent, as if they belonged to the realm of natural, not civil, law" (2015, 29).

It is this political imaginary, I have been arguing, that lurks beneath Esposito's attempt to conjugate an "affirmative" biopolitics of "life," the "common," and "community," and it is decisively subverted by close attention to *ecology* in the sense that I have developed above out of systems theory and deconstruction—an ecology in which relict species and sky islands make it clear that "there is no world," an ecology of the "not us" and the "not ours" and the "not now"—a shift in perspective that has particular resonance in the geopolitical setting of Big Bend National Park. To put it another way, fundamental to an "everything is connected" idea of nature or ecology is the sovereign presumption that there *is* some Archimedean point from which one could see that "everything is connected," a vantage from which—to reach back now to the *first*-generation systems theory of Gregory Bateson[11]—the "map" may be presumed to coincide with the "territory," which is to say the presumption that one could escape the performativity, iterability, and autopoietic self-reference for which the "island"—especially in Derrida's remarkable second year of *The Beast and the Sovereign* seminars—is an especially powerful figure.[12] This is precisely why Derrida is so resistant to the discourse of "globalization," which is "simply alleged and not even there, and where we, we who are worldless, *weltlos, form* a world only against the backdrop of a nonworld, where there is neither world nor even that poorness-in-world that Heidegger attributes to animals" (2005, 155). Indeed, he asserts, "sovereignty is a circularity, indeed a sphericity" (13). In other words, we usually think "one world/everything is connected" idea of ecology and sovereign "national boundaries" are *opposites*, but what I have tried to show is that they are, in fact, two sides of the same coin.

NOTES

Originally published as "Of Ecology, Immunity, and Islands: The Lost Maples of Big Bend," *New Geographies* 8 (September 2016).

1. Big Bend has been designated a gold-star-level International Dark Sky Park by the International Dark Sky Association. See Grebowicz (2015, 88n32).

2. This quotation occurs in the second year of Derrida's seminars on the death penalty—in the session for January 10, 2001—as yet not translated into English but quoted in Naas (2014).

3. For more on how these questions cross-pollinate with Derrida's work, see Wolfe (1998, 78–94; 2013, 60–86).

4. For a subtle and thoroughgoing critique of this confusion, one that traces some of its adventures in the history of philosophy, see Padui (2011, 89–101).

5. Namely in the last two chapters of *Before the Law*.

6. See also Esposito (2011, 205–220).

7. For Derrida's conjugation of the *munus*, see Derrida (2005, 35).

8. For more on "double finitude," see Wolfe (2010, "Introduction").

9. See Luhmann (1995, 12–58). Further references are in the text. In this connection, Esposito's characterization of Luhmann's rendering of the immunitary paradigm is entirely symptomatic, when he writes that Luhmann's model focuses on "an internal self-regulation of systems that is *completely independent and autonomous* with regard to environmental pressures," which has the effect of "breaking *any possible relationships with the outside* but also of *calling into question the very idea of 'outside'*" (40, emphasis added).

10. See Wolfe (2010, xx–xxv).

11. As Bateson (1987) puts it, borrowing from the "general semantics" of Korzybski, "the map is not the territory," and "what gets onto the map, in fact, is *difference,* be it a difference in altitude, a difference in vegetation, a difference in population structure, difference in surface, or whatever" (449, 451). That is to say, all maps are about self-referential schemata and the systems of differences they do (and do not) make available to observation.

12. For an explicitly political exploration of the question of performativity, see Derrida's well-known essay "Declarations of Independence" (2002, 46–54).

WORKS CITED

Bateson, Gregory. 1987. *Steps to an Ecology of Mind: Collected Essays in Anthropology, Psychiatry, Evolution, and Epistemology*. Northvale, N.J.: Jason Aronson.

Borradori, Giovanna, ed. 2003. *Philosophy in a Time of Terror: Dialogues with Jürgen Habermas and Jacques Derrida*. Chicago: University of Chicago Press.

Derrida, Jacques. 1982. "Différance." In *Margins of Philosophy*. Chicago: University of Chicago Press.

——. 2002. "Declarations of Independence." In *Negotiations: Interventions and Interviews, 1971–2001*, ed. and trans. Elizabeth Rottenberg, 46–54. Stanford, Calif.: Stanford University Press.

———. 2005. *Rogues: Two Essays on Reason.* Trans. Pascale-Anne Brault and Michael Naas. Stanford, Calif.: Stanford University Press, 2005.

———. 2011. *The Beast and the Sovereign.* Vol. 2. Trans. Geoffrey Bennington. Chicago: University of Chicago Press.

Derrida, Jacques, and Elisabeth Roudinesco. 2004. *For What Tomorrow: A Dialogue.* Trans. Jeff Fort. Stanford, Calif.: Stanford University Press.

Esposito, Roberto. 2011. "The Person and Human Life." In *Theory After "Theory,"* ed. Derek Attridge and Jane Elliott, 205–220. New York: Routledge.

———. 2013. *Terms of the Political: Community, Immunity, Biopolitics.* Trans. Rhiannon Noel Welch. New York: Fordham University Press.

Fritsch, Matthias. 2008. "Antagonism and Democratic Citizenship (Schmitt, Mouffe, Derrida)." *Research in Phenomenology* 38 (2): 179–180.

Grebowicz, Margret. 2015. *The National Park to Come.* Stanford, Calif.: Stanford University Press.

Hägglund, Martin. 2008. *Radical Atheism: Derrida and the Time of Life.* Stanford, Calif.: Stanford University Press.

Haraway, Donna J. 1991. "The Biopolitics of Postmodern Bodies: Constitutions of Self in Immune System Discourse." In *Simians, Cyborgs, and Women.* New York: Routledge.

Kauffman, Stuart. "The Re-Enchantment of Humanity: Implications of 'No Entailing' Laws." Unpublished manuscript.

Luhmann, Niklas. 1995. *Social Systems.* Trans. John Bednarz Jr. with Dirk Baecker. Stanford, Calif.: Stanford University Press.

Maturana, Humberto, and Francisco Varela. 1998. *The Tree of Knowledge: The Biological Roots of Human Understanding.* Ed. J. Z. Young. Trans. Robert Paolucci. Boston: Shambhala.

McPhee, John. 1981. *Basin and Range.* New York: Farrar, Straus and Giroux.

Naas, Michael. 2014. *The End of the World and Other Teachable Moments: Jacques Derrida's Final Seminar.* New York: Fordham University Press.

Padui, Raoni. 2011. "Realism, Anti-Realism, and Materialism." *Angelaki: Journal of the Theoretical Humanities* 16 (2): 89–101.

Rorty, Richard. 1991. *Philosophical Papers.* Vol. 1: *Objectivity, Relativism, and Truth.* Cambridge: Cambridge University Press.

von Uexküll, Jakob. 2010. *A Foray Into the Worlds of Animals and Humans,* with *A Theory of Meaning.* Trans. Joseph D. O'Neil. Minneapolis: University of Minnesota Press.

Wauer, Roland H., and Carl M. Fleming. 2002. *Naturalist's Big Bend.* College Station: Texas A&M University Press.

Wolfe, Cary. 1998. *Animal Rites: American Culture, the Discourse of Species, and Posthumanist Theory.* Chicago: University of Chicago Press, 1998.

———. 2010. *What Is Posthumanism?* Minneapolis: University of Minnesota Press.

———. 2013. *Before the Law: Humans and Other Animals in a Biopolitical Frame.* Chicago: University of Chicago Press.

III

INORGANIC RITES

7

AFTER NATURE

The Dynamic Automation of Technical Objects

LUCIANA PARISI

The dominance of information technologies in contemporary culture has led to the widespread use of algorithmic processing, which has transformed our conception of automation. From smartphone apps to high-frequency trading, from the use of multiagent systems in design to social-media censorship and national surveillance-control centers, automation, it will be argued, coincides with the proliferation of inhuman functions of decision making. While industrial forms of automation aimed to reproduce physical movement mechanically, the automated systems of the information age have become intelligent agents, exhibiting a degree of autonomy from their programmers through complex algorithmic interactions.

It has been argued that the information age reveals a condition in which the overlap between automation and autonomy has been triggered by the speed of algorithmic procedures of evaluation, control, and decision making. In the contemporary context of "big data," rule-based formal structures of data processing work at volumes and speeds that exceed the capacities of individual human cognition and operate beyond human awareness (Galloway 2006, Terranova 2004, Fuller and Goffey 2012). From this standpoint, this chapter discusses the paradoxical overlapping between thought and automation by questioning the assumption that technoscience—and the contemporary convergence of biotechnology, nanotechnology, information technology, and cognitive science—cannot (and should not) constitute the ground for explaining the ontological consistency of being. This chapter questions the claim that it is now, more than ever, necessary to distinguish human life from human technoculture in order to keep the ontological autonomy of being separate from the history of technical objects.

This argument is not new, and there have been many attempts at rearticulating the relation between human nature and technology. As such, this chapter will pay particular attention to Gilbert Simondon's conception of the technical object because his theory has been central to contemporary debates about the question of technology in human culture (Deleuze 1993, 2001; Brian Massumi 2009; Bernard Stiegler 2009). *On the Mode of Existence of Technical Objects* discusses Gilbert Simondon's proposition that technical objects are not opposed to human nature but are instead the result of the artificial action of human beings through which a mutual relation with machines is engendered. I want to take this proposition seriously and suggest that this rather difficult question about the mode of existence of the technical object involves a fundamental acknowledgment that artificiality can and does surpass and irreversibly modify the autonomy of being. While rejecting both the hylomorphism of form and matter and the mechanical view of the universe that runs according to the repetition of the same initial conditions, Simondon's proposition seems to embrace the technoscientific revolution of the Enlightenment, for which machines were not simply tools but became embedded in culture and used as a means of governance. However, for Simondon, while technical objects are cybernetic organisms, they are also importantly imbued with a potential to aggregate and change over time. They are responsive to the environment and constantly probed by what he refers to as human creative action (Simondon 1958). Technical objects are not transcendent structures abstracted from materiality only to reveal once again the ultimate horizon or the desire to transcend human finitude. Simondon's view of the ontogenetic capacity of technical objects to exist in a dynamic field of relational constituency, I will argue, contributes to the disentangling of the philosophy of technology from the question of human finitude. In particular, it suggests that this disentanglement is crucial to an articulation of the ontological modality of technical objects. Nevertheless, this chapter also addresses the limits of Simondon's theory—showing that this ontological modality is attached to an energetic conception of continuity between what being *is* and how being *does*, which risks overlooking the tendency of the artificial to become autonomous from a mere energetic plane insofar as artificiality is also, and importantly, a construction or a conceptual architecture of a mode of thinking. To tackle this question, I propose we revisit the concept of the digital vis-à-vis the notion of computation by looking at recent developments in information theory—for example, Gregory Chaitin's (2006b)—that expand on Alan Turing's discovery of the limit of computation (that is, the incomputable or randomness) to suggest that automated forms of algorithmic calculation are not simply absorbers of human creativity. Instead, algorithmic systems of decision making seem to suggest that technical objects are not one with human nature and, more importantly, are not simply simulations of human cognitive functions. In opposition to the hype around interactive interfaces and the affective usage of digital media, I contend that in order to challenge the critical theory of technology that maintains a separation between human nature and human production, it is necessary to acknowledge that this production involves the automation of logical reasoning and cannot be explained solely as a mutual relation between energy and information. The production involves the inhuman function of decision making, which may be understood in terms of an emerging— or rudimentary—autonomy of and for mechanized thinking. In what follows, I will first outline Simondon's philosophy of technology in order to explain how it is underpinned by a strong antiautomation view. I will then clarify the relation between digitalization and computation so as to locate my critique of Simondon's energetic view of machines within the context of the historical development of the mechanization of logic.

TECHNICAL OBJECTS

For Gilbert Simondon, the mode of existence of a technical object needs to be explained fundamentally in terms of culture, but he also argues that the technical object is culture's constitutive vehicle for artificialization (1980, 47). This means that the technical object needs to be divorced from the critique of instrumental reason and at the same time from any substantialist and essentialist ontology. Instead, Simondon claims that "culture must incorporate technical entities into its body of knowledge and its sense of values. Recognition of the modes of existence of technical objects should be the result of philosophical consideration" (1).

The mode of existence of the technical object, however, is not to be thought in terms of prosthesis or artificial extension of an always already incomplete human nature. The reproducibility of the natural into a mechanized form is not Simondon's concern—he offers instead a counterintuitive theory that the technical object is at its most concrete phase when it is able to behave as a natural object. In particular, it is the concretization of the technical object (defined by increasing levels of internal resonances and by the crystallization of its phases) that makes it independent from the external regulatory environment (laboratory, factory) in which it was ideated. By achieving an internal coherence in which its systemic functions become closed and organized, a technical object therefore becomes comparable to an object produced spontaneously—independent from the environment and defined by its capacity to have incorporated external dynamics within itself. Its concrete state thus shows conditions of operation that are regulatory and self-maintaining. In other words, by becoming concrete, the technical object loses its artificial character. The concretized technical object approximates the mode of existence of natural objects because, like a natural object, it tends toward internal coherence, closing the system of the causes and effects that operate circularly inside its boundaries. It thus incorporates and transforms parts of the natural world that intervene as a condition of its functioning (14).

But what does it mean to conceive of the process of concretization of the technical object in terms of a tendency to achieve a natural state? It is evident that Simondon is here pointing to a new notion of functionality of the technical object. Far from being the incarnation of instrumental reason, Simondon argues for a dynamic function of the technical object primarily understood in terms of an energetics of time (9–10). Here the functionality of the technical object is explained in terms of critical phases of change in which the past and the present are coextensive rather than being defined by discrete and sequential units of time. The technical object is not, however, defined by an autonomous ontology or an autonomous being of time. Instead, the technical object incorporates and is at once part of the time of the living. Its process of concretization involves a transindividuation of different fields of potentialities that allows the technical object to become individuated as an ensemble of living energies (62). This is also how a technical object is able to acquire a natural state and thus cease to be the artifice of man.

From this standpoint, to enter Simondon's philosophy of technology is also to ask: What does it mean for artificial objects to become natural? To what extent does this view offer us a nonhumanistic or nonanthropomorphic approach of the technical that sees beyond the horizon of human finitude? I want to suggest that Simondon's emphasis on the mode of existence and, thus, of being—of the what and the how of the technical object—provides us with an important point of departure from which to develop a new view of the digital and of digital automation. In particular, I will focus on how Simondon responds to and radically transforms one of the kernels of the digital in his original analysis of the relation between energy and information. Simondon's view makes us realize the importance of an entropic conception of energy for understanding the mode of existence of technical objects (9, 27–31).

However, while recognizing Simondon's novel philosophical account of the technical machine developed in view of the historical shift from a thermodynamic to a cybernetic conception of technology, I want to add that a closer analysis of information theory may show us that the digital has more than simply a regulatory function for energy. Instead, information theory presents us with a challenging view of the mathematic theory of communication. It is well known that Claude Shannon (1949) conceived of entropy not as a threat to communication but rather as an element of surprise. Extending Shannon's insight, I characterize this element of surprise in terms of entropy that is immanent to information. From my discussion, it may become evident that Simondon's energetic view of technical objects is fundamentally an ontological study of technical objects, one concerned with the being of or the becoming of technical objects. However, I suggest that the theorization of technical objects in terms of energetic accumulators of being, involving an *ontogenesis of being*, cannot explain how thought becomes formalized as it is incorporated in technical objects. By drawing on Gregory Chaitin's (2006a, 29–32) view of the relation between energy and information in terms of the tendency of information to increase in size (i.e., information is itself entropic or non–size compressible), I aim to disentangle the symbiosis of being and thought granted by the mutual relation between energy and information.

Drawing on the central role that Alfred N. Whitehead (1929, 1967) assigns to capacities of abstraction, I will tend to view the automated processing of information in terms of a mode of thought that includes both physical and conceptual prehensions, or the speculative functions of elaboration and the becoming thought of physical data. I do so not to merge being and thought but rather to explain the process of becoming as the asymmetrical continuity between the physical and the conceptual. But, to clarify these distinct levels, I will first explore Simondon's energetic philosophy.

ANTIAUTOMATION

It is well known that Simondon rejected the positivistic view of the universe predicated upon the physics of reversibility premised in classical mechanics and sustained by the figure of the Laplacian Demon and its ability to calculate infinite amounts of information without error (Longo 2008, 68–70). Here, error is seen as an external and arbitrary incident in opposition to the function and efficiency of the machine. According to Simondon (1980, 49–50), this deterministic view of the machine, defined by the constant return to its initial condition, is represented by the regulatory function of information in cybernetics. As Simondon notes, the technical ensemble of the twentieth century—thermodynamic energy—is replaced by information theory, a guarantor of stability (regulation and stabilization). The cybernetic machine augments the quantity of information, leading to the emergence of negentropy that surpasses or incorporates the degradation of energy. The cybernetic machine is opposed to disorder, the heat-death of the universe. Its aim is to stabilize the world.

However, this clear-cut reaction against cybernetics may perhaps reveal a richer history of cybernetics to which Simondon's own theory has implicitly contributed. Cybernetics (and, in large part, computation) is less an extension of positivism and more an attempt at constituting a formal language that could take into account the inevitable truth of a universe losing the equilibrium theorized by concepts of irreversibility and entropy that constitute the core of the nineteenth century's second law of thermodynamics. Also, contrary to classical determinism, which argues that all results could be known in advance, cybernetics accounts for the centrality of variations, or the impossibility of foreseeing whether previous conditions could determine future outcomes.

Similarly, we know that Simondon's philosophy is influenced by Heisenberg's Uncertainty Principle and by Luis De Broglie's study of the wavelike behavior of particles, which reveal that the exact measurement of positions and momentum cannot be known.[1] Isabelle Stengers (2003), for example, points out that in Simondon's *On the Mode of Existence* there is a shift from a notion of causal indeterminism linked to the effects of measurement to Niels Bohr's conception of indeterminism beyond measure—here conceived as being constitutive of physical reality itself. Bohr's conception of indeterminism assigns a constitutive role to the energetic fluctuation of matter and explains that randomness is not an incident of measurement but is instead the very condition against which measurement could not but fail. Indeterminism is said to be the veritable condition of microphysical reality. It seems safe to argue that it is this new condition that allows Simondon to reconceive the technical object away from mechanistic tradition and move toward embracing an energetic understanding of matter.

In *Modes of Existence of Technical Objects*, we witness a renewed notion of physicalism, no longer inscribed in the substantialist, atomistic, and monistic traditions in which the smallest part is the universal absolute principle of the individual and of the individuated universe. Instead, Simondon's thermodynamic conception of matter in terms of potential energy, of scales and degradation, is rather a radicalization of microphysical indetermination—arguing that any object is primarily inexhaustible despite the tendency toward increasing degradation. But for Simondon, decay does not correspond to the inevitable finitude of Being or beings. Instead his view embraces the quantum state of matter, revealing that the modality of being is defined by critical phases of energy in which nothing is lost and the potential reserve is never scarce. A fundamental dynamism is here sustained by the constant process of transformation of energy for which there is no finite moment or arrest.

This energetic view also characterizes his insistence on the intrinsic validity of technology for human culture. From this standpoint, what Simondon calls concretization does not simply mean that ideas become physically tangible in human culture (a sort of Aristotelian physicalism of tools). Instead, technical objects are here cast as processors of live energy that becomes crystallized, and somehow objectified, and thus rendered a mode of existence in itself. But this is not to be confused with a notion of externalization or passive prosthetics of human faculties. Simondon's argument against mechanization instead aims to keep the technical object alive, to bring back its energetic impulse. In terms of culture, this means that technical objects are living repositories of ideas, practices, forms, and innovations that are modulated by or in a transductive relation with humans. Ultimately humans and machines do not share the same ontological conditions of being. On the contrary, Simondon is more interested in matter's self-organizing tendencies, involving the assemblage or temporal synthesis of heterogeneous elements that constitute the machine.

Hence, the mode of existence, or being, of technology is not to be framed in terms of a recuperation of ontology or an appeal to an eternal state. Rather, Simondon's reconceptualization not only of the mode of, but also the condition of, the technical object crucially reveals that the process of concretization is above all part of a larger ontogenetic process of being. As such, the technical object enjoys a condition beyond the world of the constructed and exists within the very energetic field of matter.

CYBERNETICS

In this section I will home in on the ways in which Simondon's philosophy can or cannot help us sustain the theorization of the inhuman condition of digital, algorithmic computation. Contrary to much of the critique against cybernetics, for Simondon, information is neither an input nor an output—it is not to be looked at in its finite state of individuated data. Information above all instead corresponds to "more than unity," insofar as its discrete, individuated, and binary state is inseparable from the preindividual field of energetic potential through which information individuates a difference. In other words, information is understood in terms of topological formations whereby a tendency toward change, and not a binary state, defines information in terms of its dynamic form, or "*la bonne forme*," as Simondon puts it (1980, 190).[2] It is this aspect of Simondon's philosophy that provides a sense of the neutralization of information that results in a rejection of its discrete state.

To put forward his philosophy of technology, then, Simondon calls into question both Shannon's mathematical model of communication and Norbert Weiner's conception of information. For Simondon, a philosophy of technology cannot be founded on research into form, or the efficiency of form, in the transmission of information. Against the univocal concept of information and energy, according to which these are equivalent and already determined unities, Simondon points out that information is not a preexisting form that is then to be applied to matter. In a radical move, Simondon reconceptualizes information as the variability of forms, the addition of variation to a form, and the unpredictability of variation in form (13–17).

It is clear that Shannon defines information neither as form nor as meaning. Information instead denotes the uncertainty of the outcomes and/or the necessity of noise in the transmission of a signal. But the cybernetician Norbert Weiner gives information a regulatory function, ensuring that the thermodynamic tendency of a system to run out of equilibrium is reduced, modulated, and put to work. Under this interpretation, information is the measure of order and is located in opposition to entropy, which is marked as the measure of uncertainty or thermodynamic decay. Following Simondon, one could argue that not only is information univocally related to energy while energy is locked within a process of irreversible degradation but also that cybernetics establishes an equivalence between humans and machines in its assertion that any system—living and nonliving—resolves the threat of entropy through the regulatory function of information. This regulatory function can both lead to a return to homeostasis and more notably, channel energy in order to overcome the threshold of decay and thus enter a new level of order. This is the notion of *negentropy* (the negative use of entropy that results in a modulation of energy in a transmission channel): information produces more information and becomes the motor of evolution toward a more complex level of order. Simondon rejects this functionalist and instrumental view of information as the regulator of energy's tendency toward chaos and instead argues for the necessity of an evolutionary mode of being whereby information is always already in the process of formation.

This is not simply, for Simondon, the claim that emphasis on the notion of "in-formation" is not an essentialist return to a fundamental analogy. This is instead an ontological claim of a totally different nature.

He uses this notion to argue that the manner in which (or how) information becomes is constitutive of what it is. Thus not substance but quantic indeterminacy is the motor of the emergence or modification of form— information has no content, no structure, and no meaning, in itself; it is disparity, not univocity.

But what does this mean, and how does this explain the mode of existence of technical objects as involving a becoming natural of the artificial? To answer this question, first we need to bear in mind that Simondon's fundamental conception of self-differentiation or individuation is characterized in terms of a primary disparity between information and energy. The mode of existence of a technical object thus implies a resolution between two different orders of reality— information and energy—sustained by the operation of transduction. This also means that the disparity between information and energy cannot be understood in quantitative terms and is therefore free from any kind of stable formalization.

Nevertheless, it may perhaps be reductive to argue that Simondon rejects all notions of quantification as a reduction of qualities to probabilities. He is known, on the contrary, to embrace the puzzles offered by the quantum physics of his time. Rather than being reductive, then, Simondon does not simply refuse but audaciously rearticulates the notion of quantity, quantification, and measuring from the standpoint of quantum physics. Drawing from De Broglie, Simondon uses the idea that a quantum leap coincides with the discharge of a measurable amount of energy but adds that such a leap also importantly coincides with a passing of a threshold to *a qualitatively new level of existence*. This inclination to explain the relation between energy and information in terms of what could be defined as the quality of a phenomenal experience also reveals that his indebtedness to De Broglie's notion of indeterminacy is based on the observer's measuring effect on the movement of quantic particles. Nevertheless, Simondon's close reading of this scientific enterprise adds another level of understanding of indeterminacy. While he suggests that quantification is always laboring under a deficit of potential and that formalization always labors under a deficit of energy, he crucially argues for a notion of qualitative differentiation and energetic variation of form that defies the centrality of the subject and the human. Extending Bohr's conception of indeterminacy as the fundamental condition of the microphysical fluctuations of matter, Simondon endorses a microphysical understanding of the technical object. This means that he rethinks the exchange between information and energy in terms of the qualitative variations of quantity. Here quantity is no longer described in terms of measure—as external observer or tool for measure—but becomes the manifestation of energetic variation. This means that there is no information without energy and that the relational exchange between them is precisely the process that ensures the constant transmutation of quantity into quality. The exchange indeed is not determined by a system of equivalence but rather of irreversible differentiation whereby information is always already affected by the energetic impulse. Here the energetics of a system are not defined by the system's decay; rather, because of the ontogenetic force of the relation, information becomes a catalyst for novel ensembles of elements and individuals. In other words, one can suggest that for Simondon it is not measurement that involuntarily causes fluctuation but that it is information itself that becomes an energetic form of measure.

This is a key point, as it places Simondon outside the phenomenology of experience and brings his work closer to a metaphysical investigation in the mode of the existence of technical objects—to the dynamic being of technical objects. In this sense, information ceases to be seen as the regulatory form of pure chance (otherwise described as entropy), and, going back to Simondon's radical notion of *la bonne forme*, information is seen now as corresponding to a process of individuation or ontogenetic formation (Simondon 1992). Information seen in this way includes the potential field for difference that triggers and conditions the process by which individuation occurs (Simondon has in mind a form-taking informational activity that starts from preindividual energetic variations). But it is a mistake to compare this preindividual field to an amorphous pool of flows. The preindividual is not undifferentiated but is radically indeterminate; that is, it does not correspond to an ontological being but circumscribes the margins of indetermination that are always left to run with, or are consistent with, the informational operations of the technical object. One could, thus, argue that Simondon offers a new model of cybernetics that, although it is set in opposition to the mechanistic view, nonetheless *embraces* information by way of its proposal to understand measuring and quantification as defined primarily by the indeterminacy of physical reality.

So, what does it mean to rethink information in microphysical terms? As opposed to a function set to reduce energy or to sacrifice and/or use qualitative variation, information is the expression of individuation and differentiation. In its transductive process of energy exchange, information is an event. In other words, information is subtracted from scientific objectivity and positivistic measurement in order to expose the levels of crystallization emerging from its relation with energy. One could also talk here of an *affective* relation to the extent that information is the affected recipient of an uncontainable spring of energy.

What does it mean to com-penetrate quantity with quality or to suggest that the indeterminacy of physical reality is fundamental to understanding what appears to be the inhuman or inorganic dimension of technical objects? And what are the consequences of developing an energetic conception of information? Can such a microphysical conception of the technical object indicate a posthumous life in which it is possible to account for a continuous relation between the biological and the technical, human nature and human production, without falling back into binarisms and flat ontology? And finally, to what extent can this microphysical motor of machines coincide at all with the same level of order as technical systems? By merging the molecular—physical and biological—and the technical orders of function into one plane of continuous variation, information itself becomes neutralized and unexplained.

By specifically addressing Simondon's dynamic conception of information, Gilles Deleuze (1993) helps us clarify that this form-taking activity is not the result of an external relation nor the result of interactive communication between energy and information. On the contrary, this activity involves "at least two orders of magnitude, of two disparate scales of reality . . . it implies a fundamental difference, like a state of asymmetry. If it is nevertheless a system, it is only insofar as difference exists as potential energy, as a difference of potential distributed within certain limits" (246). Deleuze conceives of difference in terms of an internal principle of dedifferentiation through which he argues that quantity comprises a difference in itself that cannot be precisely measured and, thus, remains an intensive quantity. Individuation on this model reads as the organization of a solution, of a "resolution" (derived from two complementary yet asymmetric activities) for an objectively problematic system (a metastable system existing in a state of asymmetry). This means that, on the one hand, difference coincides with internal resonance, "the most primitive mode of communication between realities of different orders" (Simondon 1992, 298), and on the other hand, it is information that, in its turn, establishes communication between two disparate levels—one defined by a form already contained in the receiver, the other by the signal carried from the outside.

The relation between information and energy, however, never involves a simple breakdown. Instead it entails a temporal resolution—a crystallization or concretization in the form of critical phases of change. The relation therefore is also a self-differentiation, always a more than one, so that the relation is both superimposed above the terms and simultaneous to itself. Nevertheless, the relation as individuated is again multiple because it is "multiphasic"; it is a "phase of becoming that will lead to new operations" (Deleuze 2001, 49). From this standpoint, one can understand the digital object as a mode of existence of a metastable system, which is itself a structure (not yet a synthesis) of the heterogeneous.

At this point, it is interesting to note how Deleuze and Deleuze and Guattari follow Simondon in theorizing the technical object in terms of machinic assemblages and are therefore motivated to ask to what extent, if at all, the machinic assemblage embraces a nuanced conception of information. One could argue that it is possible that a machinic assemblage involves the capacity of the technical object to behave as a metastable system in which the artificial is no longer built out of what exists in nature. If so, the machinic—but not the mechanical or the technical—becomes a natural object only to the extent that it is placed within the larger field of metastable systems in which the relation between information and energy primarily leads to—or becomes the causal efficacy for—the concretization of preindividual potentialities. In this case, information ceases to coincide with a quantification of energy and becomes the subjective form—in Whiteheadian terms—of energetic force. Entropy can then be thought as productive and not simply regulated by information; its power is amplified rather than repressed. Far from being the measure of (deterministic) chaos, here entropy marks the margin of indetermination of any system of communication. It becomes the condition for communication rather than the moment of its deconstruction. Entropy is indeed no longer put in contrast to information, insofar as quantity cannot be divorced from difference. Similarly, noise cannot be completely silenced or expelled from

the signal. Entropy is thus amplified and distributed across both communication channels and the algorithmic infrastructure of the communication apparatus itself.

But what is entropy in the digital? Do digital systems not operate precisely by transforming energy into information, modulating variation
through an immanent mode of calculating potentialities, ergo, at base by
eliminating friction?

Like Simondon, Deleuze and Deleuze and Guattari reject the automatism of communication, and at the same time they propose a new notion
of information and of the technical machine in which energy is the motor
of computation (Deleuze and Guattari 1983, 1996). Defying the deductive
method of formalism upon which computation is based, the machinic is
envisioned as a praxis defined by the encounter, the affective and transductive production of the absolutely novel. The disparity between information and energy is, thus, resolved by means of reverse causality. Entropy
is not the effect of measuring the incident emerging in communication
but is of another order all together that constitutes information.

My initial question about what it means for the technical object to
become naturalized through concretization needs to be readdressed. In
line with Simondon, the machinic is not defined by an ontological equivalence with the natural. Instead, the machinic explains the primary process
of difference defined by the superiority of the relation, the transductive
(and not deductive or inductive) operation of the encounter. Taken in
this way, the machinic is precisely this continuous process of transduction between energy and information, not simply a naturalization of the
technical. One can locate this point in Guattari's (1995) discussion of the
emergence, within the postmedia age, of machinic heterogeneities understood as living or live machines and no longer as a subset of technology.
This is a big departure from Laplace's or Turing's universal, mechanical
models in that the machine *is* ontogenesis, a mode of existence or a modification of being, an irreversible transformation of nature. However, if the
machine acquires the status of a living being, one must reject the specific
function of information because it, thus, can only be subsumed in the field
of energy and cannot exist on its own. From this perspective, it is impossible to surpass the incumbent paradox that has afflicted the critique of
digital technologies: While technical objects cannot be reduced to their
information function because their processes of individuation coincide

with becoming alive, digital objects, with their heavy reliance on discrete states and formal axiomatics, can never achieve such a state. What we are left with is the assumption that the becoming alive—or concretization or naturalization—of technical objects involves not a posthumous life or the seeping of inhuman, inorganic functions into the biological order of the organism but a tendency to incorporate the inhuman into the organic order without recognizing the information capacities of technical objects to be more than organic.

COMPUTATION

In this section, I return to theories of information that have tackled the question of entropy in the context of algorithmic machines, and I propose that the digital function of information needs to be contextualized in the history of mechanized logic. I revisit these kernels of cybernetics and information theory, from both the computational model of quantification and from the problem of entropy in systems of prediction and control, in order to develop a view about the specific function of information in automated systems.

As already mentioned, I suggest that the interactive paradigm, which more clearly sits in second-order cybernetics, has become the dominant form of mediatic communication today by way of its expansion of interactive algorithms. This paradigm specifically involves:

1. An organization of data no longer originally inputted into the system but rather derived from the outside via the social and personal use of software and the capacities of algorithms to interact with one another in the computational architecture of digital media
2. A computational model of prediction based on an open architecture of axioms, whereby new rules can be added as a result of the adaptive response of algorithms to the environment
3. A conception and use of the incomputable, whereby real numbers or infinities of data are no longer incalculable but are part of the data landscape that algorithms transform into patterns that then become quantified as randomness, complexities, or Omega infinity algorithms

The dominance of the interactive paradigm thus requires a new level of critique that can no longer hold the algorithmic regime of communication in opposition to the qualitative form of variation of the living, the subject, and thought. A return to computational theories reveals that incomputables or randomness—noncompressible quantities of information—are part of the computational tendency to elaborate information and establish a functional order of integration of data at the limit of the incomputable. Not only are incomputable quantities of data at the epicenter of communication, but also they unravel the importance of another order of entropy.

Here entropy—the measure of disorder/chaos/noise in a system, or simply uncertainty—is no longer, or not, exclusively resolved through a negentropic process of transduction whereby the information becomes modulated as if it were energy; the problem of algorithmic randomness in the computational processing of increasing quantities of data instead denotes the tendency of information itself to increase in size—defining a fundamental complexity corresponding to patternless information. Gregory Chaitin's (2006a) algorithmic information theory shows that the incomputable is not a probability known in advance that the program is expected to perform: Incomputables are maximally *unknowable* information quantities—infinities—that are nonetheless detected within computational processing as *incomputables*. These are at once discrete and infinite quantities of information. On this interpretation, computation does not resemble the Laplacian Universal Mechanism that repeats preset conditions, since incomputables (or randomness emerging in algorithmic sequencing) define computation as an incomplete affair. Likewise, such incompleteness ceases to mark the limit of information in terms of measure, as it involves the discovery of algorithmic entropy, that is, the capacity of information to increase in size/quantity. In short, in the algorithmic age, entropy cannot exclusively be explained by recourse to the function of primary energetic order. Whereas the function of energetic fields cannot of course be denied, what happens in the computational processing of vast amounts of data is indeed the emergence of patternless information. This defines the tendency of information as increasing in size complexity and reveals the function of information as implicating an algorithmic elaboration of primitive data. As such, in the order of iteration of the algorithmic function, this continuous compression of data leads to the emergence of Omega, or patternless information—of which only few ciphers can be decoded with rational numbers. This form of information complexity, however, must not be confused with an absolute incalculability that breaks down the ceaseless patterning of data into information. This entropic or size complexity of information can instead be used to explain how information constitutes its own domain of complexity, a domain emerging from the function of integration of data and involving an algorithmic metaelaboration for which randomness cannot be reduced to algorithms yet emerges from its own function (Chaitin 2006b, 74–81).

The size complexity of algorithmic information as described above points to an even more alien (or nonreducible to the energetic ground) complexity rather than an intensive notion of information. Drawing on Simondon, we saw that Deleuze argues for forms of energetic variation against the univocal relation between energy and information. Rather than opposing discreteness and continuity, or the digital part and the differential dynamics of the whole, the incomputable exposes incompleteness and complexity both at the core of discreteness and in models of quantification. On this view, one can suggest that the extension of microphysics to the technological field problematically overlooks the algorithmic order of the function of information and its own levels of complexity (that is, the incomputable character of the discrete). One could therefore argue that information entropy is central to what could be theorized as third-order cybernetics, which also explains anew the persistence of an inhuman form of elaboration of data that cannot be explained in terms of its physical causality.

But this new centrality of the algorithmic randomness/entropy embedded into computational interactive systems implies that it is no longer possible to overlook either the discrete nature of the digital or the possibility of algorithmic randomness in digital sequencing. It is therefore also impossible to exclude the mechanization (or the becoming inhuman) of the function of reason by either remaining attached to the primacy of the being of the sensible (or of affective potentiality) or by returning to a cognitivist being of the intelligible—a formalism that imposes a logic of symbol manipulations on data.

What I am proposing here is neither a re-vision of computation attached to the ontology nor the ontogenesis of being. Computation, within the context of algorithmic machines, instead points to an ontological process by which an inhuman thought emerges from the complex elaboration of data and patterns. The notion of information entropy in this instance enables a new understanding of the nature of the digital itself as expressed from within a dynamic conception of computation, already envisioned in the nonlinear form of interactive algorithms and parallel distributed systems. But this dynamism stems not from an energetics of information but rather from within the quantitative nature of information itself by way of the computational problem of incomputable and incompressible randomness. To revisit the question of the digital from that problem implies a retheorization of the meaning of information. The quantitative attribute of information denotes not simply a mechanical view of complex computational dynamics but points to the problem of incompressible quantities within algorithmic systems: a potentiality that is properly quantitative and involves the entropic character of quantities, or the irreversible tendency of information to increase in size. Despite its limits, Chaitin's work on the problem of the incomputable produces a new notion of both the digital and information—now defined by incomputable dynamics (irreversible change) rather than a relation of exchange with energy, which is not to dismiss the epistemological and ontological importance of such a relation. I intend to unpack the layers of dynamism in both the microphysical and the computational order. In other words, far from extending the microphysics of energy to information, I revisit the digital in order to find a way to articulate a computational dynamism proper to the strata of encoded information.

Simondon's recuperation of the technical, the technical machine, and, to some extent, reason from the cognitivist and representational understandings of technology constitutes an important step in the formulation of an inhuman form of information that bestows the tendencies to the artificial. However, I argue that, while this microphysical conception of the technical object has now become central to the dominance of an energetic view of interaction in the computational paradigm, further scientific and philosophical investigation into the algorithmic nature of computation can help us challenge the mechanical view of technical objects more directly. Doing additional research would, I think, expose the irreconcilable, perhaps nonresolvable, schism defined by the incomputable dimension of quantification rather than a mutual relation between information and energy. This is not simply to delimit our engagement with the complexity of computation to the mere level of information but to zoom in to this informational level in order to amplify the levels of dynamic change and thus propose that the two orders of randomness or entropy— both at the informational and energetic levels— point to a computational process in which one order cannot be collapsed onto the other.

INHUMAN THOUGHT

Information randomness indeed cannot be resolved by energetic random- ness but instead needs to be explained and unpacked in terms of a real mode of information production, one where the precondition is not being organically bound. I borrow this idea from Whitehead's discussion of the function of reason, which is set in contrast from both pure and practical reason and from deductive and inductive methods—he insists that the function of reason remains speculative (Whitehead 1929). This nonde- ductive understanding of reason and ultimately of a function of thought that can draw consequences from the physical prehension of reality, sus- tains my view that the artificial dimension of thinking encapsulated in the computational order of information also involves a speculative function- ing that goes beyond the ontological grounding of being in the energetic impulse of living. In a sort of reversal between life and thought, being and reason, Whitehead insists that thinking is a function for living. In other words, for Whitehead, reasoning involves the conceptual elaboration of physical data, or the envisaging of the consequences of such a process, which precisely requires placing reason after nature. Instead of a bifur- cation between the complexity of nature and the complexity of thought, Whitehead insists that the function of reason is to elucidate, explain, and unpack the reasoning behind living and thus beyond the art of surviv- ing. "Reason is a factor in experience which directs and criticizes the urge towards the attainment of an end realized in imagination but not in fact" (1929, 16). The finality of reason, however, is not part of a grand plan toward a preestablished schema. Instead, it is embedded in the actual occasion of which reason has a constructive function. This means that while Whitehead recognizes that all thinking emerges from the biophysi- cal constraints of the living, he also argues that the function of reason is to elucidate and evaluate the causes through which these can be transcended. For Whitehead, the function of reason has almost a metaconceptual task, to the extent that it transcends immediate fact by relating systems of ideas and generalizing observations. In other words, the function of reason involves a passage leading from the awareness of what is observed in (sense) experience—also defined in terms of a physical prehension—to the conceptual understanding of this observation, which leads to complex related entities as presented by mathematical thought (1920, 13–15). In particular, for Whitehead, mathematical thought enlightens all thought; mathematics is an essential element in the history of thought. Whitehead's process philosophy thus leaves an important place to mathematical logic and to the importance of separating mere matter of fact from the purely abstract conditions that they indicate. The function of reason is not determined by the direct apprehension of experience but is rather a function of abstraction of the particular entities involved and crucially involves the elaboration of the general conditions of the observations that they are expressible without having to make reference to particular relations or to particular relata occurring in that particular occasion of experience. For Whitehead, the rational attainment of this condition of generality ensures that these hold for an indefinite variety of other occasions (1967, 24–25). With mathematical logic, the concreteness of entities and relations has become a variable and absolutely abstract propositional function. This does not mean, however, that pure mathematics is imposed on concrete occasions. Instead, the function of reason is precisely to explain the process by which logic or the establishment of rules is attained through a process of abstraction in which the varieties of sense awareness are conceptually prehended. Whitehead's view on the speculative function of reason thus can help us push Simondon's philosophy of technical objects in another direction. As I have suggested, Simondon's reflections on the mode of existence of technical objects inscribes the relation between energy and information into a physicalist-oriented schema concerned with the material—or even biophysical—order for which information patterns are fundamentally propelled by a preexisting energetic field of dynamic or highly perturbed level variations. In other words, this relation is defined in terms of a transductive modulation by which information ultimately coincides with *informed energy*. For Whitehead, instead, process involves various levels of abstraction, including physical, conceptual, and metaconceptual or generalizing systems of relations that highlight the importance of a rational articulation of data. While for Simondon, the conceptual, ideal, or abstract configuration of a technical object needs to become concrete in order to be an effective or *bonne forme* of a culture, society, and mentality, Whitehead instead argues that indetermination, or not knowing, is the condition by which conceptual generalization—or determination—can be achieved through distinct degrees of prehension. From this standpoint, Whitehead's insistence on the central function of reason can support a view of technology that involves not simply a mechanical function of organization whereby information is defined by the algorithmic pattern that encodes it. His view allows us to shed light on the importance of the logic by which technology functions. This means that Whitehead's processual ontology could allow us to explain the emergence of computational logic not in terms of predetermined axioms but according to an emerging abstraction by which the energetic field cannot simply be the grantor of the ordering of information into patterns and the logical abstraction of patterns into a series of determinations. While it is true to say that algorithms are above all encoders of complexity and that Turing's Universal Machine was devised to compress infinite variations into finite instructions to accomplish determinate tasks, this algorithmic procedure cannot be isolated from the problem of the incomputable nor from the entropic tendency of information to increase quantitatively and add volume to patterns. From the standpoint of algorithmic information theory, computational logic has thus become ampliative, which means that results are not directly deducted from inputs (that is, they do not reproduce premises) but instead add more information to the starting set of instructions. In other words, the algorithmic elaboration of data also involves a growth of information. However, since the incomputable is the condition of information, the fallacy of deductive logic is not simply an error in the system expressing the limit of computation; instead, the emergence of a nondeductive logic of interactive, distributive, and learning algorithms implies the partial or continuous determination of randomness. Indeed, this nondeductive logic has its own function of reason that, one can suggest, implies a process of algorithmic determination of randomness. This can also be theorized as the advance of an alien thought in the form of complex algorithmic evolutive patterns that continuously—albeit partially—incorporate randomness.

Nevertheless, in relation to my suggestion, one could easily argue that the function of reason in computational processing is strongly limited by the lack of a metaconceptual form able to accomplish anything else than a faster mechanization of decision making. My suggestion therefore may work ultimately to reinforce rather than challenge the fundamental problem of automation that Simondon took pains to avoid. In other words, while Simondon's theory of a technical object may help us unearth the inhuman field of energetic potential that challenges the instrumental view of technology as an extension or a prosthetic of human skills, it could appear to some that the form of automation that I am proposing only, and at best, manages to reproduce the mechanistic assumptions of the formulation of the inhuman in the context of a software culture.

Nevertheless, it is possible to recognize that the automated processing of information does not fulfill Whitehead's model of the function of reason, insofar as algorithmic computation cannot achieve the order of abstraction that encompasses an infinite infinity of variations without representing any particularity. In other words, this kind of processing seems to be too attached to the operations of classification and labeling of information, however complex these may be. On the other hand, however, it is also problematic to dismiss the specificity of algorithmic information and the emergence of a computational formalism that moves from indetermination to the determination of randomness—or the incomputable. My point is that this kind of formalism could instead serve us to explain the complexity of the technical object in terms of the algorithmic procedures that run it and according to a form of logic that no longer obeys its initial premises or axiomatic preestablished truths. This is a form of abstraction that can neither be ascribed to a bland mechanism nor to the force of energetic impulse.

Neither mechanism nor vitalism thus will suffice to explain how important it is to radicalize both informational and energetic dynamisms because these have both become defined in terms of their margins of indetermination. This is true not only of the living but also, significantly, of thought—both concrete and abstract, both practical and theoretical, and in both its natural and artificial dimensions. From this standpoint, it is crucial that algorithms cease to be representational forms of energy and become hosts of the order of information entropy that must be addressed from within their computational matrices. The algorithmic information theory's conception of entropy can serve as a point of departure for theorizing a third-order cybernetics in which the technical object needs to be understood in terms of a function of reasoning that is inorganic and nonbiologically bound to any particular organism. How we understand this functioning, however, involves not falling back into the representational framework of cognitivism promoted in classic AI discussions. To conclude, I want to reiterate that my reflection on the advance of the inhuman thought of technical systems is inspired by the condition of randomness at the core of the complex function of elaboration in the algorithmic processing of data, which are embedded in socialities of all sorts within distributive agglomerations of systems. The task ahead, then, is to elaborate further the distinct orders of function of energetic and information systems without grounding them in a specific field (either energetic or mechanical). The task, therefore, is to analyze the process by which algorithmic classification and evaluation of data involves functions of abstraction and how these functions map the emergence of automated reasoning and its inhuman capacities of thinking after nature.

NOTES

1. This is also to say that Simondon does not reject but reevaluates the statistical calculus in the mathematical formulation of the principle of indetermination insofar as this calculation can be useful as a diagnostic tool of the effective reality of the quantic behavior of particles and the knowledge or the measure of it. A parallelism between what is and what is possible to know allows Simondon to propose a new notion of information as being always already attached to the reality of the preindividual.

2. Simondon's critique of the concept of information proposed by Norbert Weiner's cybernetics was also a critique of the notion of totality of form proposed by Gestalt, which, contrary to the hylomorphic schema, based on the analogy between matter and form, pointed to a fundamental equilibrium of the form. However, for Simondon the concept of information had to account for an internal dynamics that sustained the process by which the unity of individuation could rather encompass the variations of form, the crucial variation emerging from the tension between energy, matter, and information. Information is thus important because it can resolve the disparity between matter and energy insofar as it is here posed as another level of condition for achieving the integrity of complexity of form. However, as I discuss in this chapter, Simondon's notion of information is profoundly attached to the microphysical explanation that precondition information (and matter) is ontologically dependent on the highly vibrating consistency of the energetic field.

WORKS CITED

Chaitin, Gregory J. 2006a. *Meta Maths: The Quest for Omega*. London: Atlantic.

——. 2006b. "The Limits of Reason." *Scientific American* 294 (3): 74–81.

Deleuze, Gilles. 1966. "Review of Gilbert Simondon's 'L'individu et sa genèse physic-biologique.'" *Pli* 12 (2001): 43–49.

——. 1993. *Difference and Repetition*. New York: Columbia University Press.

Deleuze, Gilles, and Felix Guattari. 1983. *Anti-Oedipus: Capitalism and Schizophrenia*. Minneapolis: University of Minnesota Press.

——. 1996. *What Is Philosophy?* New York: Columbia University Press.

Fuller, Matthew, and Andy Goffey. 2012. *Evil Media*. Cambridge, Mass.: MIT Press.

Galloway, Alexander R. 2006. *Protocol: How Control Exists After Decentralization*. Cambridge, Mass.: MIT Press.

Guattari, Felix. 1995. *Chaosmosis: An Ethico-Aesthetic Paradigm*. Bloomington: Indiana University Press.

Longo, Giuseppe. 2008. "Laplace, Turing, and the 'Imitation Game' Impossible Geometry: Randomness, Determinism, and Programs in Turing's Test." In *Parsing the Turing Test*, ed. R. Epstein et al., 67–83. New York: Springer.

Massumi, Brian, Arne De Boever, Alex Murray, and Jon Roffe. 2009. "On Gilbert Simondon: Technical Mentality, Revisited." *Parrhesia* 7:36–45.

Shannon, Claude E., and Warren Weaver. 1949. *The Mathematical Theory of Communication*. Urbana: The University of Illinois Press.

Simondon, Gilbert. 1958. *Du mode d'existence des objets techniques*. Paris: Aubier, Editions Montaigne.

——. 1980. *On the Mode of Existence of Technical Objects*. Trans. Ninian Mellamphy. Preface by John Hart. University of Western Ontario. https://english.duke.edu/uploads/assets/Simondon_MEOT_part_1.pdf.

——. 1992. "The Genesis of the Individual." In *Incorporations*, ed. J. Crary and S. Kwinter, 297–319. New York: Zone.

——. 2005. *L'invention dans les techniques: Cours et conférences*. Paris: Seuil.

Stengers, Isabelle. 2003. *Cosmopolitiques II*. Paris: La Découverte.

Stiegler, Bernard. 2009. "The Theater of Individuation: Phase-Shift and Resolution in Simondon and Heidegger." *Parrhesia* 7:46–57.

Terranova, Tiziana. 2004. *Network Culture: Politics for the Information Age*. London: Pluto.

Whitehead, Alfred N. 1920. *The Concept of Nature*. Cambridge: Cambridge University Press.

——. 1929. *The Function of Reason*. Princeton, N.J.: Princeton University Press.

——. 1967. *Science and the Modern World*. New York: The Free Press.

8

NONPERSONS

ALASTAIR HUNT

People are people, so why should it be that you and I should get along so awfully?

—DEPECHE MODE, "PEOPLE ARE PEOPLE"

It is important to explore the category of the non-person *separately from that of the* thing.

—BARBARA JOHNSON, "ANTHROPOMORPHISM IN LYRIC AND LAW"

I AM A PERSON

On two consecutive days in January 2010, two law courts, one American, the other European, issued announcements that neatly express the contemporary legal establishment's view on the fundamental question, Who is a person?

The first was the U.S. Supreme Court's widely reported judgment on *Citizens United v. Federal Elections Commission*, published on January 21, 2010. The court was asked to decide whether the use by corporations and other collective bodies of their general treasury funds to pay for mass communications opposing or supporting candidates in political elections is a form of speech to which they have the constitutional right guaranteed in the First Amendment's free-speech clause. As commentators from across the political spectrum agree, in holding that they do have this right, the court effectively confirmed and even amplified what U.S. law has for decades already held: Corporations are persons with some of the constitutional rights human beings have.[1]

The day after the announcement of *Citizens United*, the European Court of Human Rights decided that two appeals seeking to force the Austrian government to allow a resident of Vienna to be appointed a legal guardian were inadmissible under the terms of the European Convention on Human Rights.[2] While the court's response was considerably less widely reported, what makes it especially interesting is that the individual in question, one Matthias Pan, is, as his family name intimates, a chimpanzee. The animal had been at the center of legal struggles between the government, animal activists, and Immuno, the corporation that in 1982 had imported him, along with ten other chimpanzees, from West Africa to Austria to be the subject of medical experiments. The legal struggles climaxed in 2006 when an anonymous individual offered to donate a substantial amount of money to the Austrian animal-rights organization Union Against Animal Factories (Verein Gegen Tierfabrik), which had assumed responsibility for Matthias's care, but only on the condition that he be appointed a legal guardian to ensure that the money was spent in his best interests. Applications to secure a guardian were met with rejection in all Austrian courts, up to and including the nation's Supreme Court (Oberster Gerichtshof), and the appeals to the European Court of Human Rights aimed to overturn these judgments. That they were so unsuccessful the court refused even to hear the case is no surprise. Unlike corporations such as Immuno, which for many years was the legal owner of Matthias, chimpanzees are not recognized by the court as persons. Indeed, the court's response reiterates the reality that no legal system in any country in any region of the world currently recognizes any animals as persons with fundamental legal rights.[3]

There exists widespread public skepticism of the established legal status of corporations as persons, especially, though not exclusively, on the political left. One expression of this skepticism is a series of photographs posted on the Flickr account of the Communication Workers of America Union, portraying men and women of various ethnicities holding pieces of paper on which these printed words appear:

I

AM A PERSON

_____ IS NOT

In each photo the blank line at the beginning of the second clause is filled in with the hand-written name of a corporation, such as "BANK OF AMERICA," "WALMART," "VERIZON," "AT&T."[4] It is not just the signs held by the subjects of the photos that communicate a confident conviction that the personhood of human beings is a self-evident truth given in the nature of things, while the so-called personhood of corporations is self-evidently an artifice, even a fiction. The portrait-style format of the photos, in which the face is central, invites us both to imagine the portrayed individuals as performing the announced assertion of personhood by using their mouths to speak the words written on the signs—after all, as any number of authorities have said, nothing confirms personhood more than saying "I"[5]—and, at the same time, to ask what corporations have mouths with which they can speak or faces that can be photographed.

No similar degree of public skepticism of contemporary law's categorization of animals as nonpersons exists even on the political left on either side of the Atlantic. In fact, if people are aware of the issue at all,[6] a majority are sincerely incredulous at efforts to have animals recognized as persons. One instance of this incredulity comes from that pretty reliable barometer of U.S. left opinion, Comedy Central's satirical news television program *The Daily Show*. On an episode that aired in February 2012, reporter Wyatt Cenac profiled an attempt by the animal-rights group People for the Ethical Treatment of Animals (PETA) to sue the marine animal theme park chain SeaWorld Entertainment on behalf of five performing orca whales, on the grounds that keeping them in captivity violates the Thirteenth Amendment's prohibition of slavery. During an interview the PETA spokesperson Lisa Lange explains the legal strategy: "Because the Thirteenth Amendment doesn't use the word 'person,' or 'people' in its language, we're hoping that a court will see that it can apply to animals," to which Cenac, in a voiceover paired with an image of the U.S. Constitution centered on the words "We the People," ironically responds: "She's right. If the Constitution was written for people, not whales, it would say that somewhere" (2012).[7] The implied point is not just that the Thirteenth Amendment can be fairly read as assuming slavery to be the ownership and exploitation of "persons" or "people" but that these words in turn obviously mean human beings, not animals. In his decision on the suit, Judge Jeffrey Miller of the U.S. District Court for Southern

California underscored just how self-evident the nonpersonhood of animals is: "The only reasonable interpretation of the 13th Amendment's plain language is that it applies to persons, and not to nonpersons such as orcas" (quoted in Siebert 2014, 33).

In short, then, whereas the public's skepticism toward actually existing corporate personhood bristles with the emphatic certainty of the civil rights–era slogan "I AM A MAN" that the above photos borrow, the acceptance of established law's skepticism toward calls for the legal recognition of animals as persons updates the cartoon published in the English periodical *Punch* in 1861, in which an ape is pictured holding a written sign posing a rhetorical question that has only one self-evident answer: "Am I a man and a brother?"

In this chapter I argue that an honest and consistent radical democratic position on the legal question "who is a person?" requires challenging the capture of political subjectivity within a humanist framework. I make this argument principally by reading the work of the political theorist Hannah Arendt. I turn to Arendt not because she offers any insights about such nonhuman candidates for personhood as corporations or animals but rather because she offers a compelling way to think about personhood as such. Admittedly, Arendt is an odd figure to appeal to for anyone seeking to question political humanism. Neither she nor the theorists who take up her writing, such as Étienne Balibar, Jacques Rancière, Judith Butler, and Bonnie Honig, ever seriously consider that the subjects of democracy could be anything other than human beings. Even those who do consider this possibility believe that for Arendt the measure of political subjectivity is "the human being's capacity for speech and language" (Wolfe 2013, 7). In my reading, however, Arendt's professed humanism only draws attention to the elements in her own argument about personhood, rights, and political community that provide resources for contesting the presumption that personhood always has a human face. What she helps us see, I contend, is that personhood is a much weirder thing that we generally assume it to be, for it possesses formal dimensions that are not in any simple sense human. My thesis, then, is not merely that nonhumans, such as animals, should become recognized as persons but that the inhuman, indeed impersonal formal dimensions of the figure of the person suggest that its reduction to the human is less an ontological necessity than it is an ideological effect.

BEFORE THE LAW

One way to illuminate the perplexities of legal personhood, Arendt argues, is to turn to the etymology of the word "person," which, she notes, "has been adopted almost unchanged from the Latin *persona* by the European languages with the same unanimity as, for instance, the word 'politics' has been derived from the Greek *polis*" (2003, 12). In *On Revolution* (1965) she offers the following reading of the Roman meaning of *persona*:

> In its original meaning, it [*persona*] signified the mask ancient actors used to wear in a play. (The *dramatis personae* corresponded to the Greek τὰ τοῦ δράματος πρόσωπα.) The mask as such obviously had two functions: it had to hide, or rather to replace, the actor's own face and countenance, but in a way that would make it possible for the voice to sound through. At any rate it was in this twofold understanding of a mask through which a voice sounds that the word *persona* became a metaphor and was carried from the language of the theater into legal terminology. The distinction between a private individual in Rome and a Roman citizen is that the latter had a *persona*, a legal personality, as we would say; it was as though the law had affixed to him the part he was expected to play on the public scene, with the provision, however, that his own voice would be able to sound through.
>
> (96–97)

While many scholars note the etymological fact that in ancient Rome the word *persona* had not one but two different meanings, one literary and the other legal, few stress, as Arendt does, that the legal sense of the word was historically secondary with regard to, and indeed derived from, an older use of the term in literature.[8] Given that the literary use of *persona* came first, the law's use is, she says, a "metaphor." No doubt, it is often said that calling a corporation a "person" is, or at least initially was, a metaphorical use of a term the law otherwise uses literally.[9] But Arendt's suggestion is, more radically, that *all* uses of "person" in a legal context—no matter whether they're applied to corporations, human beings, or whatever—signify in a manner that is figurative rather than literal.[10]

What's most intriguing about Arendt's account of the etymological archive of *persona*, though, is her further claim that the figurative transfer of the term from literature to law dramatizes a figurative logic already operating in the literary sense. If a mask enables an actor to perform a character, it does so because it "hides" or "replaces" a face. As she puts it elsewhere, a mask "cover[s] his [the actor's] individual 'personal' face" (Arendt 2003, 12). And what a mask covers a real face with is not, of course, another real face (that would be gruesome!) but a representation of a face. A mask conceals a face and displays an image of another face. To be sure, in the above passage Arendt notes that the theatrical mask reveals as well as it conceals, insofar as it allows the actor's "own voice to sound through," something that echoes her earlier account of the disclosure of the agent through speech in *The Human Condition* (1958, 179).[11] However, the stress she here places on the mask's concealing function suggests that although speaking and acting may be the modes in which agents manifest who they are, what makes it possible to speak and act in the first place is the assignment of a recognized role in which to speak. To appear as a person is in this sense *not* to put in a personal appearance. The prerequisite of disclosing oneself through speech is, oddly enough, actually concealing who and what one visibly is. The term that Roman law used figuratively, then, was *already* operating in a figurative manner in the Roman theater. In borrowing from the language of literature, the law took on not any old thing but the very literary principle of taking on.

Arendt's characterization of the figurative nature of the theatrical mask is borne out by a brief consideration of the substitution of appearing for being that characterizes the activity of acting. In an episode of the British sitcom *Extras*, the British actor Ian McKellen, playing himself, interviews an actor, Andy, played by Ricky Gervais, auditioning for a part in a play. Apropos of nothing in particular McKellen asks, "How do I act so well?" While Andy would quite reasonably expect some insights on acting method from an acclaimed practitioner in theater and film, what he gets is the following discursive stoner insight:

> What I do is I pretend to be the person I am portraying in the film or play. You're confused. It's perfectly simple. Case in point. Peter Jackson comes from New Zealand to me and says, "Sir Ian, I want you to be Gandalf the wizard." And I say to him, "You are aware that I am not really a wizard?"

And he said, "Yes I am aware of that. What I want you to do is to use your acting skills to portray a wizard for the duration of the film." So I said, "OK." And then I said to myself, "Mmmm. How would I do that?" And this is what I did. I imagined what it would be like to be a wizard. And then I pretended and acted in that way on the day. And how did I know what to say? The words were written down for me in a script. How did I know where to stand? People told me. If we were to draw a graph of my process, of my method, it would be something like this: Sir Ian, Sir Ian, Sir Ian. Action. [*Loudly*] Wizard, you shall not pass! Cut. [*Normal voice*] Sir Ian, Sir Ian, Sir Ian. You see?

(*Extras* 2006)

As you can imagine, Gervais's character is incredulous that the great Sir Ian McKellen is explaining the obvious to him. And indeed, the point is in a way quite simple: The character who appears to spectators in a dramatic production, although it is manifested by the actor's performance, is not the actor who exists. Gandalf the wizard is the creation of the wizardry of drama. This does not mean that for the spectators there is no perceptible relation between the specific actor and the person she pretends to be. In various ways spectators inevitably register an actor's singular interpretation of the role she is playing. When a character in a television series is transferred from one actor to another, viewers can spot the difference, and there is no guarantee that they will take to the new actor. Similarly, people can have their favorite James Bond actor. The phenomenon of easily recognizable "star" actors, such as Meryl Streep, also means that when viewers watch a film, they are, at some level, also perceiving the actor herself and experiencing her not just as a character but *as an actor*. But none of this means that a character has to be played by a certain actor. To say no one else could do Gandalf the way McKellen does is not to say that only McKellen could play Gandalf. No matter how good actors are, no matter how much they distinguish themselves as an actor, acting involves actors simultaneously disappearing as actors and appearing as something that they fundamentally are not. Indeed, the better the actor, the less visible they are to the spectators, since the actor's actions and words appear to be those of another. This happens when the mask put on by the actor is misread as a real face. Insofar as what is real in a play is the mask or character put on, it is the actor who is a fiction. "The doer is merely a fiction added to the deed," as Nietzsche said, "—the deed is everything" (1968). Neither does the absence of physical masks in modern acting complicate this point. For while acting now seems premised on the display of the actor's face in the detail of close-up camera shots, the use of makeup, costuming, and special effects remind us that all actors wear masks in the sense that they "put on," "take on," or "assume" a person, thereby appearing as someone they are not. In this regard, Ian McKellen's stoner insight intimates something more profound than it appears to. For in delivering his monologue on acting McKellen is acting as an actor who is not, at the moment, acting. Yet only someone who is not stoned enough would ever confuse the character of Ian McKellen that Ian McKellen is playing with Ian McKellen himself.

The same substitutive principle is in play, Arendt argues, when persons appear before the law. Just as theatrical spectators take an interest in not so much the actor herself as the character she puts on, so the law concerns itself with a person that each of us naturally is not but is asked to perform. Arendt's best explanation of the point comes not in *On Revolution* but in a speech she delivered in Copenhagen in April 1975 on the occasion of her acceptance of the Sonning Prize for contributions to European culture. Early in the speech she admits feeling discomfort with being "transformed into a public figure" through the act of public recognition that the award of the Sonning Prize is. By "personal temperament and inclination," she admits, "I tend to shy away from the public realm." What helps her bear public recognition is her knowledge that recognition takes as its object our masks and not our faces. After rehearsing the account of the etymology of the Latin *persona* offered in *On Revolution*, she writes:

> The Roman mask corresponds with great precision to our own way of appearing in a society where we are not citizens, that is, equalized by the public space established and reserved for political speech and political acts, but where we are accepted as individuals in our own right and yet by no means as human beings as such. We always appear in a world which is a stage and are recognized according to roles which our professions assign us, as physicians or lawyers, as authors or publishers, as teachers or students, and so on. . . . In other words, the advantage of adopting the notion of *persona* for my considerations lies in the fact that the masks or roles which the world assigns us, and which we must accept and even

acquire if we wish to take part in the world's play at all, are exchange-able; they are not inalienable in the sense in which we speak of "inalien-able rights," and they are not a permanent fixture annexed to our inner self. . . . It is in this sense that I can come to terms with appearing here as a "public figure" for the purpose of a public event.

(Arendt 2003, 7–8, 13)

The condition of Arendt's acceptance of the public recognition that comes with the award of the Sonning Prize is her conviction that the part of polit-ical theorist affixed to her is not something that she is required to confuse herself with. Like a mask she can don, rather than a face she irremissibly has, her public persona is a role she can inhabit rather than embody. It can be "affixed" to her without thereby being a "permanent fixture." To be sure, she well understands the great temptation of recognition, namely "taking oneself seriously and identifying the woman with the author who has his identity confirmed, inescapably, in public" (Arendt 1968a, 96). But as she underscores at the end of the speech, such identification is always an error because it can only offer recognition "*as* such and such, that is, as something which we fundamentally are *not*."

Thankfully we do not need to be as talented at acting as Ian McKellen or as good at theorizing politics as Arendt to appreciate that acting is some-thing required of us in so-called real life, including whenever we appear before the law. As Arendt puts it elsewhere: "Each society demands of its members a certain amount of acting, the ability to present, represent, and act what one actually is" (Arendt 1968b, 84). In fact, even if we are terrible actors, the sheer fact of being recognized as a legal person means we are taking on a role assigned to us. This is not to say that law cannot distinguish individual persons in their unique performance of their per-sonhood, the words they speak and the deeds they perform. Just as spec-tators can evaluate an actor's individual performance, so the law can judge us for our actions. Nevertheless, what makes the term *persona* attractive for the law is the fact that it refers to a role or position that an individual, or indeed several individuals, can inhabit rather than to the preexisting properties of an individual. Even as the law addresses each of us as an individual, it sees us not for who we really are but rather insofar as we conform to an institutionalized structure of address with its own schema of intelligibility. Focalizing whatever each of us is through the figure of the person that it does not derive from us, the law deploys its own mask-recognition technology to see the persons it can see. That what the law perceives is appearance rather than being should not, Arendt would insist, be taken as a slight against the law, for in the light of publicity "more than anywhere else, we have no possibility of distinguishing between being and appearing. In the realm of human affairs, being and appearing are one and the same" (1965, 88). We are, then, the persons we play before the law. For as soon as one enters its stage, the mask becomes one's face.

DISAPPEARED PERSONS

Even as the mask of legal personhood covers my own face, it is the spectators, including legal institutions, who decide whether the mask stays on it. Arendt's account of the spectators' role in mediating the practical relation of mask to face is marked by a keen awareness that not all faces are permitted to wear masks. In the passage from *On Revolution* we considered in the previous section she reminds us that Roman law, far from assuming that everyone, or at least every human being, is a person, conceived of personhood as a legal position the enjoyment of which was limited to certain kinds of human beings. "Without his *persona* there would be an individual without rights and duties, perhaps a natural man—that is, a human being or *homo* in the original meaning of the word, indicating someone outside the range of the law and the body politic of the citizens, such as for instance a slave—but certainly a politically irrelevant being" (1965, 97; cf. 36).[12]

What makes a slave in Rome such a good example of a nonperson, according to most scholars, is that they were objects of the very property right that, as nonpersons, they themselves did not have. In Roman law *persona* referred to the bearer of legal protections, such as privileges, immunities, and, most importantly, rights, such as the right to own property and the right of redress through the courts in cases of damage to or theft of property. As means to the end of *persons* who cannot be treated exclusively as means to ends, such property could include slaves, who were assumed into the category of the nonperson in the form of the *res vocalis*, the thing with the ability to talk. Possessed rather than possessive individuals, then, slaves did not just not own property; they did not even own themselves, being instead property owned by someone who was a legal person.[13]

Without denying that slaves are violently subject to the will of legal persons,[14] Arendt offers another, quite different explanation of their predicament. As well as suffering the loss of personal freedom, slaves, she contends, are also invisible in the sense that they do not appear in the light of the public realm. She makes this argument in the same chapter from *On Revolution* containing her discussion of personhood. Interestingly, however, she does not explicitly connect the two topics of slavery and personhood, giving instead as her major example of nonpersons the impoverished masses during the French Revolution. The reason why the latter feature so heavily in this chapter is that it is here that Arendt makes the argument, infamous to some, that although the French Revolution has had a great influence on subsequent political revolutions, it nevertheless failed to found a lasting democratic political community because it was overwhelmed by the economic or, in her terminology, "social" problem of poverty. The flipside to this thesis is, of course, that by comparison the American Revolution succeeded to the limited extent that it did because the Americans, luckily enough, did not really concern themselves with the issue of poverty. And why not? Because, as she writes, "the predicament of poverty was absent from the American scene." While this assertion seems to shock some readers into not reading further, Arendt, to her credit, immediately acknowledges that this is "a sweeping statement that stands in need of a twofold qualification" (1965, 58).

Her first qualification is that it would be more precise to say that what were not present on "the American scene" was not poverty but "misery and want." On this topic Arendt doesn't always keep her terminology straight—just few pages earlier she writes: "Poverty is more than deprivation, it is a state of constant want and acute misery" (50)—and I'll admit that the distinction between mere poverty and misery isn't one of her most exact. However, what is worth our attention is her suggestion that whereas in France *les miserables* were driven by the needs of their suffering bodies onto the scene of the political revolution, which they ruined, the merely poor in America didn't pose a serious social question and hence were able to witness the foundation of a democratic nation-state. I say "witness" because according to Arendt the poor did not participate, at least fully, in the act of revolution, for the simple reason that "the 'continual toil' and want of leisure of the majority of the population would automatically exclude them from active participation in government—though, of course, not from being represented and from choosing their representatives."[15] The lesson here is that the condition of poverty brings not so much want as "obscurity" or "darkness." Arendt attributes this insight to John Adams:

> The poor man . . . feels himself out of the sight of others, groping in the dark. Mankind takes no notice of him. He rambles and wanders unheeded. In the midst of a crowd, at church, in the market . . . he is in as much obscurity as he would be in a garret or a cellar. He is not disapproved, censured, or reproached; *he is only not seen.* . . . To be wholly overlooked, and to know it, are intolerable.
>
> (58–59)

For our purposes, all of this is a setup for Arendt's second qualification, which, since it is the one that especially interests us, I will quote at some length.

> And the fact that John Adams was so deeply moved by [the obscurity of the poor], more deeply than he or anyone else of the Founding Fathers was moved by sheer misery must strike us as very strange indeed when we remind ourselves that the absence of the social question from the American scene was, after all, quite deceptive, and that abject and degrading misery was everywhere present in the form of slavery and Negro labor.

Ironically referencing William Penn's description of the United States as "a good poor man's country," she continues:

> We are tempted to ask ourselves if the goodness of the poor white man's country did not depend to a considerable degree upon black labor and black misery—there lived roughly 400,000 Negroes along with approximately 1,850,000 white men in America in the middle of the eighteenth century, and even in the absence of reliable statistical data we may be sure that the percentage of complete destitution and misery was considerably lower in the countries of the Old World.
>
> (60–61)

One might think that admitting the fact of chattel slavery in America would scuttle Arendt's claim that misery was not present on "the American scene" and thus weaken her explanation for the comparative success of the American Revolution and the failure of the French. How can she hold that the miserable French masses overwhelmed their revolution but the miserable American slaves did not? Here is Arendt's bold explanation:

> From this, we can only conclude that the institution of slavery carries an obscurity even blacker than the obscurity of poverty; the slave, not the poor man, was wholly overlooked. . . . [The Americans' lack of feeling for African slaves] must be blamed on slavery rather than on any perversion of the heart or upon the dominance of self-interest. . . . The social question, whether genuinely absent or only hidden in darkness, was nonexistent for all practical purposes.
>
> (61–62)

While Arendt does not explicitly connect this passage to the discussion of personhood that comes later in the same chapter, each illuminates the other. In the first place, the passage on the obscurity suffered by slaves makes clear the extent to which an individual is a person only insofar as she appears before spectators. Arendt highlights the requirement by presenting exemplary nonpersons, slaves, as those who do the opposite of appear. The fundamental plight of slaves in America, before their loss of personal freedom, was their categorical disappearance from the public realm. Just how thoroughly slaves disappeared in public is something Arendt underscores with her sharp distinction between their obscurity and that of the nonenslaved working poor in revolutionary America. While John Adams could claim that the poor were overlooked as actors, they did nevertheless enjoy a range of important legally protected civil rights as persons. Indeed, the not-so-poor (and not-so-feminine and not-so-black) could even vote for wealthier political actors who in principle acted on their behalf and in their interests. As Arendt points out, however, slaves, as nonpersons, were not even permitted the privilege of this obscurity. Compared to the white poor, it was *they* who were "wholly overlooked." Returning to the theatrical basis of the legal metaphor of personhood, we could say that if poor white people in late eighteenth-century America were like actors who are given nonspeaking parts as extras on the political stage, slaves were like stagehands who, unlike even minor *personae mutae*, if they happened to be dimly perceived scurrying around in the darkness shrouding set changes between scenes, remained utterly unseen in the scenes taking place on stage.

Even if we accept that slaves in revolutionary America were categorized as nonpersons, though, we might still find it difficult to accept that their nonpersonhood consists in a fundamental disappearance. Could the revolutionaries really not see these human beings as persons at the very moment when they were engaged in a project of founding an independent nation-state on "the self-evident truth" that "all men are created equal"? It's helpful here to return to Arendt's account of personhood. For as much as the passage on slaves' disappearance highlights how being a person consists in a certain visibility, so the account of personhood helps us describe with some precision the nature of the disappearance of slaves. What is crucial is the distinction between, and confusion of, mask and face upon which, we've seen, legal personhood turns. For if nonpersons such as slaves are in a radical way unable to be seen, what is invisible about them is not so much a face as a mask on their face. The American revolutionaries did not suffer a kind of mass failure of sensory perception that prevented them from seeing faces that belong to human beings. In fact, the bodies of slaves, including their faces, could be seen very well—so well that slaves could be captured, counted, appraised, assigned a financial value, bought and sold, forced to work, raped, and so on. But even when slaves were within an empirical line of sight, they were seen in the mode of not being seen as wearing the mask of legal personhood. Again, the theatrical basis of the legal metaphor is useful. Because stagehands do not play any onstage roles whatsoever, our obscure awareness of them working in the dark is fundamentally different from our observation of any actors on stage. When we happen to see them, we may certainly recognize them as human beings, just like the actors, but if we are to enjoy the play, even to understand it, we must not see them as characters in the play. In a similar way, when the civil institutions founded by the revolutionaries gazed upon the faces of white human beings, they saw the claims of legal personhood. But when they gazed upon the faces of black human beings, they may have seen a human face, but without the mask of personhood such faces were legally and politically irrelevant.[16]

One way in which the appearance of persons and disappearance of nonpersons in the theater differs from that in the law bears highlighting. In the theater the lead actors, the nonspeaking extras, and the stage crew all fill their roles because of their talents and interests as individuals. On the political stage of late eighteenth-century America, by contrast, the highly visible political actors, the impoverished people who they in principle represented, and legal nonpersons such as slaves were believed to fill their roles because they were fitted for them by virtue of being born as members of certain groups. Meryl Streep, the minor actors supporting her, and the sound engineer all get their jobs largely because they possess the skills necessary to perform their jobs. But what qualified political actors, the silent people, and enslaved nonpersons for their roles was not anything they did but what kind of human beings they irremissibly were by nature. What decided whether they got their various jobs was not their résumé but their birth certificate.

In *On Revolution* Arendt highlights this breakdown of the metaphor when she notes that one naturally given fact about the individuals on and around the political stage in revolutionary America was that the rich and poor who appeared as legal persons were white, while the slaves who existed as nonpersons were black. She leaves implicit, however, the exact contribution of race to the formation of legal personhood. The most she says, as we've seen, is that the obscurity of slavery is "even blacker than the obscurity of poverty." But even this says a lot. For the sheer fact that the racial category of blackness *figures* a difficulty of perception implies something important about the specific operation of the substitution of legal mask and natural face at this time. It implies that appearance as a person was experienced as legible in the white skin with which some people were born, while the obscurity of nonpersons was experienced as legible in the black skin with which others were born. The problem for slaves, then, was not so much a painful misfit between black skin and white masks but that the legal mask of personhood, which strictly speaking is colorless, was understood only to fit, was even confused with, a face with white skin. The categories of race that divided free persons from enslaved nonpersons were, after all, taken to be self-evident at birth. American slavery, thus, exemplifies the kind of slavery Arendt describes in *Origins* as "an institution in which some men were 'born' free and others slave," in which "it was forgotten that it was man who had deprived his fellow-men of freedom, and . . . the sanction for the crime was attributed to nature" (1968b, 297). Thus Arendt herself helps us see what, given her effort to contrast the French and American revolutions, she herself may not wish to see, namely that the Americans displayed just as much contempt as their French counterparts for the distinction between the face and the mask. For in hallucinating the presence of the mask of personhood in their white faces, and the absence of this mask in black faces, they also attempted to "reduce politics to nature" (1965, 99).[17]

But the racially organized distribution of legal masks did not just naturalize slavery in the United States; it also excluded a large part of the population from membership in the "people." In *Origins* Arendt argues for a strict conceptual separation between the state and the nation, even as she tells the story of the widespread conquest of the former by the latter. The state, she explains, comprises a centuries-old set of legal institutions inherited from monarchical and despotic governments. By contrast, the nation emerged only in the eighteenth and nineteenth centuries when peoples began to experience themselves as culturally and historically bounded entities associated with a particular territory. With the formation of the nation-state during this time, the state, as the supreme legal institution, was tasked with the responsibility to protect everyone who happened to be living in its jurisdiction. What happened, however, was that the state increasingly came to "grant full civil and political rights only to those who belonged to the national community by right of origin and fact of birth. This meant that the state was transformed from an instrument of the law into an instrument of the nation" (Arendt 1968b, 230). Interestingly, Arendt elsewhere seems to exempt America from this narrative. "The United States is not a nation-state in the European sense and never was. The principle of its political structure is, and has always been, independent of a homogenous population and of a common past" (1959, 47). But however true the comparative heterogeneity of the nation in the United States was after the Civil War, beforehand it was very not true in one important aspect. Indeed, so long as slavery existed, "the people" in America comprised persons who, at least in one fundamental way, all looked alike: They were not black. Indeed, the radical disappearance of slaves from the public sphere in revolutionary America suggests that they rank among those non-European victims of imperialism whose radical exclusion from community anticipated the production of stateless peoples in Europe after World War I. Black slaves were, in other words, effectively stateless.[18] Although slaves were for statistical purposes counted as physically present on the sovereign territory of the United States, they were not counted as members of the community entitled to the legal protection of the state. They were not merely denied particular rights that members of a community enjoy. More fundamentally they were denied the membership in a community that is the very condition of having any rights.[19] Their predicament is, as Arendt puts it in reference to the example of the stateless flung up by World War I, "not the loss of specific rights" but "the loss of a community willing and able to guarantee any rights whatsoever" (1968b, 297). This is why the position of slaves is not one *before* the law but *outside* the law—or "outside the range of the law and the body politic of the citizens" (1965, 97).[20] It was by drawing upon the self-evident assumption that appearing as a person before the law was the same as appearing as racially white that the American revolutionaries, far from permitting black people to be agents or beneficiaries of their project, to number among "We the People" by whom and for whom the revolution occurred, could ignore them as politically irrelevant nonpersons.

THE RIGHT TO HAVE RIGHTS

While it might strike us as inexplicable that the law sanctioned racially organized slavery at the very moment it also gave expression to a democratic revolution, the law seems already to have corrected its error by extending the recognition of legal personhood to all human beings without exception. In the United States this correction took the form of legislation passed in the wake of the end of slavery (specifically the Civil Rights Act of 1866 and the Fourteenth Amendment of 1868).[21] On the more panoramic stage of international law it arrived with the sixth article of the Universal Declaration of Human Rights of 1948: "Everyone has the right to recognition everywhere as a person before the law." Of course, not all nation-states in the world recognize all human beings within their jurisdictions as persons, and even within those that do, serious questions remain about the extent to which such recognition improves the lived experiences of real people and about the ability and willingness of national and international institutions to enforce the law's own claims in this regard. But even contemporary theories of radical democracy basically accept that the law has achieved a clear-eyed perception of all persons. That is why they explicitly focus not on understanding how democracy can include those not already recognized as persons but on enriching the ways those already recognized as persons can participate in democratic life. Jacques Rancière, for example, writes that the denial of "political rights to certain parts of the population on sexual, social, or ethnic grounds . . . seems outdated in the West," while the restriction of "citizenship to a definite set of institutions, problems, agents, and procedures" is "a contemporary issue of major importance" (2010, 57).[22] What everyone accepts, then, is not that all persons already enjoy equality before the law or equality of political action but more minimally just that legal institutions, insofar as they articulate a principled claim that all human beings, and not just a privileged few, must be recognized as persons before the law, now understand when a face deserves the mask of legal personhood.

For Arendt, any confidence that we have resolved the epistemological problem of recognizing a person when we see one is misplaced. To be sure, Arendt has nothing but praise for the recognition of all human beings as persons before the law. However, she is very critical of the reasoning most often invoked to support this recognition. In one way or another the reasoning is that the legal concept of the person, while not the same as the human being, just as a mask is not a face, is nevertheless held to have a motivated relation to it. Personhood is the legal expression of the nature of human being. As legal terminology has it, only human beings are "natural persons," individuals whose personhood, far from being something they are *given* by any legal institutions, as is the case for "artificial persons" such as corporations, is *a given*.[23] For Arendt the belief that personhood comes already built into the human species does not just amount to a fantastic appraisal of the political valence of the fact of being biologically human; it amounts to a kind of biopolitics—a scientifically updated version of the ideological appeal to nature used to authorize the restriction of personhood to privileged groups of human beings. And indeed the idea that human beings, and no other living creatures, are "natural" persons eerily echoes the assumption of race-based slavery that some are naturally free and some are naturally slaves.[24]

However, Arendt's warning against a biopolitical model of personhood is but a version of her larger critical claim that no act of affixing the mask of personhood to faces can unfold as the expression or imitation of a nature that exists in faces prior to the acquisition of a mask. The basis of this argument comes in her reading of the Declaration of Independence in *On Revolution*. The preamble to this document, she notes, contains two appeals to what she calls "a transcendent source of authority for the laws of the new body politic": an appeal to "Nature's God" and an appeal to self-evident truths, as in the famous words, drafted by Thomas Jefferson, "We hold these truths to be self-evident that all men are created equal, that they are endowed by their Creator with certain inalienable rights." A self-evident truth, Arendt argues, is one

> that needs no agreement since, because of its self-evidence, it compels without argumentative demonstration or political persuasion. By virtue of being self-evident, these truths are pre-rational—they inform reason but are not its product—and since their self-evidence puts them beyond disclosure and argument, they are in a sense no less compelling than "despotic power" and no less absolute than the truths of religion or the axiomatic verities of mathematics.
>
> (1965, 184)

What's interesting about the self-evident truths in the Declaration of Independence, however, is that they are presented not as simply beyond agreement but as the professed beliefs of those collaborating on the revolutionary project of founding an independent nation-state. And indeed, Jefferson writes not "These truths *are* self evident" but rather "*We hold* these truths to be self-evident." This phrasing is important because it indicates that the quest for an absolute authority for the revolution against British sovereignty, demanded by the natural-law philosophy that informed the Declaration at the level of ideas, was given an unexpected answer by the text's own demonstration that the real motor of its claims is the performative act of declaring. In Arendt's words, a declaration is not "so much 'an argument in support of action'" as "the perfect way for an action to appear in words" (121). Far from being simply a means for justifying a preexisting political community, the declarative act is a structural part of this community's appearance. To declare the existence of an independent nation-state is not to make an argument for this community as if it already exists; it is to enact this community. To be sure, the act of publishing the Declaration of Independence did not by itself secure the existence of the United States against the military response of Britain. However, as the first public performance of the United States, the Declaration did not just certify the prior existence of the country but posited its existence as a pure act.[25]

The performativity that structures the founding of the United States is also on display in the laws of the country that recognize individuals as persons. Indeed, if no persons appear before the law without the elementary act of having masks affixed to their faces, then in a very real sense persons *only* appear by virtue of such acts of affixing. The obvious example of this is the corporation, which cannot in any simple way be said to have a face to which one could attach a mask. To be sure, several of the familiar legal theories of corporate personhood—concession theory, group theory, and real theory—would all dispute so-called creature theory, as exemplified in Chief Justice John Marshall's characterization of the corporation as a "mere creature of law . . . an artificial being, invisible, intangible, and existing only in contemplation of law," in *Trustees of Dartmouth College v. Woodward* (1819). The group theory of the corporate person, for instance, would say there are faces to which the mask of legal personhood can be attached and would even say that these faces are human, for they belong to the multiple individual human beings whose collective will forms the animating basis of a corporate project. However, the idea that corporations have faces that the legal act of incorporation merely recognizes is not universally compelling, something that is clear from the power of the photographs, considered earlier, of human beings emphatically displaying their faces while denying that corporations are persons. This does not mean, however, that the artificial claims of corporations to personhood can be easily contrasted to the supposedly natural claims of human beings, as these photographs seem to assume. Arendt suggests as much when she presents corporate and human personhood as merely two forms personhood may take:

> The *persona*, in its original theatrical sense, was the mask affixed to the actor's face by the exigencies of the play; hence, it meant metaphorically the "person," which the law of the land can affix to individuals *as well as*

to groups and corporations, and even to "a common and continuing purpose," as in the instance of "the 'person' which owns the property of an
Oxford or Cambridge college [and which] is neither the founder, now
gone, nor the body of his living successors."

(1965, 97, emphasis added)

In fact, drawing upon Arendt's analysis of the Declaration of Independence as an active verbal deed rather than a statement of truth, we could
say that the legal personification of the corporation, far from being a
mimic version of human personhood, is the exemplary model for all acts
of affixing personhood to faces, including those instances where the faces
are human.[26]

If an entity can have a mask attached to it when it does not even have
a face, then in the case of those entities that do have faces, nothing about
that face by itself decides whether the mask fits. Far from being a determining ground for the decision to assign or not assign the mask of personhood, the face drops away. The point is not that personhood does not
just appear as a representation, as opposed to presence, but that such
appearances require a performative power to posit personhood where no
nature can supply it.[27]

Befitting her accent on the active character of words, Arendt's account
of the plight of slaves in revolutionary America offers a practical demonstration of the performative process by which persons appear. As we
have already seen, she observes that John Adams sees something that his
fellow revolutionaries do not, namely that poor white people are "overlooked" by the institutions that regulate political participation. Yet Arendt
also observes something that even Adams does not, namely that black
slaves are *even more unseen* than poor white people, for they are not just
excluded from full participation in the management of public affairs but
categorically invisible to the law as persons. His insight about the blindness of others is offered in the mode of a blindness, one to which he is
himself blind, moreover.[28] And clearly Arendt's insight here has immediate political implications, insofar as she presents slavery as an injustice
"incompatible" with revolutionary republicanism, which it is correct to
outlaw by reforms to the U.S. Constitution.[29] What Arendt here performs
is a concrete declaration of what in her account of the plight of stateless
people in *The Origins of Totalitarianism* she calls "the right to have rights."

As others have noted, this right is best understood not as a moral right but rather as a practical political principle. In Étienne Balibar's reading, it refers to the "hyperbolic" process of actively claiming new rights as opposed to merely enjoying the rights one already has (2002, 166–167).[30] But given that Arendt never names the subject of the right to have rights, we may read it with another emphasis as the claim of those whom a community currently sees as nonpersons to recognition as persons before the law. That is, this right unfolds not just as the acquisition of new rights by those already viewed as persons but as the emergent appearing of new rights-bearing persons. The point is that the apprehension of new persons, such as Arendt's perception of the injustice of the position of slaves in revolutionary America, takes the form not of a positive identification of persons but rather as a critical awareness that the law's recognition of persons commits an error. The emphatic tone of the assertion "I *AM* A PERSON" is not epistemological certainty. It is an agonistic response to the denial of personhood. Any truth that such photos declare is fundamentally conditioned by the performative nature of the declarative act.

What I want to stress, however, is that the critical perception of the law's inability to see individuals as persons cannot guarantee that it is not itself a performance of the very blindness it critiques. I mean this not so much in the sense that even as Arendt decries race-based slavery in America, her own views of Africans are not without their shortcomings as affirmations of democracy, though this is undoubtedly true (Gines 2014). I mean it rather in the sense that the performative process by which persons appear before the law has its own formal logic, and that logic is radically impersonal in the sense that it resists our attempts to master it. Arendt suggests this in her repeated assertions of an "inadvertence," a disjunction between intention and results that structures all forms of acting in public, especially the political action that makes new persons appear before the law (1965, 19, 23, 27, 32, 34, 37, 164). But the idea is most sharply articulated by Paul de Man, whose images of blindness and insight I have already been employing. In a series of essays he wrote on personhood in lyric poetry, de Man (1984) suggests that the experience of persons is enabled by impersonal figurative operations that cannot themselves be reduced to the recognition of natural persons. However, as he himself points out, the very claim to identify something deeply impersonal about personhood inevitably ends up imposing itself as something persons may understand and control, and hence as a repetition of the very delusion being critiqued. To repurpose what Barbara Johnson has said in a brilliant essay on one of de Man's essays on persons, we can "explain" the difference between persons and nonpersons, but we cannot "anticipate" it. The difference "could only exist as a disruption" of what we think we know (Johnson 2008, 212). If our blindness to persons is something we can neither fully see nor be fully blind to, then even Arendt's assertion that all human beings, but only human beings, deserve recognition as persons before the law cannot pretend to enjoy certain knowledge that it does not misrecognize itself.

WHO, WE THE PEOPLE?

I want to conclude by briefly explaining how I think Arendt helps us respond first to theoretical critiques of the figure of the person and second to the practical debates about persons in the law. The rhetoric of personhood, along with the accompanying rhetoric of rights, has lately received a fair amount of criticism, especially within animal studies, biopolitical studies, and new materialism. The basic charge is that the very concept of the rights-bearing person distorts our experience of ethics and politics in two ways. First, it is held that the act of attaching the mask of personhood offers a constrained form of recognition because it effaces the unruly materiality of the entities to which it is attached. Within animal studies, for instance, this claim takes the form of an assertion that, since the person is paradigmatically human, personhood can be affixed to animals only on the condition of humanizing them (Braidotti 2011, 89). The second way that personhood is said to distort our experience of ethics and politics is that it is said to perpetuate a violent hierarchy between those recognized as persons and those disappeared as nonpersons. Roberto Esposito (2011) articulates the point sharply: Not only is personhood "valuable exactly to the extent that it is not applicable to all," but "*persona* means whoever can reduce others to the condition of the thing" (209).[31] What Arendt allows us to see is that the figure of the person distorts our ethical and political perceptions only to the extent that we have distorted personhood by not attending to its impersonal formal dimensions.

For persons emerge through a performative identification of mask and face that is as irresistible as it is impossible to control. So while I appreciate the wariness toward mainstream deployments of personhood, I believe that an outright dismissal of the figure as humanist or exclusionary concedes too much to the figure's self-authorizing proprietors and risks missing an opportunity to frame democratic action in the dominant language of law and politics.

If the person is not what we tend to think it is, then perhaps the right to inhabit this figure is not so easily usurped by human beings to the exclusion of everything else except the corporations that terrorize them. I have not offered direct reasons why corporations should not be recognized as persons or why animals should be recognized as legal persons. I think the strengths and the limitations of both the case against corporate personhood and the case for animal personhood have already been well demonstrated, and I have nothing to add (Crouch 2011, Francione 2011). The insight that Arendt offers us is purely formal, though not without normative implications. All claims to personhood are epistemologically equal in the sense that they cannot be resolved simply by an appeal to the nature of those who bear these claims, whether they are corporations, animals, or human beings. Indeed, insofar as contemporary democratic theory tries to frame personhood as a legal expression of human nature, it avoids acknowledging that what generates the perception of an individual as a person is not knowledge but an anxiety of ignorance that no knowledge can control. If this is true, then the discussion of corporate and animal personhood is mistaken to focus so heavily on the nature of corporations (the priority of the profit motive, their possible immortality) or the nature of animals (their cognitive abilities, their species membership). I'm not suggesting that arguments about who, or what, is person should not take into account various facts about candidates. But I do think it is naïve to think that naturalistic descriptions alone can decide for us. Decisions about how to distribute the mask of personhood is political, and our public deliberations about them must involve analyses of the figure of the person in all its formal weirdness. Indeed, our appreciation of the artificial, contestable, and unknowable character of claims to personhood can be the basis for a more coherent and honest radical democratic response to these claims.

NOTES

An earlier version of this essay was presented at a symposium hosted by Tom Tyler at Oxford-Brookes University in September 2013. My colleagues at Portland State University, Bishupal Limbu, Sarah Ensor, Sarah Lincoln, Anoop Mipuri, and Bill Knight all read an even earlier draft and offered immensely helpful comments. Once again, Cassia Gammill provided invaluable research assistance.

1. I say that the court "effectively" affirmed that corporations are persons because the Court's majority decision does not explicitly reason that corporations are persons. James Marc Leas and Rob Hager make the strong argument that, although *Citizens United* is having the practical effect of "turning elections into the high-return investment vehicles they are today," the judgment is not a defense of the constitutionally protected personality of corporations. Rather it synthesizes two earlier Supreme Court judgments: *Buckley v. Valeo* (1976), which asserts that money is a form of speech, and *First National Bank of Boston v. Bellotti* (1978), which overturned a law regulating money in politics by asserting that speech (that is, money) is speech, no matter the identity of the speaker. Leas and Hager conclude that *Citizens United* merely extends these two ideas into the realm of electioneering by arguing not that corporations have the right to express their opinions but that people have the right to hear all the information available. See Leas and Hager (2012).

2. The decision came in the form of two letters from the European Court of Human Rights, both of which are dated January 22, 2010. Thank you to Paula Stibbe for sharing these with me.

3. While no one disputes that legal systems around the world classify animals as nonpersons, not everyone agrees that they therefore comprehensively lack all rights, since anticruelty statutes provide animals with protections against certain forms of mistreatment. As Gary Francione (1995) has argued, though, within the laws that affect most animals, the only "right" that is provided animals is the demand to have rather important interests, such as the interest in not being killed, balanced against less important interests of its human owners, such as the interest in making money. And it is deeply questionable whether that is a right at all. Underlying Francione's argument is a recognition of the law's distinction between persons as objects of direct duties and nonpersons as objects of indirect duties, the upshot of which is that while one can physically injure a nonperson such as an animal, one cannot harm it in a legal sense. In "Force of Law" Jacques Derrida (1992) puts it this way: "An animal can be made to suffer, but we would never say, in a sense considered proper, that it is a wronged subject, the victim of a crime, of a murder, of a rape or a theft, of a perjury. . . . What we confusedly call 'animal,' the living thing as living and nothing else, is not a subject of the law or of law. The opposition between just and unjust has no meaning in this case" (18).

4. A version of the phrase "I am a person. Corporations are not" is also found on the website of Move to Amend, a coalition of organizations, formed in response to the *Citizens United* decision, that seeks the legal abolition of corporate personhood or at least the legal recognition of corporations as holders of certain basic rights (2014).

5.   Immanuel Kant's (2006) version of this claim resonates deeply with our topic: "The fact that the human being can have the 'I' in his representations raises him infinitely above all other living beings on earth. Because of this he is a *person*, and by virtue of the unity of consciousness through all changes that happen to him, one and the same person—i.e., through rank and dignity and entirely different being from *things*, such as irrational animals, with which one can do as one likes" (15).

6.   Two leading national newspapers, one on either side of the Atlantic, recently featured cover articles on the question of legal personhood for animals (Siebert 2014; *Die Zeit* 2014).

7.   It's worth noting that Comedy Central's other satirical news television program, *The Colbert Report*, also regularly mocks corporate personhood. While the Supreme Court deliberated the *Citizens United* case, Stephen Colbert, tongue firmly in cheek, pontificated thus: "I want to stand up for an oppressed minority whose free speech is being infringed: corporations. Other minorities have had their breakthrough, but corporations are still forced to the back of the bus" (2009). Colbert's activism on behalf of corporations only increased after August 2011, when presidential candidate Mitt Romney, in response to heckling from a crowd in Iowa, blurted out: "Corporations are people, my friend."

8.   Current scholarly opinion supports Arendt's assertion that in ancient Rome *persona* was used in the theater before being used in law. Marcel Mauss (1985, 1–25) stresses the priority of the *persona* as mask in his classic panoramic reading of the anthropological and etymological archive of "person." Not all insightful readings of the literary and legal meanings of *persona* in ancient Rome discuss the figurative process involved in the transfer of the word from literature to law (e.g., Rorty 2000, 542–543; Slaughter 2007, 17–18). According to the *Oxford English Dictionary, persona* is perhaps a loanword from the Etruscan *phersu*, "mask," which itself likely comes from the Greek *prosopon* (πρόσωπον), "face, countenance, mask, dramatic part, character."

9.   "When a jurist first said, 'A corporation is a person,' he was using a metaphor," conceded Arthur W. Machen (1911, 262) in his apology for corporate personhood. More recent legal theories deny that there is anything figurative in saying corporations are persons (Schane 1987, 563–609).

10. Going slightly beyond what Arendt actually says, we could add that the whole reason why the law borrows "person" from literature is not in order to substitute for an ordinary, literal legal term, and thereby to achieve a certain rhetorical effect, but rather because no literal legal term exists to refer to the kind of thing that a person is. Hence it would be more technically correct to say that person in law is a *catachresis*. To give an ordinary example of this extraordinary trope, the vertical things that prevent the flat top of a table from falling to the ground have no natural or proper name. In order to refer to them in a familiar way, we have to borrow from our nomenclature for the body parts of living animals and call them "legs." The same goes with the "hands" of a clock, the "mouth" of a river, and the "foot" of a mountain, all of which are both figures of speech and the ordinary ways to refer to the referents in question.

11. In a footnote to the passage in *On Revolution*, Arendt calls into question the revealing function of the voice. Although she is "tempted to believe that the word carried for Latin

ears the significance of *per-sonare*, "to sound through," she realizes that "the etymological root of *persona* seems to derive from *per-zonare*, from the Greek ξωνη, and hence to mean originally 'disguise'" (1965, 283n43).

12. In contemporary English, "person" can still signify a man or woman of high rank, distinction, or importance.

13. Slaves could have possessions of a sort, called *peculium*. This could in some cases even be quite a lot of money and could even include other slaves. But unlike the property of persons, these possessions were held by the slaves by the permission of their owner at his pleasure. For a fascinating account of the details of these aspects of Roman law, see Watson (1967, 1987).

14. Arendt (1958, 31) argues the subjected position of slaves insightfully stresses their being forced to bear the burden of laboring to sustain the lives of others. Thus, the "unhappiness" of Greek slavery consisted in subjection to necessity (the body's poor health and poverty) and subjection to manmade violence.

15. In other words, they did possess the "rights of subjects to be protected by their government in the pursuit of private happiness" (Arendt 1965, 118) but did not enjoy "the right to become 'a participator in the government of affairs,' the right to be seen in action" (121).

16. Arendt's account of those foreclosed from the political is one important source of Giorgio Agamben's theorization of "bare life" (1998). Indeed, bare life, or *nuda vita*, "naked life," implies the kind of privation of a mask and clothing in general that Arendt discusses.

17. Arendt here anticipates Jacques Rancière's (1999) concept of "the partition of the sensible" ("le partage du sensible"), an organization of the social sensorium that literally enables some beings to appear as political actors while constraining others to appear as, by nature, nonactors.

18. To some extent Arendt anticipates contemporary work in critical race studies exploring how being black in white supremacist societies is itself a form of statelessness. See Sexton (2010, 2011) and Sexton and Copeland (2003).

19. My use of Arendt to claim that slaves are excluded from community is one that calls into question her own contrast in *Origins* between the marginalization of slaves and that of the stateless: "Yet in the light of recent events [the experiences of the stateless], it is possible to say that even slaves still belonged to some sort of human community; the labor was needed, used, and exploited, and this kept them within the pale of humanity. To be a slave was after all to have a distinctive character, a place in society—more than the abstract nakedness of being human and nothing but human" (1968a, 297). Her description of slavery here is disputable, but her contention that slaves belong to "some sort of community" is already uncertain, and I am not surprised that Arendt herself revises this assessment of slavery in her later work. Indeed, in her Sonning Prize speech she describes nonpersons such as slaves in the very terms that in *Origins* she reserves for the stateless: "In Roman law *persona* was somebody who possessed civil rights, in sharp distinction from the word *homo*, denoting someone who was nothing but a member of the human species, different, to be sure, from an animal but without any specific qualification or

distinction, so that *homo*, like the Greek *anthropos*, was frequently used contemptuously to designate people not protected by any law" (2003).

20. In *Origins*, Arendt uses the similar phrases "out of legality altogether" (1968a, 294) and "outside the pale of the law" (1968a, 277, 295, 297, 302).

21. Especially decisive is not just the assertion, in both pieces of legislation, that no state shall deny any "person" in its jurisdiction "the equal protection of the laws" but also, and more basically, the assertion that all those "born or naturalized" in the United States, and not just white people, are entitled to be "citizens"—and hence in possession of the rights of person-hood, such as the right to make and enforce contracts, sue and be sued, give evidence in court, and inherit, purchase, lease, sell, hold, and convey real and personal property.

22. That politics is a matter of action is, of course, a quintessentially Arendtian thesis, and her influence is everywhere detectable in not only the work of Rancière but also that of Balibar (2002, 2014), Agamben (1998), and Honig (1993).

23. Thom Hartmann (2010, 11) explains natural personhood as follows: "Only humans in-herently have rights. Every other institution created by humans—from governments to churches to corporations—has only privileges, explicitly granted by government on behalf of the people with the rights." The ontological obviousness of natural human per-sonhood also underwrites the commonsense appeal of the various recently proposed amendments to the U.S. Constitution that seek to limit the personhood of corporations. The amendment published on the website of Move to Amend reads: "The rights pro-tected by the Constitution of the United States are the rights of natural persons only. Artificial entities established by the laws of any State, the United States, or any foreign state shall have no rights under this Constitution and are subject to regulation by the People, through Federal, State, or local law. The privileges of artificial entities shall be determined by the People, through Federal, State, or local law, and shall not be construed to be inherent or inalienable" (2014).

24. Arendt makes this connection most explicitly in the German version of *Origins* when she writes that in the idea of human rights extends racism's fiction of "natural-born free-dom . . . to everyone, even slaves," along the way ignoring the fact that "freedom and unfreedom are products of human action and have nothing to do with 'nature'" (1986, 615, my translation). I have explored this issue further in "Just Animals" (Hunt 2016). I will just add here that the emphasis on the natural character of human personhood in the popular response to corporate personhood also pushes it not just toward figures of birth but toward the biologism that animates "human rights" or "personhood amendments" to the U.S. Constitution proposed by that country's pro-life movement.

25. Arendt's reading of the performative character of the Declaration of Independence is worth reading beside Jacques Derrida's own reading in "Declarations of Independence" (2002). For an excellent discussion of the similarities and differences between the two readings, see Honig (1993).

26. This is the position taken by Ernest Barker. Barker's "very illuminating" essay is the only source Arendt cites in her discussion of personhood (1965, 283n43). Barker writes: "It [legal personality] is a mental construction, or juristic creation. It has, as we have seen, a certain artificiality; and it has this character both when it is ascribed, as in the vast major-

ity of cases, to an individual, and when it is ascribed, as it is in other cases, to the purpose in pursuing which a number of individuals are joined. We cannot say that legal personality in the one sort of case is artifice, and in the other not" (Barker 1950, lxxv–lxxvi).

27. Introducing the theatrical metaphor of personhood, Arendt says that it can be contrasted with metaphors drawn from organic phenomena. Her emphasis on an expressivist logic in the latter—"explosion of an uncorrupted and incorruptible inner core through an outward shell of decay and odorous decrepitude" (1965, 96)—suggests that the substitution of a mask for a face is not structured in this way.

28. My figures of insight and blindness here allude to Paul de Man's claim that the critical demystification of epistemological claims itself takes the form of a repetition of mystification. Interestingly de Man (1983) offers a version of this claim that resonates with our topic in a reading of Edmund Husserl's unwittingly ethnocentric critique of ethnocentrism.

29. In "Reflections on Little Rock" she calls slavery "the one great crime in America's history" (Arendt 1959, 46).

30. See also Balibar's other notable engagement with Arendt (Balibar 2014).

31. For an equally compelling criticism of the figure of the person, see Wolfe (2013).

WORKS CITED

Agamben, Giorgio. 1998. *Homo Sacer: Sovereign Power and Bare Life.* Trans. Daniel Roazen-Heller. Stanford, Calif.: Stanford University Press.

Arendt, Hannah. 1958. *The Human Condition.* 2nd ed. Chicago: University of Chicago Press.

——. 1959. "Reflections on Little Rock." *Dissent* 6 (1): 45–56.

——. 1965. *On Revolution.* New York: Penguin.

——. 1968a. "Isak Dinesen." In *Men in Dark Times*, 95–110. San Diego, Calif.: Harvest.

——. 1968b. *The Origins of Totalitarianism.* New ed. San Diego, Calif.: Harcourt.

——. 1977. *Eichmann in Jerusalem: A Report on the Banality of Evil.* Rev. ed. New York: Penguin.

——. 1986. *Elemente und Ursprünge totaler Herrschaft.* München: Piper.

——. 2003. "Prologue [Sonning Prize Acceptance Speech]." In *Responsibility and Judgment*, 3–14. Ed. Jerome Kohn. New York: Schocken.

Balibar, Étienne. 2002. *Politics and the Other Scene.* Trans. Christine Jones et al. London: Verso.

——. 2014. "Hannah Arendt, the Right to Have Rights, and Civic Disobedience." In *Equaliberty: Political Essays*, trans. James Ingram, 165–186. Durham, N.C.: Duke University Press.

Barker, Ernest. 1950. "Translator's Introduction." In *Natural Law and the Theory of Society, 1500 to 1800*, by Otto Gierke, ix–xci. Cambridge: Cambridge University Press.

Braidotti, Rosi. 2011. "Animals and Other Anomalies." In *Nomadic Theory: The Portable Rosi Braidotti*, 81–97. New York: Columbia University Press.

Crouch, Colin. 2011. *The Strange Non-Death of Neoliberalism.* London: Polity.

de Man, Paul. 1979. *Allegories of Reading.* New Haven, Conn.: Yale University Press.

——. 1983. "Crisis and Criticism." In *Blindness and Insight: Essays in Contemporary Criticism*, 2nd ed., 3–19. Minneapolis: University of Minnesota Press.

——. 1984. *The Rhetoric of Romanticism*. New York: Columbia University Press.

Derrida, Jacques. 1992. "Force of Law." In *Deconstruction and the Possibility of Justice*, trans. Mary Quaintance, ed. Drucilla Cornell et al., 3–67. New York: Routledge.

——. 2002. "Declarations of Independence." In *Negotiations: Interventions and Interviews, 1971–2001*. Trans. Elizabeth G. Rottenberg, 46–54. Stanford, Calif.: Stanford University Press.

Esposito, Roberto. 2011. "The Person and Human Life." In *Theory After Theory*, ed. Derek Attridge and Jane Elliot, 205–219. London: Routledge.

Francione, Gary. 1995. *Animals, Property, and the Law*. Philadelphia: Temple University Press.

"Get Corporate Money Out of Politics." Communication Workers of America Union. https://www.flickr.com/photos/cwaunion.

Gines, Kathryn T. 2014. *Hannah Arendt and the Negro Question*. Indianapolis: Indiana University Press.

Hartmann, Thom. 2010. *Unequal Protection: How Corporations Became "People" and How You Can Fight Back*. 2nd ed. San Francisco: Berrett-Koehler.

Honig, Bonnie. 1993. *Political Theory and the Displacement of Politics*. Ithaca, N.Y.: Cornell University Press.

Hunt, Alastair. 2016. "Just Animals." *South Atlantic Quarterly* 115 (2): 231–246.

"Ich bin wie du: Brauchen Tiere Meschenrechte?" *Die Zeit*. May 14, 2014.

Johnson, Barbara. 2008. "Anthropopmorphism in Lyric and Law." In *Persons and Things*, 188–207. New Haven: Yale University Press.

Kant, Immanuel. 2006. *Anthropology from a Pragmatic Point of View*. Ed. Robert B. Louden. Cambridge: Cambridge University Press.

Leas, James Marc, and Rob Hager. 2012. "The Problem with *Citizens United* Is Not Corporate Personhood." *Truthout*. January 17. http://www.truth-out.org/news/item/6095:the-problem-with-citizens-united-is-not-corporate-personhood.

Machen, Arthur. 1911. "Corporate Personality." *Harvard Law Review* 24:253–267.

Mauss, Marcel. 1985. "A Category of the Human Mind: The Notion of Person, the Notion of Self." In *The Category of the Person: Anthropology, Philosophy, History*, ed. Michael Carrithers et al., trans. W. D. Halls, 1–25. Cambridge: Cambridge University Press.

Nietzsche, Friedrich. 1968. *On the Genealogy of Morals*. Trans. Walter Kaufmann and R. J. Hollingdale. New York: Vintage.

Rancière, Jacques. 1999. *Disagreement: Politics and Philosophy*. Trans. Julie Rose. Minneapolis: University of Minnesota Press.

——. 2010. "Does Democracy Mean Something?" In *Dissensus: On Politics and Aesthetics*, trans. Steven Corcoran, 45–61. London: Bloomsbury.

Rorty, Amélie Oksenberg. 2000. "Characters, Persons, Selves, Individuals." In *Theory of the Novel: A Historical Approach*, ed. Michael McKeon, 537–553. Baltimore, Md.: Johns Hopkins University Press.

Schane, Sanford A. 1987. "The Corporation Is a Person: The Language of a Legal Fiction." *Tulane Law Review* 63:563–609.

Sexton, Jared. 2010. "People-of-Color-Blindness: Notes on the Afterlife of Slavery." *Social Text* 28 (2): 31–56.

——. 2011. "The Social Life of Social Death: On Afro-Pessimism and Black Optimism." *InTensions* 5.

Sexton, Jared, and Huey Copeland. 2003. "Raw Life: An Introduction." *Qui Parle* 13 (2): 53–62.

Siebert, Charles. 2014. "The Rights of Man . . . and Beast?" *New York Times Magazine* (April 27): 28–33, 48–50.

Slaughter, Joseph. 2007. *Human Rights Inc.: The World Novel, Narrative Form, and International Law*. New York: Fordham University Press.

Trustees of Dartmouth College v. Woodward. 17 U.S. 518 (1819).

Watson, Alan. 1967. *The Laws of Persons in the Later Roman Republic*. Oxford: Clarendon.

——. 1987. *Roman Slave Law*. Baltimore, Md.: Johns Hopkins University Press.

——. 1997. "Rights of Slaves and Other Owned Animals." *Animal Law* 3 (1): 1–6.

Wolfe, Cary. 2013. *Before the Law: Animals in a Biopolitical Frame*. Chicago: University of Chicago Press.

TELEVISION PROGRAMS CITED

Cenac, Wyatt. "SeaWorld of Pain." *The Daily Show*. Comedy Central. February 15, 2012.

Extras. Directed by Ricky Gervais and Stephen Merchant. BBC and HBO. October 12, 2006.

9

SUPRA- AND SUBPERSONAL REGISTERS OF
POLITICAL PHYSIOLOGY

JOHN PROTEVI

Many posthumanist analyses focus on the present, but we need not so limit ourselves in time like that. Nor need we adopt the naïvely technoutopian "me and my Google Glass, how cool is that!"; the portentously futurist "Google Glass? Hah! You ain't seen nothing yet! We're just now on the cusp of changing human nature with our technological breakthroughs!"; or even the body-abjecting and singularity-mongering "to hell with the meatsack, let's upload!" perspectives of too much of the popular takes and some semischolarly versions of posthumanist thought. (For some of the usual suspects, see Kelly 1995, 2010; Kurzweil 2005; and for criticism of such, Hayles 1999. For solid work, see Ansell Pearson 1997, Wolfe 2009, Braidotti 2013.)

What I want to do in this chapter is provide some case studies to show some fifth-century BCE experiments in geo-bio-techno-politics. While not as far back in time as some "originary technicity" people will go (for an overview of Derrida, Steigler, and others, see Bradley 2011, or for a philosophy-of-mind take, see Clark 2003), these case studies should show that tying the body to material-affective networks is nothing new and that this interlacing is formative rather than being caught in a dichotomy of liberation versus repression.

We can use the term "political physiology" to name the production of bodies politic, that is, the way in which the human subject is formed by linking the social and the somatic. To understand political physiology we must go above and below the subject; that is, we must see the production of the human subject via the inhumanly social and via the somatic (neither societies nor bodies are restricted to the human, nor can they be understood with categories restricted to human subjectivity). In this chapter I will examine the supra- and subsubjective registers in ancient Greek politics and philosophy. First, I will look at how suprasubjective circuits of carbon-mediated solar energy explain the fifth-century BCE Athenian empire's expansionism. Then I will look at subsubjective registers of rhythm and entrainment as thematized in Plato's *Laws*.

THE SOCIAL OR SUPRASUBJECTIVE REGISTER: BIO-SOLAR-POLITICAL PHYSIOLOGY

The sun and war are linked in the suprapersonal register of political physiology. One need not credit all of Bataille's more melodramatic statements to recognize that looking at circuits of solar energy as the basis of life is a staple of introductory biology textbooks (e.g., Mader 2009). And, certainly, looking at the ways societies waste excess on wars or monuments via Bataille's notion of *dépense* or radical expenditure is the key to Deleuze and Guattari's grappling with the Marxist questions about capitalist crises of overproduction and the "realization of surplus value" (Deleuze and Guattari 1984, 4n, citing Bataille 1991). Scarcity is not natural but is produced, via *dépense*, Deleuze and Guattari claim, to instill a fear of starvation/isolation (friendship can be made scarce by shunning—and friendship is as physiologically necessary as food, even if you die of loneliness more slowly than you starve) that will reinforce social structures that code flows (28). Concerning war and "political physiology," finally, I define that term as the study of the construction of "bodies politic," that is, the interlocking of emergent processes that link the patterns, thresholds, and triggers of affective and cognitive responses of somatic bodies to the patterns, thresholds, and triggers of actions of social bodies. Political physiology has a wider application than consciousness studies, since political institutions interlock with individual physiology in emotional responses to commands, symbols, slogans, and images; such responses at least strongly condition actions, through unconscious emotional valuing, but sometimes—even if rarely—provoke behavior that completely eludes conscious control, as in depths of panics and rages (Protevi 2009). Political physiology also asks us to look at the act of killing and its relation to political sovereignty (Protevi 2008). The traditional definition of sovereignty is that it is vested in the political body that holds the monopoly on the legitimate use of force within a clearly defined geographical territory. Thus, at the limit, a political body must be able to control the triggering of killing behavior in the bodies of its "forces of order" (army and police).

In principle we could look to the articulation of political physiology and bio-solar politics in contemporary life, for example in the two Iraq wars and their bizarre American offspring, the Hummer, where global petroleum wars meet the anxious individual driving an armored car around suburbia. But here I will focus on the ancient eastern Mediterranean. We can begin with some geopolitical basics. Consider some factors in the genesis of ancient Mesopotamian empires involving irrigation-intensive agriculture in flat river valleys. Keeping warnings against geographical determinism in mind, we can nonetheless point to the traditional idea that *poleis* benefitted from mountainous terrain to maintain independence, each mountain range enclosing a farming region supporting small farmers who were able, by forming a phalanx of armored hoplites, to overcome aristocratic dominance and demand *isonomia* or equality before the law (Sacks 1995, 101, 190). I am of course here treating very complex matters with a very broad brush; nonetheless the hoplite–equality connection is, for all its simplicity, well attested to in even the most technical discussions aiming to eliminate the too-crude notion of a "hoplite revolution" and pin down the role of hoplites in supporting tyrants who break aristocratic dominance (Raaflaub, Ober, and Wallace 2007, 34–36, 121, 133–136, Ste. Croix 1981, 260, 282; 2004, 126–127).

It may seem odd at first, but from a solar-politics perspective we can claim olive oil as a key factor in the genesis of Athenian democracy. Olive oil is a storage form of solar energy burned for light in lamps and for energy in human bodies. One of the "tipping points" in the democratization process in Athens occurs when Solon forbids debt slavery and debt bondage (Raaflaub, Ober, and Wallace 2007, 59; Ste. Croix 1981, 137, 282) as well as all agricultural exports except olive oil. This last provision stabilizes the middle class of small farmers who were threatened by aristocratic dominance (Milne 1945, Molina 1998). These farmers produce olive oil as a cash crop—a small part of their total production, to be sure, for it is the large farmers who dominate the oil market; nonetheless, it is a crucial money source. This stabilization of a mass olive oil–export market creates demand for work by urban artisans who produce jars for olive oil and manufactured goods for export (and also arms for hoplites to forestall aristocratic reconquest). A growing urban population needs grain importation, however, and protecting the import routes requires a naval force. In turn, what we can call the "military egalitarianism" thesis retains its force; it claims that a dependence on a naval force pushes the regime toward urban democracy, that is, to expanding the political base beyond that of the hoplites, for rowers are drawn from ranks of urban masses unable to afford hoplite gear (Raaflaub, Ober, and Wallace 2007, 119–136; Gabrielsen 2001).

Democratic rowing in the Athenian navy (leaving aside the question of seaborne marine troops) was relatively low intensity, at least compared to the hand-to-hand fighting depicted in Homer and the phalanx clashes of the classical age. (Actually, to respect the Deleuzean emphasis on assemblages, we should note that "hand-to-hand" is a misnomer, for shield and sword/spear is itself quite a bit less intense than just one-on-one with hands.) Thus for rowers there is less necessity for the high-intensity training needed for noble single combat. To produce such a single-combat-warrior body you need to traumatize its members via lots of intensive hunting and fighting as boys; here we can think of Odysseus's scar from his adolescent rite of passage, the boar hunt (*Odyssey* 19.390–395, [Fagles 1999]). Phalanx training was intermediate between aristocratic single combat and naval rowing; it is less intense than single combat, thanks to teamwork, that is, emergence. In the phalanx, you stand by your comrades rather than surge ahead. Recall Aristotle's definition of courage as the mean between rashness and cowardice; in comparative terms, rashness for the hoplite, the phalanx soldier, is standard behavior for the Homeric warrior, while phalanx courage—staying with your comrades— would be mediocrity if not cowardice for the aristocratic warrior (*Nicomachean Ethics* 2.8.1108b20–27 [Barnes 1984], Ward 2001). The discrepancy between Aristotelian and Homeric courage is an excellent example of the Deleuzean distaste for essentialism: You will never come up with a set of necessary and sufficient conditions to define "courage," so it is much better to investigate the morphogenesis of warrior and soldierly bodies and see if there are any common structures to those production processes. The question we want to ask from this perspective is: How are the warrior and the soldier different actualizations of the virtual multiplicity linking political physiology and hydro-solar-politics? And this standing together is the key to the *erôs*, the ecstatic union with a social body, of the phalanx. As we will see in a moment, William McNeill's *Keeping Together in Time* (1995) allows us to account for this human bonding in terms of collective resonating movement provoking the entrainment of asubjective physiological processes supporting emotional attachment (117).

But before we go below the subjective level to entrainment and political physiology, we should note its complement in the suprasubjective materialist explanation of Athenian foreign policy by the noted Marxist historian G. E. M. de Ste. Croix. In *Origins of the Peloponnesian War* (1972) and again in *Class Struggle in the Ancient Greek World* (1981), Ste. Croix points out that the geopolitical key to the transition from Athenian democracy at home to the "Athenian Empire" after the Persian Wars is the threshold of human energy production from grain ingestion. Ste. Croix (1981) uses this to undercut ideological explanations of Athenian foreign policy: "I have . . . explained why Athens was driven by her unique situation, as an importer of corn on an altogether exceptional scale, toward a policy of 'naval imperialism,' in order to secure her supply routes" (293). The singularities in the Athenian actualization of the geo-hydro-bio-political multiplicity are what gets us out of ideological condemnations of a supposed Athenian "lust for power," an analysis that is stuck on the level of the human subject. As Ste. Croix (1972) points out, rower-powered war ships had a much shorter range than sail-driven merchant ships, which are able to capture solar energy in form of wind power—itself generated from a multiplicity of temperature differentials of landmass/sea/water currents producing wind currents (47–49; see also Gomme 1933). So the Athenian democrats needed a network of friendly regimes whose ports could provide food and rest for the rowers of their triremes. That is, to use our terminology, to replenish the biological solar energy–conversion units of the triremes qua "man-driven torpedo[es]" (Gabrielsen 2001, 73). Bringing the geo-solar-hydrological dimensions of the multiplicity together with biotechnical and more traditionally sociopolitical dimensions, a recent scholarly article puts it this way: "The concept of *thalassokratia* [sea power] implies intense naval activity, primarily in order to defend existing bases and to acquire new ones, and intense naval activity, in its turn, requires command over enormous material and financial resources" (Gabrielsen 2001, 74).

Adopting this viewpoint allows us to understand the suprasubjective and anti-ideological materialism of a key passage from Ste. Croix: "Athens' whole way of life was involved; and what is so often denounced, as if it were sheer greed and a lust for domination on her part, by modern scholars whose antipathy to Athens is sharpened by promotion of democratic regimes in states under her control or influence, was in reality an almost inevitable consequence of that way of life" (1981, 293). But we should not be content with only going above the subjective; we should go below it as well. McNeill's reading of the political consequences of entrainment-provoked military solidarity takes us below the subject, complementing Ste. Croix's suprasubjective geomaterialism. McNeill writes: "The Athenian fleet developed muscular bonding among a larger proportion of the total population than ever fought in Sparta's phalanx" (McNeill 1995, 117). Furthermore, "feelings aroused by moving together in unison undergirded the ideals of freedom and equality under the law. . . . The muscular basis of such sentiments also explains why the rights of free and equal citizens were limited to the militarily active segment of the population" (118).

To understand McNeill's move, we have to go below the subject to understand how rhythm and entrainment work in subjectification.

THE SOMATIC OR SUBSUBJECTIVE REGISTER: BIORHYTHMIC POLITICAL PHYSIOLOGY

SCIENCES OF RHYTHM

In an informative obituary of his mentor, Steven Strogatz (2003) writes of Arthur T. Winfree's work on biological oscillators. First he examines the way Winfree poses the question of entrainment, the synchronization of oscillators in organismic behavior.

> How is it that thousands of neurons or fireflies or crickets can suddenly fall into step with one another, all firing or flashing or chirping at the same time, without any leader or signal from the environment? Winfree

studied the nonlinear dynamics of a large population of weakly coupled limit-cycle oscillators.

Strogatz then shows how a critical point or threshold is reached in the coupling of the oscillators, that is, the rates at which the fireflies fire or crickets chirp:

> The oscillators remain incoherent, each running near its natural fre-
> quency, until a certain threshold is crossed. Then the oscillators begin to
> synchronize spontaneously. (The effect is somewhat like the outbreak of
> synchronous applause after a magnificent concert.)

The next step is crucial. The models of these biological phenomena have a mathematical similarity to physical systems: "Winfree pointed out that this phenomenon is strikingly reminiscent of a thermodynamic phase transition, but with a twist: The oscillators align in time, not space." And here Strogatz describes the "deep analogy" that allows the multiple registers of dynamic-systems modeling:

> This deep analogy has since been explored by many statistical phys-
> icists interested in non-equilibrium phase transitions. . . . Winfree's
> work on coupled oscillators provided one of the first tractable exam-
> ples of a self-organizing system. It began as a problem in biology but
> has had a major impact on dynamical systems theory and statistical
> physics.

I would add that this "deep analogy," which is the heart of dynamic-systems modeling's ability to work across multiple registers by exploiting deep mathematical similarities between self-organizing behaviors of systems in widely varying fields, is not the result of transposing biology into physics; it's not a biological metaphor. But that's not to say it's completely irrelevant that Winfree discovered the deep analogy through biology. It might be that this is just a historical accident about the order of discovery, while it's the similarity in the order of being that is what's important. Or, if biology is particularly close to the virtual, it might be no accident that Winfree was working in biology. To explore this question, let us turn to *A Thousand Plateaus* (Deleuze and Guattari 1987).

A PHILOSOPHY OF RHYTHM

Recent secondary literature (Toscano 2006, Welchman 2009) suggests that the virtual/actual distinction of *Difference and Repetition* is attenuated if not forgotten in *A Thousand Plateaus* (1987), where we find the emphasis on the intensive, on multiple material systems forming haecceities or spatiotemporal interactive dynamisms precisely when a threshold of synchronization in multiple rhythms is realized, triggering self-organization.

Given the complexity of the conceptual network Deleuze and Guattari use to discuss living systems in *A Thousand Plateaus*, we can only sketch some of the relations among the key terms: milieus and codes, strata and territories. We begin with "milieu," which is a vibratory, rhythmic, and coded material field (313) for bodies (strata) and territories (assemblages). Heterogeneous milieus are "drawn" by rhythms from chaos, while territories form between ever-shifting milieus. Now milieus are coded—the "code" is the repetition of elements such that milieus are a "block of space-time constituted by the periodic repetition of the component"—but the rhythm is always shifting in "transcoding" (313). Thus "rhythm" is the difference between one code and another: "There is rhythm whenever there is a transcoded passage from one milieu to another, a communication of milieus, coordination between heterogeneous space-times" (313).

Deleuze and Guattari's notion of rhythm is differential: "Rhythm is critical; it ties together critical moments" (313). "Critical" here means a threshold in a differential relation, a singularity in the linked rates of change of a living system in its ecological niche. Coming back to McNeill and the Athenian trireme rowers, we can see their entrainment as a limit case of rhythm: a dynamic interaction of multiple processes converging on but never reaching a pure cadence. In the approach, affective group-bonding effects are produced. In colloquial language, groups share emotions by getting on the same wavelength.

We should not think these notions are limited to Deleuze and Guattari. In fact, the philosophical literature on entrainment is developing. A recent study (Tollefsen and Dale 2012) shows how philosophical accounts of joint agency have been top down: Using adult human models, they have shown complex intentional alignment. The authors call attention to research demonstrating the alignment of low-level behaviors such as body postures and conversational rhythms, so that they aim for a "process-based, dynamic account of joint action that integrates both low-level and high-level states" (385; they refer to Shockley, Richardson, and Dale 2009 among many other fascinating studies of low-level alignment).

After looking at the suprasubjective or social register in historical accounts of Greek war, we can now look at the way in which we can use "political physiology" to explore the subsubjective register as it appears in Greek philosophy.

RHYTHM AND THE GREEKS

Richard Shusterman (2007) has some interesting reflections on the social and corporeal roots of moral intuitions:

> Much ethnic and racial hostility is not the product of logical thought but of deep prejudices that are somatically expressed or embodied in vague but disagreeable feelings that typically lie below the level of explicit consciousness. Such prejudices and feelings thus resist correction by mere discursive arguments for tolerance, which can be accepted on the rational level without changing the visceral grip of the prejudice.
>
> (25)

After David Hume and Friedrich Nietzsche, as Jesse Prinz (2007) reminds us, we are used to the idea that one needs to consider bodily experience and emotional reactions in order to have a complete view of moral intuitions and moral judgments as they occur in flesh-and-blood humans. But there are also ancient philosophers who propose close links between emotions, bodies, and moral intuitions, and here we will briefly discuss Aristotle and Plato.

Both Plato and Aristotle agree on the emotional core of character development and on the shared emotional dispositions of people raised under one political regime or another. For Plato, ethical development entails an emotional reaction prior to any rational justification ("he will rightfully object to what is ugly and hate it while still young before he can grasp the reason": *Republic* 402a [Hamilton and Cairns 1961]). For Aristotle the ethical virtues are constituted by the right disposition of emotions, and such dispositions are attained by the consistent training of children's emotional relation to pleasure and pain (*Nicomachean Ethics* 2.3.1104b10–13 [Barnes 1984]). The widest context for the habitual development of ethical virtues is the customs and laws of the city, such that the character of the citizens is the most important task of the legislator (*Nicomachean Ethics* 2.1.1103b2–5, *Politics* 8.1.1337a10 [Barnes 1984]).

For Aristotle ethical behavior is not simply a matter of having controlled appetites; ethical excellence is not simply the psychic control of the corporeal. The intuitive faculty of the soul is understanding, *nous*, which can be both practical and theoretical; *nous* involves the perception, the immediate seeing (*aisthesis*) of particulars (*Nicomachean Ethics* 1143b5). The undemonstrated practical intuition of the properly trained person, his immediate grasp of the right course of action, is the standard in ethics, for experience has given him the eye with which to see correctly (b14). While intuition is a faculty of the soul, developing practical intuition is a matter of the body politic. While it may look like simply natural development (b6), the development of practical intuition depends on embodied political experience, the enmeshing of the social and the somatic, for the quality of practical *nous* achieved by the body politic is appropriate to one's age (b8).

Let's look at musicality in Plato for a look at such "political physiology" (Protevi 2009). Book 7 of the *Laws* (Hamilton and Cairns 1961) begins with the Athenian saying that despite its importance the nurture and education of children can only be a matter of advice to heads of household rather than law (788b–c), even though habits of transgression from petty misdeeds can ripple up to bad effects in a polity (790b, 793c). So it can be hoped that citizens will take the advice to them on these matters as a law to them and to their households (790b). Political affect is of the utmost importance to Plato, but he must describe rather than prescribe its genesis.

The reason why Plato must describe, though he cannot prescribe, is twofold. First, in the *Laws* he forgoes the blank slate he gives himself in constructing the ideal city in the *Republic*. So he has to describe child-drearing that is realistically constrained by real geography and custom; he can't just prescribe what should happen. Second, he cannot prescribe in detail the singular intercorporeal rhythms that lie at the root of emotional moral development; he can only describe the irreducible singularity of relation between nurse and infant rhythms, a singularity that blocks rational description.

Now the concern with reproduction begins before pregnancy. The matrons engaged by the State supervisors can investigate marital sexual relations—presumably frequency, timing, and so on—and what tips them off is the denunciation of a married but childless couple who is "paying regard to aught else than the injunctions imposed amid the sacrifices and rites of matrimony" (784a–d). Once pregnancy occurs, the Athenian recommends that pregnant women take walks so that the external shaking of the fetus helps its body grow into robust health (789b–790b). And with regard to the soul we must pay the same sort of attention to imposed movement; analogous to the way dancing prescribed by priestesses will help those afflicted with "Corybantic troubles" (see Dodds 1951, 78–80, for a social and somatic functionalist/cathartic reading of this passage), so too will rocking and singing calm an infant (790d).

Continuing the discussion, the Athenian explains that "fright is due to some morbid condition of soul. Hence, when such disorders are treated by rocking movements the external motion thus exhibited dominates [*kratei*] the internal, which is the source of the fright or frenzy" (790e). The lawgivers must rely on custom for the most efficacious selection of these songs and on the caregiver's sensitivity and skill in delivering them at the proper time, with proper intensity and with proper rhythm. The lawgiver can set the context for their use but cannot discuss the details of the lullaby or its somatic/psychic effects. Now why is the Athenian so concerned here? It's because temper (the proper relation to fear) and moral excellence are so closely connected (791b–c). But then comes the admission at 792a that the harmonizing of the soul of the infant with regard to the placidity of its temper must rely on the "guesswork [*tekmairontai*]" of nurses, who are able to discern the proper course of action—the right rocking motion, the right lullaby—in placating a screaming child.

Once children are born, there is also supervision of the collective games of children in the public setting of the "local sanctuary" between the ages of three and six (794b–c). But note the difference between recommendations by officials to citizens for the citizens to oversee the lullabies of the nurses of infants at home and the direct supervision by public officials of nurses as they accompany the public games of children. The key point is that in the lullaby there's a singularity of bodily rapport between nurse and infant that is resistant to rational supervision, so that the nurses must resort to guesswork. But that guesswork is of fundamental importance to the corporeal development of proper emotional balance and hence moral intuition. Again, political affect is of the utmost importance to Plato, and the linchpin of the system described in the *Laws* is the guesswork of slave women.

CONCLUSION

Let me conclude with this speculation that brings together Gabrielsen's adjunct-subjective biotechnical assemblage (the trireme as "man-driven torpedo"), Ste. Croix's suprasubjective geopolitics, and McNeill's subsubjective entrainment-provoked emotional solidarity. Let us see these as the dimensions of a hydro-solar-bio-techno-political multiplicity whose democratic naval actualization in Athens may help explain why Plato in the *Laws* put the ideal city away from the sea and why in the same dialogue hoplite victory in the land battle of Marathon is praised over democratic rowers' victory in the sea battle of Salamis. "We . . . insist that the deliverance of Hellas was begun by one engagement on land . . . Marathon, and completed . . . at Plataea. Moreover, these victories made better men of the Hellenes, whereas [Salamis] did not." Plato concludes in words that at least point to our concern with geo-hydro-solar-bio-technical-political multiplicity: "In our investigations into topography [*chōras phusin*] and legislation [*nomōn taxin*] [we focus on] the moral worth of a social system [*politeias aretēn*; "political excellence/virtue"]" (707c–d, cited in Hamilton and Cairns 1961). Here I think we can see an implicit posthumanism in Plato: the virtue of a political regime put in the context of investigations into the interplay of geography and law. That Plato has a different reading of the events than Ste. Croix should not detract from our recognition of the common matrix of their thought: human affairs and the details of the earth thought together in mutual implication.

NOTE

Portions of this chapter originally appeared in John Protevi, *Life, War, Earth* (Minneapolis: University of Minnesota Press, 2013).

WORKS CITED

Ansell Pearson, Keith. 1997. *Viroid Life: Perspectives on Nietzsche and the Transhuman Condition.* New York: Routledge.

Barnes, Jonathan, ed. 1984. *The Complete Works of Aristotle.* 2 vols. Princeton, N.J.: Princeton University Press.

Bataille, Georges. 1991. *The Accursed Share.* Vol. 1: *Consumption.* Trans. Robert Hurley. Cambridge, Mass.: MIT Press.

Bradley, Arthur. 2011. *Originary Technicity: The Theory of Technology from Marx to Derrida.* London: Palgrave Macmillan.

Braidotti, Rosi. 2013. *The Posthuman.* Cambridge: Polity.

Deleuze, Gilles, and Félix Guattari. 1984. *Anti-Oedipus.* Trans. Robert Hurley, Mark Seem, and Helen R. Lane. Minneapolis: University of Minnesota Press.

——. 1987. *A Thousand Plateaus.* Trans. Brian Massumi. Minneapolis: University of Minnesota Press.

Dodds, E. R. 1951. *The Greeks and the Irrational.* Oakland: University of California Press.

Fagles, Robert, trans. 1999. *The Odyssey.* New York: Penguin.

Gabrielsen, Vincent. 2001. "The Social and Economic Impact of Naval Warfare on the Greek Cities." In *War as a Cultural and Social Force: Studies in Ancient Warfare,* ed. T. Bekker-Nielsen and L. Hannestad, 72–89. Copenhagen: Royal Danish Academy of Sciences and Letters.

Gomme, A. W. 1933. "A Forgotten Factor of Greek Naval Strategy." *Journal of the Hellenic Society* 53:16–24.

Hamilton, Edith, and Huntingdon Cairns, eds. 1961. *The Collected Dialogues of Plato.* Princeton, N.J.: Princeton University Press.

Hayles, N. Katherine. 1999. *How We Became Posthuman.* Chicago: University of Chicago Press.

Kelly, Kevin. 1995. *Out of Control: The New Biology of Machines, Social Systems, and the Economic World.* New York: Basic Books.

——. 2010. *What Technology Wants.* New York: Penguin.

Kurzweil, Ray. 2005. *The Singularity Is Near: When Humans Transcend Biology.* New York: Penguin.

Mader, Sylvia. 2009. *Biology.* 10th ed. New York: McGraw Hill.

McNeill, William. 1995. *Keeping Together in Time: Dance and Drill in Human History.* Cambridge, Mass.: Harvard University Press.

Milne, J. G. 1945. "The Economic Policy of Solon." *Hesperia: The Journal of the American School of Classical Studies at Athens* 14 (3): 230–245.

Molina, Luis. 1998. "Solon and the Evolution of the Athenian Agrarian Economy." *Pomoerium* 3. http://pomoerium.eu/pomoer/pomoer3/molina.pdf.

Prinz, Jesse. 2007. *The Emotional Construction of Morals.* Oxford: Oxford University Press.

Protevi, John. 2008. "Affect, Agency, and Responsibility: The Act of Killing in the Age of Cyborgs." *Phenomenology and the Cognitive Sciences* 7 (2): 405–413.

——. 2009. *Political Affect: Connecting the Social and the Somatic.* Minneapolis: University of Minnesota Press.

——. 2013. *Life, War, Earth: Deleuze and the Sciences.* Minneapolis: University of Minnesota Press.

Raaflaub, Kurt, Josiah Ober, and Robert Wallace. 2007. *Origins of Democracy in Ancient Greece.* Berkeley: University of California Press.

Sacks, David. 1995. *A Dictionary of the Ancient Greek World.* New York: Oxford University Press.

Shockley, Kevin, Daniel C. Richardson, and Rick Dale. 2009. "Conversation and Coordinative Structures." *Topics in Cognitive Science* 1:305–319.

Shusterman, Richard. 2008. *Body Consciousness: A Philosophy of Mindfulness and Somaesthetics.* Cambridge: Cambridge University Press.

Ste. Croix, G. E. M. de. 1972. *The Origins of the Peloponnesian War.* Ithaca, N.Y.: Cornell University Press.

——. 1981. *The Class Struggle in the Ancient Greek World.* Ithaca, N.Y.: Cornell University Press.

——. 2004. *Athenian Democratic Origins and Other Essays.* New York: Oxford University Press.

Strogatz, Steven. 2003. "Obituaries: Arthur T. Winfree." *Society for Industrial and Applied Mathematics News.* http://www.siam.org/news/news.php?id=289.

Tollefsen, Deborah, and Rick Dale. 2012. "Naturalizing Joint Action: A Process-Based Approach." *Philosophical Psychology* 25 (3): 385–407.

Toscano, Alberto. 2006. *The Theatre of Production.* Basingstoke: Palgrave Macmillan.

Ward, Lee. 2001. "Nobility and Necessity: The Problem of Courage in Aristotle's *Nicomachean Ethics.*" *American Political Science Review* 95 (1): 71–83.

Welchman, Alistair. 2009. "Deleuze's Post-Critical Metaphysics." *Symposium* 13 (2): 25–54.

Wolfe, Cary. 2009. *What Is Posthumanism?* Minneapolis: University of Minnesota Press.

10

GEOPHILOSOPHY, GEOCOMMUNISM

Is There Life After Man?

ARUN SALDANHA

Our recalcitrance won't last forever, of course. Sooner or later events will break through even our carefully buttressed denial.

—BILL MCKIBBEN, *THE END OF NATURE*

Will all these microrevolutions, these profound challenges to social relations, remain contained within restricted areas of the socius? Or will there be a new interconnectedness that links one with another, without thereby setting up any new hierarchy or segregation? In short, will all these microrevolutions end by producing a real revolution? Will they be capable of taking on board not only specific local problems but the management of the great economic units?

—In other words: are we going to get away from all the various Utopias of nostalgia—getting back to our origins, to nature, to the transcendent?

—FÉLIX GUATTARI, *MOLECULAR REVOLUTION: PSYCHIATRY AND POLITICS*

MAN AND MATERIALISM

Something in the world forces us to think. Deleuze's statement in *Difference and Repetition* (1994, 139) is fast becoming a rallying cry for a manner of thinking that brings itself into existence and relevance without the support of a belief in a transcendent logic. His critical project is to rid philosophy of the eternally returning dogmatic beliefs that continue to drive even so-called modern philosophy, wherein Truth, Meaning, Logic (including the dialectic), Laws (including those of "the market"), Consciousness, Technology, and Man (including human rights) take the metaphysical place that God used to inhabit. An immanent project must return to sensing how it is *this* world in all its ceaseless becoming and incomprehensibilities that summons the collective effort of thinking. To the ancient question "what is philosophy?" Deleuze and Guattari (1994) answer: a return to earth. Their geophilosophy is the name of thinking when it learns to stop looking at the heavens for answers. But what if we were to take the *geo* more literally, as the physical planet that thinking has to confront?

Often inspired by Deleuze, what calls itself *new materialism*—traversing fields such as feminism, queer theory, and technoscience studies—similarly seeks to reorient the conceptualization and study of the material world without such older categories as essence, meaning, and law. New materialism's main target does not seem to be positivism but Marxism, both in its philosophical (dialectical materialism) and sociological (historical materialism) avatars. After all, Marxism has from Marx and Engels themselves consistently and persistently claimed to carry the torch of materialism. Marx continues the legacies of Epicurus and Galilean science against those of Plato and Christianity, but they are rendered revolutionary by way of Hegel's dialectical method. If materiality for the "old" materialism revolves around labor, class, the technological interaction with the earth, the globalization of capital, commodity flows, and organizing for international revolution, the materiality of the new materialism is usually more prosaic. Suspicious of the transcendences that they presume slip into Marxism, new materialisms are at their best when theorizing the singularity of bodies and objects in their local interactions and thickness. When they do scale up, they skip the materialities of the state, money, migration, and so on, rising to the level of that which animates all bodies, *life* as such. It is this Victorian and early twentieth-century concept of life that has returned to serve as the theoretical focal point. The return to vitalism affirms the ever-present if mostly latent openings for creative practice that, it is hoped, can reform socioecological assemblages from below. Instead of the Marxist dialectics that presupposed a constitutive man/nature difference setting thinking and history into motion, difference inheres within the field of bodies and things themselves insofar as they intermesh.

A second current has called itself *posthumanism*. Its project to overcome "man" draws on Derrida's deconstruction of the man/nature dichotomy, Foucault's archeology of philosophical anthropology, and their varying indebtedness to Heidegger's theorization of technology and language and his critique of humanism. This current lays bare modern philosophy's residual dogmatic images of thought, which are ultimately anchored in the belief of man as thinking being standing separate and sovereign above the world of things. On this analysis philosophical modernity as inaugurated by Descartes supposes that only "man" perceives and represents and that nature is the mute outside. But not only has humanity never been the pure starting point apart from nature, but there is no way of knowing nature as one universal chain of being waiting to be discovered and described. This post-Cartesian confidence in the power of science is correlated with the aggressive subjugation of not just women and non-European populations but animals, plants, and environments.

Sometimes inspired by science fiction, posthumanism argues that critical thinking is barely keeping up with the biotechnological and cybernetic reconfigurations of what the human species is, or is supposed to be, and how profound changes to the rest of life is complicating traditional concepts of man. Posthumanism joins up, like much of the new materialism, with transdisciplinary trends such as systems theory, which have since the 1960s studied human and biophysical systems on the same plane of being. Though the relation is seldom discussed, posthumanism is also opposed to historical materialism. To the extent that he is heir to the Enlightenment's belief in progress and science, Marx is part and parcel of the humanist constellation that posthumanism puts into question. Moreover, as the disastrous environmental record of state communism shows, many posthumanists and new materialists would point out that Marxism as an ontology has willy-nilly been complicit with the nefarious aspects of the productivist and objectivist desire to know and govern life completely.

This chapter will claim that Deleuzo-Guattarian geophilosophy does not break with historical materialism but reworks it. Deleuze would count himself neither as a new materialist nor as posthumanist, and he has little to do with preoccupations regarding how technology is making the world posthuman. In making some openings toward an exchange among geophilosophy, Marxism, and earth-systems science, I will question the revitalization of vitalism that has been occurring in Deleuze's name (following, in part, Colebrook 2010). While it is crucial to analyze the philosophical and political pitfalls of humanism, it is the earth and not just life or bodies that an antihumanist alternative needs to start from. There have been crucial critiques of both the bias toward Eurocentric scientism and the universalist politics in the "old" materialism, often indirectly, by conceptualizing materiality otherwise. But by turning one's theoretical gaze and viscera away from the material configurations of financial collapse, global inequality, war, and climate change, there is a risk of becoming unprepared for when the world will *really* force one to think. For as environmental scientists and activists like Bill McKibben have been showing for decades, the public's recalcitrance about the possibility of global catastrophe won't last forever. Marxism, the "old" materialism, will be necessary for thinking global catastrophe. Despite its limitations it is the only discourse that has made the question of what should be done in the face of global catastrophe its central tenet.

Deleuze knew how disastrous capitalism was, but he barely could have foreseen the extent to which the earth itself is demanding that philosophy radically reframe itself in the twenty-first century. It was rather Guattari who was involved in environmental and media activism, one of the very few "French theory" intellectuals to have had a keen sense of how consumerism, militarism, and environmental destruction are linked (Guattari 2000). But that human extinction because of global warming could become part of mainstream discussion a mere two decades after their deaths would have seemed to both Guattari and Deleuze to be Malthusian scaremongering. Even then, their ontology allows for mapping the earth-shattering threshold that the earth has recently passed.

THE ANTHROPOCENE: WHO WOULD HAVE THOUGHT?

Uncontrollable global warming, rising sea levels, nuclear accidents, deforestation, desertification, myriad kinds of pollution, unimaginable waste, dead zones in oceans, network failure, cybercrime, the depletion of minerals, the sixth mass extinction: these are just some of the most discussed catastrophic trends in the world today. But they are not simply ecological problems and injustices of the present. As Jan Zalasiewicz (2010) writes in his learned speculation about the state of the planet a hundred million years from now, these processes are objective materials sedimenting into signs that tell of the unprecedented intensity with which a minority of one species has lived and affected its planetary milieu. Unlike any previous geological epoch the Anthropocene is directly ethical and political—nobody can change the Jurassic. After all, what science plainly tells us is that if it continues on the same pathway it has over the last several centuries, the Anthropocene is likely to be the end of most existing species.

The amount of CO_2 industrial capitalism has already pumped into the atmosphere and the fact the global human population is sustained only by exponential growth—that is, burning ever more fossil fuels and mining ever more desperately to produce ever more smartphones—means runaway climate change is not a theoretical "scenario" but an uneven distribution of disaster already here to stay. In contradistinction, nuclear holocaust, asteroid impact, pandemics, or an attack by extraterrestrials are mere potential risks, hence easier for thought to absorb. Unsurprisingly, therefore, the Anthropocene has fast become central to public debate, activism, and the arts and humanities. There is a veritable *geological turn* cascading through the domains of thought that cannot but address new vernacular notions such as carbon footprint, environmental racism, climate justice, extreme events, business-as-usual, and geoengineering. Who would have thought geology would become central to our lives?

What forces thinking is a reality far more disturbing, therefore, than the technological and embodied trends that inspire much of the turn toward life and the posthuman. It is easy to see why many in the humanities until recently focused on bodies and local instances of interaction. Hurricane Katrina, the Great Recession, the Haiti earthquake, the failure of the 2009 Copenhagen climate negotiations, the Syrian civil war and refugee flow, Australian drought and wildfire, fracking, and many more disasters had to occur for the scale of overall crisis to become evident. Even for physical scientists the planetary scale has long been obscured by hyperspecialization and an imposed lack of imagination. Only recently has the massive corroborated evidence from the earth and life sciences converged into the realization that European industrialization and colonization have irreversibly affected the planet to the extent the new situation deserves a name, the Anthropocene.

The Anthropocene concept can be called the most important concept ever. It would seem we are experiencing a more decisive Copernican turn than the ones associated with Copernicus, Kant, Darwin, Marx, and Freud. That is, the Anthropocene concept does not only further decenter the taken-for-granted human perspective by pushing it even further away from anthropocentrism—from the belief the universe was created *for* humans and is as such in principle fully legible, from the certitude there is a divine hand or Aristotelian logic guiding the order of being, while an absolute difference is posited between thought and nature—but simultaneously *re*centers the human species as vastly more consequential than these previous thinkers could have imagined. In the Anthropocene the human species is thought of no longer as only that which carries consciousness but as a biomass relaying energy and continuously shaping its future conditions of possibility. Though there are hints in these previous Copernican turns that human extinction could occur as a direct result of present decisions (the secular answer to the problem of evil), they almost never addressed the planet as part of that ethical and political dynamic.

It is no wonder stratigraphy (geologists dating layers of the history of terrestrial life), until very recently one of the most arcane pursuits, is suddenly sending shock waves through the media. As Elizabeth Kolbert (2011) writes in *National Geographic*:

> [Stratigraphers] take the long view—the extremely long view—of events, only the most violent of which are likely to leave behind clear, lasting signals. It's those events that mark the crucial episodes in the planet's 4.5-billion-year story, the turning points that divide it into comprehensible chapters.
>
> So it's disconcerting to learn that many stratigraphers have come to believe that *we* are such an event—that human beings have so altered the planet in just the past century or two that we've ushered in a new epoch: the Anthropocene.

It is singularly disconcerting to learn the human species is capable of a violence akin to the collision event that made the earth inhabitable for almost all dinosaurs 66 million years ago. Since industrialization took off in the early nineteenth century, the backdrop for the human subject is turning out to have started a new chapter. In a geological instant, "man" finds his powers are far more destructive than he thought but simultaneously that he is not at a Promethean pinnacle of nature or creation. He discovers he is more enslaved and befuddled than ever before by the geophysical and agroecological strata that for millennia he was attempting to master. What Kolbert and Zalaciewicz and most environmentalists are *really* disconcerted by, however, and are constitutively unable to acknowledge is that the Anthropocene's collision course is nothing but capital's feedback loop spiraling ever outward.

Deleuze and Guattari would broadly agree with the various Marxist analyses of the planetary crisis of the twenty-first century that one finds in David Harvey (2015) or John Bellamy Foster and colleagues (Foster, Clark, and York, 2010). The one vector changing the composition of the atmosphere and the earth's crust, pushing socioecological systems over the brink, fantasizing about colonizing Mars, all the while mostly escaping critical scrutiny and democratic transparency, is plainly capital. As the most deterritorializing and deterritorialized agent driving globalization, the self-augmenting nature of capital is the only explanation for the famous ice hockey–stick graphs of the Anthropocene. Few question the fact that the exponential growth of everything from food production to cities to information to emissions since 1750 (even more since 1945, still more since 1980) has been because of competitive markets. But only Marxist analyses can demonstrate that capital is not only exploitative but uncontrollable and ultimately self-destructive. It is *inhuman* not in the moral but the ontological sense.

Jason Moore (2015) therefore suggests we should name our epoch the "Capitaloscene." It isn't man or the innumerable generations of human beings that should be blamed for catastrophe but a highly peculiar coercive force, which was conceived in northern Italy during the fifteenth century, came to fruition in northern England during the eighteenth, and conquered most of the rest of the globe in the late twentieth. Moore's suggestion, as relayed by Donna Haraway (2015), has created a lively and necessary debate, not the least about the ideological concealments in bourgeois science. But there are two key reasons we should stick with "Anthropocene." First, other stratified societies have been equally ideological and destructive, and capitalism in fact subsumes them to function. Second, it is precisely because mainstream stratigraphers came up with it that this concept of concepts carries such force. From the arcane dating of rocks and fossils, they suddenly stumbled on the frightening prospect that they needed to date the longevity of the human species. It is now up to critical thinking to alert scientists of the radical political consequences their findings logically lead to.

THE MECHANOSPHERE OUTSIDE OF VITALISM

Deleuze and Guattari's critique of anthropocentrism starts to meet the challenge of the twenty-first-century world precisely because they remain committed to analyzing and dismantling the crazy axiomatic system that capital bequests. In contrast, much posthumanist and new-materialist work has until recently avoided any mention of catastrophe or inequality. Their *Capitalism and Schizophrenia* project (1983 and 1987) provides a sophisticated conceptual diagram for understanding how the Anthropocene came about. One of the facets of their ontological method they call *stratoanalysis* (1987, plateau 3). Human formations are territorializations (cuts through flows of money, populations, ideas, and so on), which are always also stratifications (layers of material homogeneity). Hence machinic assemblages (spatiotemporal systems), abstract machines (their virtual organization), and collective assemblages of enunciation (language in its pragmatic dimension) remain attached through machinic phyla (of technology, metal, minerals) as umbilical cords to the earth. A school's strata are its physical building, subsidies, pupils, teachers, staff, books, regulations, class exercises, reputation, and so on. Calling these strata instead of components—which they also are—emphasizes the fact that their materialities are layered and have varying temporalities. A school's abstract machine is what holds it together, its essence, which never stays the same.

The mechanosphere is the immense web of all those machines intertwined across the surface of the globe, whose abstract machine is the plane of consistency. "What we call the mechanosphere is the set of all abstract machines and machinic assemblages outside the strata, on the strata, or between strata" (1987, 71). *A Thousand Plateaus* closes with the very term: "Every abstract machine is linked to other abstract machines, not only because they are inseparably political, economic, scientific, artistic, ecological, cosmic—perceptive, affective, thinking, physical, and semiotic— but because their various types are as intertwined as their operations are convergent. Mechanosphere" (1987, 514). It seems the mechanosphere concept is fundamental to Deleuze and Guattari's theory of strata, territories, and assemblages. But why invent another name for this ultimate horizon, for the confluence of all processes on earth?

A clue emerges when Deleuze and Guattari write, "there is no biosphere or noosphere, but everywhere the same Mechanosphere" (1987, 69). Here appears the terminological context to the mechanosphere neologism. The suffix *-sphere* became crucial as empirical science incorporated systems thinking to study the planet in its entirety: atmosphere, stratosphere, lithosphere, hydrosphere, magnetosphere, cryosphere, biosphere, and so on. As globalization becomes conscious of itself a new series of spheres is triggered: technosphere, infosphere, blogosphere. . . . Mechanosphere supplants two terms associated with vitalist philosophy, biosphere, theorized in the 1920s by the Soviet geochemist Vladimir Vernadsky (1998), and, following him, noosphere, theorized in the 1940s by the French Jesuit and paleontologist Pierre Teilhard de Chardin (2004). Though Deleuze and Guattari do not cite either author, Teilhard was widely read in France. With the mechanosphere concept Deleuze and Guattari seem to be suggesting a return *within vitalism* if not to mechanics, at least to an ontology that rethinks the sphericality of earthly processes.

Vernadsky's biosphere concept is now widely recognized as the key precursor to earth-systems science. He was the first to alert science and philosophy that life is not only tightly interwoven by way of ecosystems and genetic evolution but is, as biomass, an integral force interacting with mountains, oceans, and the air. In fact, life has over eons shaped the planet so that evolution could continue. Teilhard presaged much New Age thinking when he argued there was a global stratum of consciousness. The noosphere is a next step of the evolution of the planet come under human production. No longer read for his science, Teilhard has become a classic author in New Age spirituality. Both Vernadsky and Teilhard engaged the vitalist theories prevalent in their times to challenge the reductionist dogma whereby life would be secondary to geophysics. But they and many others also made the more mystical suggestion that there is a planetary epidermis of conscious activity that is the culmination of the course of the physical universe. Teilhard presaged science fiction by speculating that the noosphere will ultimately diffuse throughout the rest of the universe.

In making consciousness and religion the apex of the evolution of matter itself, vitalism tends to forsake the basic antianthropocentric principle of Copernicus, Galileo, and Darwin that states that there is no direction or impetus to physical things apart from the forces that propel them to move and change. Vitalism deliberately seeks to mystify this aleatory nature of force itself. For example, Hans Driesch (1914), a key proponent of modern vitalist metaphysics, argues that when science reduces biology to physics it loses sight of what Aristotle called *entelechy*, the principle making organisms organized and intentional wholes in harmony with their surroundings. The vitalism of Driesch is not against Darwin but seeks to restore purposefulness as indispensable to the workings of life. Though a mechanistic realism remains dominant in science, vitalism lives on. Michael Ruse (2013) explains why Gaia theory, straddling neopagan spirituality and earth-systems science, is the most important current in vitalism today.

Deleuze shares with vitalist philosophies an appreciation of the crucial role interdependence plays in the evolution of species. But his ontology, especially in *Difference and Repetition*, categorically rejects Aristotelian entelechy. There is no spark that would trigger or guide the nitty-gritty of biological, technological, and mental processes toward it. Species or strata are irreducible, and none can claim perfection. Life only emerges contingently and locally *from* geophysical processes. It can never be one seamlessly unified sphere as implied by the biosphere concept. Despite an indebtedness to Bergson, the most sophisticated of vitalist philosophers, to be consistently immanentist and atheist Deleuzian ontology has to deny there is one ineffable *élan vital* or life impetus underneath all things. Life is not independent from the strata that compose it. Meanwhile, instead of purposefulness and harmony, Deleuze's more Nietzschean and psycho-analytical leanings lead him to emphasize such inevitable somber facts of life as entropy, dissipation, the death drive, dying, illness, war, wounds, and exhaustion, themes elided in Driesch and Teilhard. Deleuze's central concept of the body without organs is an attack on early twentieth philosophy's infatuations like Driesch with the wholeness of the organism.

Yet Deleuze is no pessimist. Like Nietzsche, he chooses life over the void. Guattari's work likewise remains focused throughout on exposing the oppressive features of the mechanosphere and creating concepts toward its radical reconstruction. Altogether it is possible to use Deleuzian metaphysics and Guattarian politics to replace the celebration of life in vitalism with a stratoanalytical examination of the earth and the role social formations play in altering it. It is possible, or rather necessary, to affirm both the continuing importance of science for critical thinking and the requirement to beware transcendence doesn't slip back in, whether of religion or scientific reductionism.

THIS GLOBE WHICH IS NOT ONE

How global is globalization? An important question about the biosphere and in extension the mechanosphere is whether their sphericality converges onto some kind of *totality*. Denis Cosgrove (2003) argues in his Foucauldian history of the Western geographical imagination that an intrinsic tendency in the idea of the earth as globe is a desire to totalize and control. Cosgrove shows that the imagination of the earth as self-contained and orderly whole as seen from a particular perspective (Greco-Roman, Christian, Renaissance-European, imperialist, satellite, Google) prepares its social formation for colonization. Does the mechanosphere concept repeat this totalizing vision? Not if it can remain vigilant to the eternal return of transcendence.

On a technical level, just as there is no set of sets in mathematics, there can be no ultimate oneness to globalization, the earth, or the universe. The mechanosphere is a multiplicity of elements that *in principle* cannot be exhaustively enumerated. This is because, first, as assemblage of assemblages it cannot be fully delineated from the rest of spacetime, and, second, its elements are in continuous variation and becoming. Even the cosmos cannot be a whole because "it" infinitely differs from itself, and its holeyness is essential to it: black holes, dark matter, not to speak of other or parallel universes. The reason why Deleuze and Guattari like fractal mathematics so much (1987, 486–488) is presumably because it enables a thinking of space as a distribution of singular becomings forever staving off unity and identity.

The mechanosphere is not what unifies the earth, therefore, but the name for the temporary interlocking and stratification of its flows insofar as they produce a planetary thinking. It is crucial to understand the interconnectedness of life at the planetary scale, and one can be grateful for Gaia theory and the biosphere and ecosystem concepts for combating the reductionism that plagues science. But interconnectedness is not oneness. With a critical mechanosphere concept we can emphasize the uneven, contested, fragmented, frictional, disaster-prone, and mechanical (Newtonian) aspects of the terrestrial life. Social machines change continuously and remain full of cracks and black boxes. Globalization does not add up to a unified sphericality with its own impetus toward an apex of complexity or harmony.

As we can infer from Cosgrove, the discourses of globality have become ever more totalizing and cryptoimperialist since Deleuze died, especially thanks to the Internet. The technoutopian articulations latent in the term "globalization" itself have become enmeshed with New Age philosophies and environmentalisms and are driven by a yearning for wholeness and finality. What this holism belies is a highly particular white bourgeois masculinity that needs to smooth over the fractured nature of the "spheres" it is claiming are becoming more unified. While the unprecedented intensity and extensiveness of the interconnecting of assemblages is obvious, stratoanalysis shows how globalization *increases* the fractal, antagonistic, and potentially catastrophic nature of the human species, not in the least because of the supremely abyssal nature of capital. Exposing the ideological workings of various forms of holistic globalism is a first step toward a mechanosphere concept no longer beholden to transcendence.

It is true that Deleuze and Guattari's work is replete with references to total-sounding terms like cosmos, the Plane of Immanence, and so on. But they are meant to piggyback on an existing rhetorical force—and only in order to undermine them. Like all philosophers they need generalizing terms. More specifically, in order to reach a renewed concept of immanence against idealism, it is critical to put everything on the same ontological level. Deleuze revives a principle of heretic medieval metaphysics called *univocity*, in which all beings *are* in the same manner, de jure so to speak, even if they are de facto distributed and layered differently (1994, 35ff.). What Deleuzian univocity means for thinking reality is that every being posits itself in the same way, utterly contingently and singularly. There is no serene overall plan to which every being would answer, and beings do not emanate from one divine substance or perspective.

Deleuze's materialist univocity might seem to be mystical, something like the Neoplatonic One-All through which all things and minds communicate with one another. But we should remember he makes difference his first principle and equates being with becoming. Deleuze's univocity is a post-Kantian method more than a metaphysical return to ancient pantheism. Even when he does inherit certain pantheistic tendencies running through Spinoza (God-or-Nature within which all things participate) and Bergson (cosmic duration as that which the mind participates in), if difference is originary, then oneness can never be a fact or finality. Univocity is simply that feature of every being that makes it capable of change. Although Deleuze could have perhaps more clearly distanced himself from his monistic precursors, his ontology is only consistent if it understands reality as always already split by a fractal autodifferentiation.

Here we arrive at the most important tension in materialist philosophy today, that between Deleuze (or Deleuzians) and Alain Badiou (see, for example, Tarby 2005). Badiou (2000) is probably correct that Deleuze needed to engage the mathematical theorization of infinity and multiplicity in Georg Cantor to claim a fully modern, that is, critical and atheist, conception of immanence. Pure multiplicity—as Badiou's formula has it, the one is not—is necessary for a radical response to hegemonic holisms. When Cantor discovered the foundations of mathematics through devising set theory he eliminated once and for all the recourse to a transcendent set of sets or infinity of infinities. With a rigor no domain other than mathematics can achieve, set theory succeeds in thinking multiplicity all the way "up" and "down," never pinned to an overarching oneness. The Cantor-event is another Copernican turn, and it is crucial for the overcoming of anthropocentrism and transcendence. Analytical philosophy has from Russell's paradox onward realized this, but not so the continental tradition from which Deleuze hails. The main problem, of course, is that so few of us are able to continue the study of mathematics after our school years. The Deleuze-Cantor encounter is still to happen.

To repeat, the obtuse debate about the one and the multiple that Badiou has revived has direct implications for politics and ethics because the question is whether Deleuze, if he is a post-Kantian materialist of some sort, truly overcomes spiritual transcendence as he claims to. The defense of Deleuze by Deleuzians occurs by and large at the metaphysical level and demonstrates Badiou's willful misreading (see, for example, Roffe 2012). They argue Deleuze's univocity is not substantial and mystical but a transcendental aspect presupposed by thought to accompany every thing. All beings (humans, words, insects, demons) are to be posited on the same conceptual plane even if that doesn't mean they *are* "the same," much less have the same value. The Deleuzian countercritique misses the opportunity, however, to state categorically why Badiou is wrong to doubt Deleuze's fidelity to the broadly Marxist project. Deleuze and Guattari's critique of inequality and oppression clearly takes precedence over any mystical oneness and sharply distinguishes their materialism from all prior monist and immanentist philosophers.

Still, it is useless to disagree there is a minor tendency toward vitalism and even mysticism constitutive of Deleuze's thought. But compare that with the deliberate borrowing from Christian dogma in Badiou and Žižek. What's a worse tendency, vitalism or theology? The fundamental question for materialism today is whether ontology is able to avoid both the relapse into religion and the intolerant reductionism in science, so as to create new practices for averting more catastrophe. Materialism has to remain autonomous from all-too-individualist versions of Buddhism and neopaganisms such as Gaia theory and ecofeminism, on the one hand, and the viciously self-assured and overfunded discourses of cybernetics, economics, neuroscience, and genetics on the other. Badiou's critique of Deleuze's vitalism and his radically different set-theoretical ontology are precious for materialism because the debate about how thought relates to reality cannot end with any one system.

CAPITAL IN UNIVERSAL HISTORY

What most discussions of the Anthropocene cannot confront, because it raises questions even more intractable than environmental problems themselves, is that such problems are the perfectly logical results of just one very particular economic logic, the competitive appropriation of surplus value. Marxism is the name of that intellectual and political field which offers the most longstanding and exceedingly lively (that is, sometimes deadly) debates regarding capitalism. As such it is an obligatory passage point for explaining the self-propelling becoming-spherical of capitalism. While capital overcomes barrier after barrier, penetrating ever more into fatefully named "emerging markets," no government, nonprofit organization, or "corporate ethics" is capable of slowing it down. Marx shows meticulously in *Capital* (1887) how it is precisely the essence of our mode of production to be unstoppable yet full of contradictions. Profits must be reinvested, or one cannot compete. Hence capital grows exponentially and drags the entire world into its psychotic logic.

> The bourgeoisie, during its rule of scarce one hundred years, has created more massive and colossal productive forces than have all preceding generations together. Subjection of Nature's forces to man, machinery, application of chemistry to industry and agriculture, steam-navigation, railways, electric telegraphs, clearing of whole continents for cultivation, canalization of rivers, whole populations conjured out of the ground. What earlier century had even a presentiment that such productive forces slumbered in the lap of social labor? . . . Modern bourgeois society with its relations of production, of exchange and of property, a society that has conjured up such gigantic means of production and of exchange, is like the sorcerer, who is no longer able to control the powers of the nether world whom he has called up by his spells.
>
> (Marx and Engels 1888)

The Communist Manifesto lays responsibility for industrialization and colonization (and possibly genocide) with a class subject, the bourgeoisie. One detects the profoundly ambivalent attitude of Marx and Engels toward the capitalist system. Though capital necessarily entails exploitation, crisis, war, famine, and so on, the bourgeoisie deserves thanks for lifting humanity out of feudal and primitive "backwardness" in place after place. Communism will require the infrastructures and scientific knowhow amassed by the middle classes. Indeed, Marx and Engels in all their work sharply distinguish themselves from any moralistic or populist denunciation of the system.

On a consistently antihumanist reading taking its cue from Althusser, the problem is not that the bourgeoisie is the motor of history but, on the contrary, that it has no control over the productive forces and suicidal contradictions it has unleashed. There is certainly unimaginable greed in the tiny minority of humans who are stealing the future from future generations. It is not just that the richest sixty-two individuals possess as much wealth as the 3.5 billion poorest (Elliott 2016) but that this obscene inequality corresponds to lopsided responsibility for globalization's carbon footprint. There is certainly a "false consciousness" among the masses, who believe their only political role is to vote and hope recycling makes a dent in the mountains of waste. However, greed and ideology are only the tip of the iceberg. The most inconvenient truth is that the driver of the Anthropocene is not *anthropos* or human activity at all but the "productive forces" Marx and Engels say are like evil magic. And those forces are expressions of what in *Capital* would be theorized as the self-augmenting but crisis-prone axiomatic of capital.

Deleuze and Guattari strictly follow Marx in *Anti-Oedipus* when they argue that capitalism has continuously to create, then displace its own limits. In other words, planetarity is inherent to capital. This and only this means that *universal history* has become possible in hindsight: "If we say that capitalism determines the conditions and the possibility of a universal history, this is true only insofar as capitalism has to deal essentially with its own limit, its own destruction" (1983, 140). Though it enforces the same basic *dispositif* across the globe—the imperative to compete and create surplus—globalization does not mean homogenization. Capital comes in institutional variants that include plantation slavery, Stalinist industrialization, the Nazi war machine, kleptocratic dictatorships, and Saudi Wahhabism. China's explosive growth shows again how capital voraciously incorporates all ecological and cultural difference and is outgrowing its initial white-imperialist matrix. Capitalism thrives on disparity, coercion, and segregation, and it certainly has no intrinsic relationship to democracy.

Deleuze and Guattari's *Capitalism and Schizophrenia* project is therefore explicitly universalizing in scope. It abounds with terms like mechanosphere, plane of immanence, the universal schizo, universal minoritization, and absolute deterritorialization, and they do seem to apply to the human species as such. However radical, Deleuze and Guattari follow an old European convention that presents a theory of three broad moments or stages of societal development (savage, barbarian, civilized). At radical odds with other poststructuralisms, they posit that "it is correct to retrospectively understand all history in light of capitalism, provided that the rules formulated by Marx are followed exactly" (1983, 140). *In retrospect* the becoming-geological of humanity along the supremely aggressive vector of capital *now seems* to have been waiting there all along: "Capitalism has haunted all forms of society, but it haunts them as their terrifying nightmare, it is the dread they feel of a flow that would elude their codes" (1983, 140). Once it has contingently surfaced, capital is like an undead parasite enhancing its own conditions of possibility.

Deleuze and Guattari make the controversial argument that capitalism has always and everywhere been human society's virtual shadow, insofar as it incarnates planet-encompassing production as no other system ever could. "If capitalism is the universal truth, it is so in the sense that makes capitalism *the negative* of all social formations" (1983, 153). A real tendency in all exchange and divisions of labor, capitalism is the only conceivable system capable of globalization. Capitalism is not, of course, the telos of the human species as apologetic economists claim. Again: "The only universal history is the history of contingency" (1983, 224). Capitalism is a *limit* that all production tends toward. Social formations have been unconsciously yet actively *staving off* the formation of capitalism (just as they did the state, the Oedipus complex, and general schizophrenia) because they were quite content with their nomadic lifestyles and not interested in becoming dependent on the obscure circuits of money. But one by one, these resistances (or barriers from capital's perspective) fell.

The resulting economic system is what Guattari (1984, 263ff.) in the 1970s called integrated world capitalism. It is integrated and worldwide but remains dynamic precisely because of its differentials and innovations. A major new wave of integration, simultaneously geographic, subjective, and logistical, came with computers.

All of [capitalism's] "mystery" comes from the way it manages to articulate, within one and the same general system of enrolment and equivalence, entities which at first sight would seem radically different: of material and economic *goods*, of individual and collective human activities, and of technical, industrial and scientific processes. And the key to this mystery lies in the fact that it does not content itself with standardizing, comparing, ordering, informatizing these multiple domains but, with the opportunity offered by these diverse operations, it extracts from each of them one and the same *mechanical surplus-value* or *value of mechanical exploitation*.

(Guattari 1984, 275)

There is an intrinsic push for universality in capital, but it is for entirely different reasons than neoliberals think. Guattari and Deleuze abhor the bourgeois dogma about the rational individual, and there is for them no animal instinct for utility and ingenuity. Global interconnectedness is but a side effect of a "system of enrolment and equivalence" and the self-propulsion of capital, which operates through inevitable destruction and what Guattari calls machinic enslavement. What is capable of confronting this universality other than another universality?

UNIVERSALIZATION

The Anthropocene forces politics to evoke a new scalar imagination. Traditionally politics has been defined as the collective practice aiming to improve the life of the city (*polis*), the state, or the nation. For modern thought like that of Kant and Marx, in contrast, politics finds its truth at the level of universal history, or "world history." Pointing out how world history has always been structured around Eurocentric, patriarchal, and biopolitical biases, since the 1960s it has often been communities and bodies that have instead formed the basis for politics, especially in the United States. However, real-world developments such as the nuclear-arms race, climate change, and genomics have raised the problem of universality as never before.

Naomi Klein (2014, 16ff.) explains that just when the buzzword of globalization was coming into its own as market fundamentalism was successfully promoted to governments by billionaires and millionaires in the late 1980s, the scale of the problem of global warming caused by greenhouse emissions was already becoming apparent to policy makers. It was ignored, however, so as not to stand in the way of the elite's project of dismantling the welfare state. It is not simply neoliberal ideology but the need to make profit from reinvestment that comes first, a self-perpetuating mechanism that had simply mutated in the reformist axioms of the postwar welfare state (on axioms, see Deleuze and Guattari 1987, 453ff.). Lowering carbon emissions is hampered not simply by the selfishness of the global oligarchy but by liberal democracy itself insofar as it is an axiom of capitalism. Pseudosolutions are offered through certain buzzwords that Deleuze and Guattari might call order-words, enunciations literally ordering activities to the profit of power (1987, 75ff.). Hence "adaptation," "resilience," and "sustainability" all hide their territorial logic. Just like man before, if rather less grandiosely, such buzzwords are often based on presuppositions with subtly racial undertones. They aren't just fig leaves for continued structural oppression and pollution globally but prevent imagining real alternatives to the very system that is increasing the likelihood of disaster.

There can be little doubt it is Badiou (2010) who provides the most rigorous defense of the universalist legacy of the French and Russian revolutions today. The formation of political subjectivity occurs for Badiou in the wake of a revolutionary *event* that instigates a line of escape from all identities and state institutions, including the revolutionary party. Badiou conceptualizes universality not as given but as an ongoing process. In the case of politics, universalization is entirely dependent on ordinary people coming out onto the streets to make demands from power and on the militants who continue organizing for the next mass event. Such universalism is almost the opposite of the universalisms of the church and liberal democracy, insofar as these are based on state-enforced or presumed natural rights and thus hide their particularism. A romanticized proletariat finding its true meaning in struggle and the Communist Party can also no longer be the locus of generic subjectivity. In short, there is no positive content to the subject of universalization. The temporality of revolutionary politics is immanent to the continuous collective construction of a break with the past and with what is. Consisting of nothing but self-constitutive enunciation and struggle, communist subjectivity is driven by a *hypothesis* that in stratified society real equality and justice *are* always already a possibility, especially when they don't seem to be.

Strictly speaking, politics for Badiou is agnostic about materialities such as global warming or financialization. More catastrophic injustice does not in itself increase the probability of a political rupture. In fact, he provocatively argues that environmentalism is the most deleterious ideological ploy of our times, today's opium for the masses, a theme Slavoj

Žižek (2007) has also picked up. The near consensus that "nature," "life," or "the planet" must be "saved" can be seen as an obscenely narcissistic fantasy, lulling people into assuming that "going green" is the most significant effort humans can undertake to avert civilizational collapse, instead of ending capitalism. It is also true that recent revolutionary junctures such as the overthrow of Tunisia's and Egypt's regimes, the protests against police brutality in the United States, and Syriza's electoral-governmental experiment in Greece show that popular mobilization usually proceeds without any need of addressing the earth.

Still, what the theory of the revolutionary event can learn from the climate-justice movement is that the Anthropocene is an unprecedented basis for demanding equality not just nationally and in the present but globally and for future generations. By limiting his concept of "ecology" to conservation, Badiou does not recognize that there has never been a more compelling reason for communism and universalization than the Anthropocene. In fact, justice is already being redefined by the more radical environmentalists as intrinsically planetary and revolutionary and by what the law and corporations are making impossible, especially those of the United States. But most environmentalists do not see the full extent of the capitalist crisis, as ecosocialist theorists have long complained (Foster, Clark, and York 2010). To be consistent, the climate-justice movement will probably have to become communist.

CONCLUSION: REVOLUTION

Though it appears only a handful of times in the corpus of Deleuze and Guattari, the concept of mechanosphere can be considered their prescient term for an age where the human is no longer in control of a realm supposed outside it. The mechanosphere concept allows addressing the Anthropocene without the tendency toward mystical oneness that remains hidden in the concepts of life, earth, and globality. The geological turn puts the turn to life in the humanities in perspective. Vitalism can be understood still to be beholden to an ultimately anthropocentric perspective. There cannot be an ultimate vital impulse if the biosphere creates its own collapse. To summarize, for geophilosophy: (1) deep time is a rupturing force internally affecting not just life but the planet at large; (2) the possibility of human extinction has to be confronted as a possibility, and it is already informing decisions in the present; (3) local ecological problems cannot have conceptual preference above the overall geophysical impact of the human species, which will be detectable into the very far future; (4) there is no such thing as nature, a unified harmonious whole onto which social processes would be inscribed, but only variously intermeshing assemblages in which collective thought attempts to intervene; (5) capitalism is the matrix for thinking the earth because the Anthropocene is the materialization of the logic of capital.

Of course, if biophilosophy were to be superseded by geophilosophy, the question can be raised why the next logical step away from man should not be toward an astrophilosophy, or the opposite way, toward a quantum philosophy. But to decenter for the sake of decentering is not what thinking is about. The prerogative of all ancient philosophy (including outside Europe) is to provide tentative answers to the question "how to live?" Similarly, the critical turn of Kant is to make thinking matter to the continuous reinvention of collective rational living. Though philosophers certainly have been responding to astrophysics and quantum mechanics, as they should, these scales aren't subject to multilayered crisis as the earth is. Furthermore, the crisis is induced by accumulated selfish choices of a minority of human beings and hence directly an ethical and political imperative.

Naomi Klein (2014) correctly exclaims about climate change: *This changes everything.* One chooses either capitalism with continuing catastrophe or a livable global climate regime under an egalitarian economic system. Since capitalism and climate affect almost everyone, this political question has become quasi-universal. As Klein notes, the most hopeful aspect of the Anthropocene is that an egalitarian future remains not just possible but necessary so as to avoid breakdown and war. The question is what change is conceivable. In this global-historical moment, while fascisms are rising everywhere, there are also increasing numbers of young people interested again in revolution. A century ago Rosa Luxemburg (1908) said the choice was between socialism and barbarism, immortalizing a phrase from Engels. And within socialism, she said, one must choose reform or revolution. By burrowing into the everydayness of human-nonhuman assemblages and by shunning world-historical questions and intercontinental injustices, some posthumanisms and new materialisms have resolutely chosen reformism. Many in the humanities prefer not to confront the critical threshold that the anthropic sphere has passed and the stark decisions forced on thinking. To be a materialist, one must respond to the world. If man is dead, one has to reinvent life after him.

Deleuze and Guattari are sympathetic to revolutionary and anticapitalist politics, especially that of May 1968. Even if the stakes have become infinitely bigger than they were then, this chapter has suggested their earthly ontology be reenergized by plugging it into an older enthusiasm for radical societal invention, for starting from scratch. What the machinic ontology of stratification could do is provide the material support for Badiou's austerely formalist universalism. With the resulting framework for tackling the Anthropocene, what can be seen as simmering could well be a *geocommunism*. Such a geocommunism would be capable of dismantling the present mode of production and its disastrous ways of life, first of all in the rich world. As an ever more absolute ultimatum comes crashing down on critical thinking in the twenty-first century, pretending nothing has changed with the Anthropocene is tantamount actively to bringing that ultimatum forward. The continuing livability of the planet for humans is entirely up to the political choices made in the present.

WORKS CITED

Badiou, Alain. 2004. "One, Multiple, Multiplicities." In *Theoretical Writings*. Trans. Ray Brassier and Alberto Toscano. London: Continuum.

——. 2003. *Deleuze: The Clamor of Being*. Trans. Louise Burchill. Minneapolis: University of Minnesota Press.

——. 2010. *The Communist Hypothesis*. Trans. David Macey and Steve Corcoran. London: Verso.

Colebrook, Claire. 2010. *Deleuze and the Meaning of Life*. London: Continuum.

Cosgrove, Denis. 2001. *Apollo's Eye: A Cartographic Genealogy of the Earth in the Western Imagination*. Baltimore, Md.: Johns Hopkins University Press.

Deleuze, Gilles. 1994. *Difference and Repetition*. Trans. Paul Patton. London: Athlone.

Deleuze, Gilles, and Félix Guattari. 1983. *Anti-Oedipus: Capitalism and Schizophrenia*. Trans. Robert Hurley, Mark Seem, and Helen R. Lane. Minneapolis: University of Minnesota Press.

——. 1987. *A Thousand Plateaus: Capitalism and Schizophrenia*. Trans. Brian Massumi. Minneapolis: University of Minnesota Press.

——. 1994. *What Is Philosophy?* Trans. Graham Burchell and Hugh Tomlinson. New York: Columbia University Press.

Driesch, Hans. 1914. *The History and Theory of Vitalism*. Trans. C. K. Ogden. London: Macmillan.

Elliott, Luther. 2016. "Richest 62 People as Wealthy as Half of World's Population, Says Oxfam." *Guardian, January* 18.

Foster, John Bellamy, Brett Clark, and Richard York. 2010 *The Ecological Rift: Capitalism's War on the Planet.* New York: Monthly Review.

Guattari, Félix. 1984. *Molecular Revolution: Psychiatry and Politics.* Trans. Rosemary Sheed. Harmonsworth: Penguin.

——. 2000. *The Three Ecologies.* Trans. Ian Pindar. London: Athlone.

Haraway, Donna. 2015. "Anthropocene, Capitalocene, Plantationocene, Chthulhucene [Cthulhucene]: Making Kin." *Environmental Humanities* 6:159–165.

Harvey, David. 2015. *Seventeen Contradictions and the End of Capitalism.* Oxford: Oxford University Press.

Klein, Naomi. 2014. *This Changes Everything: Capitalism vs. the Climate.* New York: Simon & Schuster.

Kolbert, Elizabeth. 2011. "Age of Man." *National Geographic.* http://ngm.nationalgeographic.com/2011/03/age-of-man/kolbert-text.

Luxemburg, Rosa. 1908. *Reform or Revolution.* https://www.marxists.org/archive/luxemburg/1900/reform-revolution.

Marx, Karl. 1887. *Capital: A Critique of Political Economy.* Vol. 1: *The Process of Production of Capital.* Ed. Friedrich Engels. Trans. Samuel Moore and Edward Aveling. http://www.marxists.org/archive/marx/works/1867-c1.

Marx, Karl, and Friedrich Engels. 1888. *Manifesto of the Communist Party.* Trans. Samuel Moore with Friedrich Engels. http://www.marxists.org/archive/marx/works/1848/communist-manifesto.

McKibben, Bill. 2006. *The End of Nature.* New York: Random House.

Moore, Jason. 2015. *Capitalism in the Web of Life: Ecology and the Accumulation of Capital.* London: Verso.

Nietzsche, Friedrich. *Thus Spoke Zarathustra: A Book for None and All.* Trans. Walter Kaufmann. New York: Viking.

Roffe, Jon. 2012. *Badiou's Deleuze.* Montreal: McGill-Queens University Press.

Ruse, Michael. 2013. *The Gaia Hypothesis: Science on a Pagan Planet.* Chicago: University of Chicago Press.

Tardy, Fabien. 2010. *Matérialisms d'aujourd'hui. De Deleuze à Badiou.* Paris: L'Harmattan.

Teilhard de Chardin, Pierre. 2004. *The Future of Man.* Trans. Norman Denny. New York: Doubleday.

Vernadsky, Vladimir. 1998. *The Biosphere.* Ed. M. A. S. Menamin. Trans. D. B. Langmuir. New York: Copernicus.

Zalaciewicz, Jan. 2008. *The Earth After Us: What Legacy Will Humans Leave in the Rocks?* Oxford: Oxford University Press.

Žižek, Slavoj. 2007. "Ecology, a New Opium for the Masses." Talk for *Lacanian Ink*, November 28. https://www.youtube.com/watch?v=fi57r_JByNE.

IV

POSTHUMOUS LIFE

11

PROLIFERATION, EXTINCTION, AND AN
ANTHROPOCENE AESTHETIC

MYRA J. HIRD

A TWENTY-FIRST-CENTURY ANXIETY

In "The Psychotherapy of Hysteria" (1891), Sigmund Freud characterized modernity as the expression of an anxiety toward the extinction of "the normal" and the proliferation of what we might now term queer; this anxiety is a reaction to the trauma of repressing our original polymorphous perversity (Hird 2002). Among other things, this trauma precipitated an enduring political and philosophical concentration on the question of sexual difference (Hird 2013); the association between sex, life, and death; and, as Foucault (1976, 1984a, 1984b) liked to remind us, a proliferating and dispersed heteronormative governance.

If the twentieth century was marked by a hysterical articulation of our repressed polymorphous perversity that transfigured into the cooptation of heteronormativity and queer within neoliberal capitalization, we may now be naming a new hysteria—this time of global scale and planetary deep time: the Anthropocene. Where once particular individuals—poor, queer, Jewish, Indigenous—were the targets of eugenic dreams of societal purity (see Raffles 2007), now pundits speak of the need to limit the human population tout court in order to save our own (and other charismatic) species from a "sixth mass extinction" (Novacek 2007). Anxieties now center, in other words, on the consequences of human proliferation.

Take, for instance, Paul Ehrlich's hugely popular *The Population Bomb*. Published in 1968, Ehrlich warns of the consequences of over-population: "a minimum of ten million people, most of them children, will starve to death during each year of the 1970s. But this is a mere handful compared to the numbers that will be starving before the end of the century. And it is now too late to take action to save many of those people" (3).[1]

Contemporary writing about the Anthropocene renders Ehrlich's pre-dictions banal, as now even our breathing has become problematic: "The breathing of human beings and their domestic pets and livestock accounts for 23 percent of all planetary carbon dioxide production, and when we include that caused by fossil fuels involved in food production we are up to half of all global carbon dioxide production" (Aravamudan 2013, 14). Indeed, this new breed of "environmental Jeremiahs" (Egan 2013) proph-esy apocalyptic extinction through its opposite, a cacophony of proliferа-tions: "Too many cars, too many factories, too much detergent, too much pesticide, multiplying contrails, inadequate sewage treatment plants, too little water, too much carbon dioxide—all can be traced to *too many peo-ple*" (Ehrlich 1968, 66–67). In what has now become the rather formu-laic reiteration that set all this contemporary Anthropocene writing in motion, Paul Crutzen and Eugene Stoermer's article in the 2000 *Interna-tional Geosphere-Biosphere Programme* (IGBP) *Newsletter* provides what has become standard statistics decrying the proliferation of humans and the consequences of our actions:

> The expansion of mankind [*sic*] both in numbers and per capita ex-ploitation of Earth's resources has been astonishing. To give a few exam-ples: During the past 3 centuries human population increased tenfold to 6000 million, accompanied e.g. by a growth in cattle population to 1400 million (about one cow per average size family). Urbanization has even increased tenfold in the past century.
>
> (17)

Proliferation and extinction mutually subtend each other. Bulleted-list lessons in the probability of species extinction, along with a clas-sification of "vulnerable," "endangered," or "critically endangered" are offered:

- The rapid loss of species we are seeing today is estimated by experts to be between 1,000 and 10,000 times higher than the *natural extinction rate*.
- These experts calculate that between 0.01 and 0.1% of all species will become extinct each year.
- If the low estimate of the number of species out there is true—i.e., that there are around 2 million different species on our planet—then that means between 200 and 2,000 extinctions occur every year.
- But if the upper estimate of species numbers is true—that there are 100 million different species co-existing with us on our planet—then between 10,000 and 100,000 species are becoming extinct each year.

(World Wildlife Federation 2014)

And,

In a slow extinction, various balancing mechanisms can develop. No one knows what will be the result of this extremely rapid extinction rate. What is known, for sure, is that the world ecological system has been kept in balance through a very complex and multifaceted interaction between a huge number of species. This rapid extinction is therefore likely to pre-cipitate collapses of ecosystems at a global scale.

(Shah 2014)

And: "human activity has increased the species extinction rate by thou-sand to ten thousand fold in the tropical rain forests. . . . Coastal wet-lands are also affected by humans, having resulted in the loss of 50% of the world's mangroves" (Crutzen and Stoermer 2000, 17). Indeed, we might say, the discourses of proliferation are themselves proliferating.

While the "cosmopolitanism" of the planet during the billions of years before the continents were separated by oceans is well documented (see, for example, Diamond 1999), writing on the Anthropocene emphasizes the acceleration of unwanted species proliferation that humans have pre-cipitated. Within a general biopolitical framework, the science of "inva-sion ecology" focuses on "ecological explosions" affecting humans such as the bubonic plague, medieval Europe's Black Death, and the pandemic of influenza during World War I, as well as outbreaks affecting flora and fauna, such as the invasive spread of the European starling *Sturnus vulgaris* into Canada and the United States in the late 1800s and early 1900s (Barrow 2009). Writing in 1958, Charles Elton argued that invasion ecology unnerves colonial and scientific assumptions of pest control:

> The real thing is that we are living in a period of the world's history when the mingling of thousands of kinds of organisms from different parts of the world is setting up terrific dislocations in nature. We are seeing huge changes in the natural population balance of the world. . . . We are artificially stepping up the whole business, and feeling the manifold consequences.
>
> (369, 376)

The military rhetoric of "invasion" of "foreign" species invites a correspondent response focused on conquest, containment, and management. Within this context, we are "no longer documenting species in the wild, but rather performing triage for bigger emergencies and on bigger scales than ever before" (Robin 2013, 406). As Kathryn Yusoff (2011) notes, this is loss and extinction not even registered as such but rather as "an achievement of eradication and control" (580).

Yet what appears to be a new recognition of our own and other species' vulnerability within invasion-ecology writing specifically, and Anthropocene writing generally, turns out to follow an Enlightenment script insofar as it is humans who have precipitated the movement of species *and* humans who are capable of controlling the continued spread of species through science, technology, and governance. Humans, in other words, have mastered nature to such extraordinary degrees as to have produced the mass global movement of species, with the implication that humans also have the capacity to alter this mass migration through more control, more governance, and further rationality. There may be proliferation, and this proliferation may be leading to extinctions, but with the further articulation of science and engineering, we may steer in another direction, put ourselves and the planet back on course, and redeem ourselves (Hird forthcoming). Take, for instance, Crutzen's "Geology of Mankind," a brief notice published in the pages of *Nature* in 2002. Crutzen exhorts (and nominates) scientists and engineers, as the helmsmen of the Anthropocene, "to guide society towards environmentally sustainable management during the era of the Anthropocene" (23).

It just may be that the Anthropocene is naming an anxiety about our own vulnerability to planetary forces, and a discourse of proliferation—of too much muchness—may be articulating this anxiety. Both proliferation and its extinction dialectic articulate what I suggest is an emerging *Anthropocene aesthetic* that describes a Western perception of our contemporary environmental and political crisis and suggests its solution. Crucially, this Anthropocene aesthetic relies upon a Western Enlightenment ideology of mastery and control over our environment and ourselves and salvation through increased reliance on science, engineering, and technology.

In this way, I suggest, while writing on the Anthropocene purports to identify a new epoch that demands new ways of thinking about ourselves and the environment, an Anthropocene aesthetic fundamentally reinforces familiar ideologies and discourses, ones that indeed circle back to the Anthropocene's inception. In this chapter I will explore this Anthropocene aesthetic through a focus on waste repositories. Dumps and landfills enjoy an ambivalent relationship with the Anthropocene: As evidence of our massive consumption, waste is something we prefer out of sight and out of mind. Yet this does not mean that waste has escaped capitalism's long reach, as its management attests. Moreover, waste's management illuminates both society's heavy reliance upon engineering and science to deal with the fallout of our rampant capitalist consumerism but also waste's imprescriptibility, waste as an interacting planetary force that exceeds the Anthropocene aesthetic.

AN ANTHROPOCENE AESTHETIC: SUBDUING NATURE TO CAPITAL

In 2000, Paul Crutzen and Eugene Stoermer proposed the term Anthropocene to capture an idea that had been circulating for years; the term signaled the end of the Holocene and set the point at which human activity intersected, in its significance and magnitude, with planetary, geophysical forces. To take the Anthropocene seriously is to entertain a radically different ontological regime—a materialist takeover bid that retroactively reassembles the planet and, paradoxically, challenges any lingering sense of human exceptionalism. In other words, the Anthropocene marks a collision of geologic and human chronologies of history (Chakrabarty 2009).

If the Anthropocene, as Nicholas Mirzoeff (2014) suggests, is a "human-created machine that is now unconsciously bent on its own destruction, a purposiveness without purpose," it may be said to mark a particular aesthetic born of a "last moment of human agency" through which we determine what we can and will perceive, find beautiful, and deem worthy of salvation (213). This paradox of the Anthropocene— that we have exceptionally created the conditions of our own existence as unexceptional—is expressed in, I suggest, a particular aesthetic. This aesthetic makes the Anthropocene *safe*: safe to articulate, to identify, and to discuss. It makes the Anthropocene resolvable. This aesthetic emerges from an Enlightenment ideology that conceptualized nature as a largely pliable, if stubborn, resource available for man to subdue and utilize (see, for example, Arrighi 1994, Blackbourn 2007). And as Mirzoeff (2014) explains, "the theory and practice of the conquest of nature has become integrated into Western aesthetics throughout the Anthropocene" (219). The Enlightenment cleaved to the beautiful as a science of the sensible and to the disclosure of the world through this new science. Aesthetic pleasure, then, "discloses to us rational order," with its variations of the beautiful as the true and the true as an objective rational order (Bristow 2011). Beauty, in other words, has come to be equated with what we can control, what is safe to articulate: what we can, and choose, to save.

As aesthetics scholars have noted, visualization is a key aspect of aesthetics and leads to well-trodden debates about subjectivity (who perceives), empiricism (what is perceived), and epistemology (perception becoming aesthetic) that are beyond the scope of this chapter. For my purposes here, an important aspect of visualization is that it is dependent upon perceiving not just things but their relationality with other things.[2] Visualization takes on a particular significance within Anthropocene writing in part because the naming of the Anthropocene is predicated on the paradoxical assumption of a future geologist casting her gaze back to the current epoch defined as the suicidal destruction of humankind and in part because the Anthropocene, *as* a biological and geological epoch, requires the impossible task of visualizing globality itself (geosphere to stratosphere). An Anthropocene aesthetic, like all aesthetics, is only achieved through selected visualization, which the Enlightenment provides.

The Anthropocene aesthetic is capitalist venture, planetary conquest, colonial authority, and limitless resource extraction. This aesthetic "allows us to move on, *to see nothing* and keep circulating commodities, despite the destruction of the biosphere" (Mirzoeff 2014, 217, my emphasis). To illustrate, Ramachandra Guha (1997) writes poignantly about the implications of such nonvisualization, or, rather, capital-directed visualization, for conservation efforts in the global South that in many cases are effected through the removal of local people from their land and the prohibition of gathering and hunting traditions. As Bonner observes:

> As many Africans see it, white people are making rules to protect animals that white people want to see in parks that white people visit. Why should Africans support these programs? . . . Africans do not use the parks and they do not receive any significant benefits from them. Yet they are paying the costs. There are indirect economic costs—government revenues that go to parks instead of schools. And there are direct personal costs (i.e. of the ban on hunting and fuel collecting, or of displacement).
>
> (qtd. in Guha 1997, 422–423)

What Guha refers to as "green imperialism" is made possible through an Anthropocene aesthetic that visualizes tigers, polar bears, or other charismatic animals roaming freely within contained landscapes, protected from the perceived threat posed by local peoples (see also Zahara and Hird 2015).

WASTE AND AN ANTHROPOCENE AESTHETIC

Lively discussions debate the proper placement of the golden spike—as much a proverbial as a material marker of the Anthropocene. Some say it was the late seventeenth century—the point at which the Industrial Revolution and the accelerated extraction and burning of fossil fuels began to take place. Others place it some eight thousand years earlier, in the Neolithic, with the clearing of forests for agriculture. Fossil fuels are a form of necro waste (Olson 2012) formed from the mainly anaerobic decomposition of buried dead organisms. Here, bacterial metabolism engages with life's remains, heat, pressure, and deep time to form solid and gaseous geologic strata. Fossil fuels speak to the longue durée of life and non-life's secular transubstantiation. Until recently, Crutzen had been of the opinion that anthropocentric time began roughly two hundred years ago, but he recently changed his mind. He now places "the real start of the Anthropocene" on a specific date: July 16, 1945—the Trinity detonation—and its signature, invisible radioactive decay. Whether in the form of mining, nuclear, industrial, hazardous, sewage, or municipal, and whether it is landfilled, incinerated, or buried deep underground, waste constitutes perhaps the most abundant and enduring trace of the human: a major human-instantiated planetary de- and restratification.

This Anthropocene aesthetic, which allows us to "move on and see nothing," also explains why rampant pollution in great urban cities such as New York, London, and Paris during and since the Industrial Revolution was not viewed as a catastrophe or even an unintended but necessary consequence of capitalism's expansion but was rather understood as a material visualization of the power, scale, acceleration, and *success* of capitalism's conquest of nature. Monet and other artists depicted industrialism's flourishing pollution—London smog, soot-covered houses, blackened sludge-filled rivers—in paintings such as *Impression: Sun Rising* (1873). In his analysis, Mirzoeff (2014) draws our attention to Monet's depiction of his own hometown of Le Havre in Normandy, which was at the time France's central port for transatlantic passenger shipping. In this painting we see the long-established rowing boats of premodernity dwarfed by industrial cranes and coal smoke billowing from steamers. Indeed, suggests Mirzoeff, it is the unique color of the yellow coal smoke combined with the blue morning light and red rising sun that makes this painting so visually remarkable.[3] This pollution, like the soot covering buildings in early twentieth-century British cities like Bradford was viewed as a verification of society's progress in industrializing.

Similarly, the middens of premodernity that grew in scale and diversity to vast trash heaps encroaching on urban landscapes were perceived as a testament to the progress of capitalist venture in what would become an emerging Anthropocene aesthetic. Take for instance E. H. Dixon's famous watercolor, the *Great Dust Heap* (1837), at King's Cross in London, which depicted an enormous garbage dump in the shape of the Alps, surrounded by urban slum dwellings, a few allotments, and a smallpox hospital. Carts loaded with more detritus are being cast onto the burgeoning heap. So well known was this garbage dump that it appeared in Charles Dickens's *Our Mutual Friend* (1865) and also in R. H. Horne's essay "Dust, or Ugliness Redeemed" published in *Household Words* (1850). Horne wrote:

> About a quarter of a mile distant, having a long ditch and a broken-down fence as a foreground, there rose against the muddled-grey sky, a huge Dust-heap of a dirty black colour—being, in fact, one of those immense mounds of cinders, ashes, and other emptyings from dust-holes and bins, which have conferred celebrity on certain suburban neighbourhoods of a great city. . . . A Dust-heap of this kind is often worth thousands of pounds.

The essay illustrates an Enlightenment aesthetic in which pollution signifies economic prosperity and industriousness. One household's discards were the scavenger's spoils, to be picked over, collected, and reinserted into the circuits of capital.[4] *Where there's muck, there's brass.*[5]

This Enlightenment aesthetic falls to engineers and scientists to actualize. Determining and controlling waste is a highly complex process. Certainly, landfills, nuclear repositories, incinerators, and other waste-management techniques have become more technically sophisticated as engineers and scientists develop better liners, gradient specifications, barriers, and so on (which doesn't mean these techniques are always or most often adopted). This "so on" is wide ranging: For instance, modern landfills are lined with a complex layering of rocks, clay, sand, and geotextile membranes and/or liners and are physically structured in waste-cell layers, which are stacked to form a number of waste columns (Yildiz and Rowe 2004). Estimating landfill settlement depends upon, among other things, the "type of waste, moisture content, compaction density, porosity, compressibility, biodegradation rate (level of nutrients available for biological activities, presence of enzymes, sludge addition, pH, temperature), and mode of landfill operation" (Elabroudy et al. 2008).

Besides these complexities, engineers and scientists must determine a landfill's integrity, expecting that they will contain certain objects—diapers, metals, plastics, Styrofoam, wood, liquids, refrigerators, pet shit and litter (as well as the dead pets themselves), batteries, food, chairs, fabrics, and so on—and the rather less expected, such as products of common industrial processes like coal fly ash, of which over 50 percent ends up landfilled (Chertow 2009). More than 308 million tons of plastics are consumed worldwide each year, most of which still end up landfilled (Plastics-Europe 2009). In addition, over 97 percent of food waste is landfilled in the United States (Levis et al. 2010). Landfills also mix waste designated hazardous with that defined as nonhazardous, including over seven million known chemicals, eighty thousand of which are in commercial circulation, with a further thousand new chemicals entering into commercial use each year (Wynne 1987, 48). Add to this the approximately 14,000 food additives and contaminants added to landfills when we waste food. The engineering and science of waste disposal, in other words, must contend with a vast heterogeneous mix of known, unknown, and unknowable objects.

Moreover, landfills assemble billions of heterogeneous bacteria whose "variations may be cyclical, directional, stochastic, or chaotic" (Collins, Micheli, and Hartt 2000). Aerobic bacteria metabolize a landfill's early life, which produces material that is highly acidic and toxic to surface water. Anaerobic bacteria do the bulk of the metabolizing work deeper in the landfill's strata, producing leachate. Leachate is a heterogeneous mix of heavy metals, endocrine-disrupting chemicals, phthalates, herbicides, pesticides, and various gases including methane, carbon dioxide, carbon monoxide, hydrogen, oxygen, nitrogen, and hydrogen sulfide. Factors affecting leachate production rate and composition include the

> characteristics of the waste (initial composition, particle size, density and so on), the interaction between the percolating landfill moisture and the waste, the hydrology and climate of the site, the landfill design and the operational variables, microbial processes taking place during the stabilization of the waste, and the stage of the landfill stabilization. Most of these factors change during the operational period of the landfill as the landfill is developed causing significant changes in leachate quality and quantity.
>
> (Yildiz and Rowe 2004, 78)

Leachate travels vertically and horizontally within landfills, and it continues to travel when it leaks beyond landfill cells and sometimes through geological strata. That is, leachate may percolate into soil and groundwater (the engineer's job is to prevent this to the largest degree possible), where it moves into and through plants, trees, animals, fungi, insects, and the atmosphere. Via leachate, bacteria create well-known, little-known, and new biological forms. Elsewhere I have described bacteria's complex and ubiquitous relationality (see Hird 2009) within and between bacterial "kinds" as well as with all other forms of life (and indeed, nonlife).

So, eventually, whatever we stash underground comes into contact with the bacterial life that dwells in the soil, or rather, given a populace of some forty million per gram, we might say they *are* the soil. Bacteria do what they have been doing since the Eoarchean era: they figure out ways of metabolizing whatever matter-energy they encounter. Each landfill is, in its own way, a unique bundle of materials, at once an ancient and a novel challenge to bacterial communities (Clark and Hird 2014).

The point is that when it comes to what bacteria ultimately make of these and other ingredients, and what in the process they make of themselves, we simply have very little idea. Neither landfill nor waste more generally is the only incitement that human activity provides for the proliferation and transformation of bacterial life. However, the unfathomably rich and complex feedstock that we are pumping underground has a special significance in the magnification of the insensible and the unknown, and its unintended consequences comprise one of the deepest and darkest ecologies of the current material-historical juncture. Containment is ultimately imprescriptible, given a long enough timeline, raising the issue of how microorganisms in the surrounding soil will respond to the leachate that sooner or later seeps from landfills. Which bacterial taxa are present, which populations will be deleteriously impacted by the specific mix of chemicals they are exposed to, and which will adapt and proliferate under novel conditions are queries of almost unfathomable complexity— questions that are effectively unanswerable (ibid.).

But what bacteria do with the substances to which we expose them, or what this exposure does to bacterial populations, may have profound consequences for humans and other organisms. In an empirical sense, we lack access to the vast majority of bacterial losses, gains, and transformations, dynamics that are obscured by the scalar mismatch of bacteria and ourselves, by the immensity of their numbers, strangeness of their forms, and the difficulty of accessing many of the environments in which they thrive (ibid.).

Today's Anthropocene aesthetic, with its roots in the Enlightenment conquest of nature and territory, colonization, biopolitics and hygiene, and invasion ecology visualizes landfills' teeming microbial masses safely stowed and contained within engineered megalandfills: nature destratified, utilized, and restratified. The earth gives up its resources—fossil fuels, minerals, rocks, water, and so on—and, through geo-bio relations humans imagine and engineer, welcomes back our detritus and indeed turns it into a resource for human use.[6] Microbes are now enlisted in relations of calculation, control, and, increasingly, services in the form of oil spill–cleanup operations, carbon sinks, and the consumption of waste to produce methane. Capitalist enterprise is even exploring the potential of landfill mining (Wallsten 2013). Through an Anthropocene aesthetic, we visualize our waste as not there: invisibly restratified, covered over, and reutilized into "natural spaces" while at the same time enlisted into capitalist regimes of profit and circulation (Hird forthcoming).

ANTHROPOCENE ANTI/AESTHETIC

A number of scholars working on Anthropocene studies are concerned with what Robin, Sörlin, and Warde (2013) refer to as creating "an archive of the future" (3). This singular archive marks humanity as a geological force, and it is primarily known to us through science and increasingly geoengineering. Where once poststructuralism, feminist theory, and postcolonial studies persuasively critiqued the universalizing "one-world" Empire (Hardt and Negri 2000) and its subjects so key to the Enlightenment, the Anthropocene appears at least to return us to, if not amplify, the obfuscation of difference. As Chakrabarty (2012) observes, the Anthropocene not only posits a single planet but also a shared humanity whose will—a "collective human agency"—constitutes a singular geologic force (13). Therefore, at the same time that the Anthropocene presents an interesting challenge to the Enlightenment separation of nature and culture—humanity as a force of geological change is only possible within a thoroughly social nature concomitant with a human agency as an "unintended consequence of our entanglements with myriad non-human forces" (Lehman 2014, 6)—it requires a way of thinking and talking about globality that clarifies and illuminates differences.

A second, related point is that moving forward and averting at least the direst predictions of contamination, proliferation, and extinction requires that "we" be able both to register and label geological changes as such. These organisms, these relations, must be caught within our field of vision, within our ken, within our classificatory nomenclature. We must, in other words, recognize them. One of the challenges, as Yusoff (2011) observes, is to "make present" those species and their relations with other in/humans that only vaguely (or not at all) populate our field of vision. In other words, Anthropocene writing requires an "imperceptible engagement" (579) with flora and fauna whose extinction makes no (as yet) perceptible mark, no list of charismatic animals moving from "vulnerable" to "critically endangered" to "gone forever." What Yusoff calls an "aesthetics of loss" I suggest is part of a wider Anthropocene aesthetic.

And it is precisely this aesthetic—inherited from the Enlightenment aesthetic of rationality, faith in science and technology, and the conquest of nature—that universalized humanity and reduced life and nonlife to nomenclature that we now primarily depend upon to navigate the future. Our Enlightenment-inherited aesthetic practices, in other words, constitute what proliferation and extinction *are* and their disclosure through science and engineering discourses. This creates, as Aravamudan (2013) notes, a particular Anthropocene nomenclature in which "scientific prediction about the future recalibrates the chronology of the present" (13).

Yet perhaps the grand "we" of the Anthropocene—provided by science and geoengineering—dissolves as popular culture adapts to and resists, transitioning "from apocalypse to way of life" in antiaesthetic visualizations (Buell, qtd. in Egan 2013, 61). After all, as Chakrabarty (2009) points out, the same Enlightenment promise that once asked us to place our faith in science and technology to bring about a new planetary society based on rationalism now asks us to recommit our faith in science and technology to excavate ourselves from our own planetary conquest.[7] *Quis custodiet ipsos custodes?* Moreover, there is a certain banality to the slow march of environmental and human destruction. Carbon emissions escalate, fossil fuels run out, coral reefs acidify, and we are somewhere on an uncertain timeline, one marked and archived in retrospect. So too, I argue, we must add the banality of landfilling, incineration, and nuclear waste, whose repositories are quotidian sites of the temporary in perpetuity, left for future generations to resolve (Hird 2015). This is waste covered up, covered over, and reimagined as the never-having-existed infrastructural base of children's play parks in middle-class suburbia.

This dissolution of the "we" brings me to my final point, which is that at the limits of the Anthropocene, the future cannot be visualized: It is an unknown aesthetic in excess of scientific prediction, human agency, and good will. It is indeterminate (Hird 2016). For this reason, and as Aravamudan (2013) notes: "There is . . . a gulf of difference separating an ecological melancholy that anticipates mass extinction and a techno-utopian optimism for solutions that include the geo-engineering of solar parasols and carbon sequestration technologies" (24). Put another way, there is an imprescriptibility to the Anthropocene—a reckless quality. An unintended stumbling into something we don't understand, are not able to predict, and are not able to control. Just as there are losses and extinctions of unclassified organisms and their relations with their environments that we have not registered, there are proliferations of microbiota deep within landfills beyond our ken.

What could define an Anthropocene antiaesthetic is an acknowledgment of the profound implications that bacteria may be generating new life forms not registered by humans *and* that this is what the microcosmos has been doing for the past 3.5 billion years. Landfills are not simply standing reserves for humans to dump their waste into and then exhume. And this, I think, is one of the major points about the Anthropocene— that humans are now contributing to, making their mark, as a geological force, and that this force is not so much from a godlike positionality of consciousness and control but must be added to the list of proliferations and extinctions with unintended and unknown consequences. And further, that the power relationship—if there is one—is squarely in favor of the microcosmos.

Thus, at the same time that waste exposes an Anthropocene aesthetic of capitalist venture, planetary conquest, and unlimited resource destratification and utilization, it further exposes an antiaesthetic or "countervisuality" (Mirzoeff 2014, 226) in which the unfathomable diversity and enormity of the noosphere's microcosmos portends capital, science, and technology's limits. That is, the antiaesthetic is as much about the failure of capitalist venture, the failure of the Enlightenment to separate us from brute nature, the failure of nomenclature, and the failure to define an "us" of common will and agency. The microorganisms metabolizing our detritus may well be creating new life forms that may equally escape taxonomic containment. As Aravamudan (2013) writes of this Anthropocene exposure: "The human is by no means the only subject or object. Endings are also mutations. The end of a singular species would still not be the end of all genres. There will be a post-ontological future of unnamable others, still new swarms that, once conceived, could fill many Chinese encyclopedias" (25). This antiaesthetic draws us, with alarm, to the imprescriptibility of anticipating, calculating, or controlling either proliferations or extinctions at a planetary scale (Hird 2012). There is a runaway quality to this Anthropocene aesthetic: that nothing—no patch of land, sea, or air—escapes the potentiality of capital's perpetual reach. Waste "management," then, signifies not only attempts to contain capital's fallout but to refuse it. And it is precisely in this refusal that the Anthropocene's antiaesthetic manifests. This antiaesthetic is not subsumed within the aesthetic; it is not an aesthetic subset. It occurs when the sensorium "can no longer make sense of what is presented to it. We cannot articulate what we perceive, namely, that the climate is wrong—too hot, too dry, too wet, or all of the above. Any suggestion to this effect is at once challenged" (Mirzoeff 2014, 214). Thus, perhaps what characterizes an Anthropocene aesthetic is a heightened anxious slippage between visualizations of the conquest of nature and the realization of our vulnerability *as* nature within a volatile and agentic environment: an aesthetic of solar panels orbiting the earth and an antiaesthetic of leaking, exploding, seeping, and combusting landfills.

NOTES

1. When interviewed by a journalist in 1970, Ehrlich followed his apocalyptic forecast with unadorned hedonism: "When you reach a point where you realize further efforts will be futile, you may as well look after yourself and your friends and enjoy what little time you have left. . . . That point for me was 1972" (in Bailey 2004).
2. For Mirzoeff (2014), visualizing may be traced back to eighteenth-century military theorists who imagined the military leader as one who could visualize in his mind an entire battlefield from the locally contextualized notes, images, and ideas supplied by his troops. Visualizing thus meant being able to perceive and act upon history as it happened, which would necessarily include the history the military leader himself created through his authority to lead troops in battle.

3. Such depictions, of course, were part of the Western discourse about the condensation of aesthesis (the full sensorial apparatus) into art as aesthetics (Buck-Morss 1992) and the possibilities art offers as an opening to different ways of experiencing and interpretation. Nevertheless, argues Mirzoeff, "the Anthropocene is so built into our senses that it determines our perceptions, hence it is aesthetic" (2014, 223).

4. According to the Welcome Collection, in 1848 the mountainous dump was moved to make way for the contemporary railway station. The dump was exported to the Soviet Union, where the ash became an ingredient in the bricks used to rebuild Moscow after the war. http://www.wellcomecollection.org/whats-on/exhibitions/dirt/image-galleries/the-street-soho-1854.aspx.

5. I thank Nigel Clark for reminding me of this famous Yorkshire saying. My father grew up in Bradford and said that the blackened buildings of his childhood were counted as a sure sign of jobs and prosperity. When my father returned to Yorkshire many years later, he was dismayed to see the freshly cleaned buildings, interpreting this as a sign of lost jobs and a city in decline.

6. As industrialization's castoffs grew in scale, the proclivity of these nineteenth-century dumps to smell, seep, spill, combust, and infect groundwater aligned with the rise of Pasteurian biopolitics. In *The Pasteurization of France* (1988), Latour argues that "Pasteur transformed our relationship with microbes" into one of pathogeneity and a threat not just to humanity but to life in general. Pasteur's entrepreneurialism fit well within the emerging biopolitics so key to modernity's approach to life, health, and capital through the connected discourses of hygiene and securitization (see Foucault 2007, 2008; Esposito 2008).

7. Thus, Crutzen and Stoermer (2002) write that we must put our faith in science "to guide society towards environmentally sustainable management during the era of the Anthropocene" (23).

WORKS CITED

Ali, S. H. 2004. "A Socioecological Autopsy of the *E. coli* O157:H7 Outbreak in Walkerton, Ontario, Canada." *Social Science and Medicine* 58:2601–2612.

Aravamudan, S. 2013. "The Catachronism of Climate Change." *Diacritics* 41 (3): 6–30.

Arrighi, Giovanni. 1994. *The Long Twentieth Century: Money, Power, and the Origins of Our Times.* London: Verso.

Bailey, R. 2004. "What Doom Will Look Like This Time Around." *Wall Street Journal*, May 20.

Barrow, M. 2009. *Nature's Ghosts: Confronting Extinction from the Age of Jefferson to the Age of Ecology.* Chicago: Chicago University Press.

Blackbourn, D. 2007. *The Conquest of Nature: Water, Landscape, and the Making of Modern Germany.* New York: Norton.

Bristow, W. 2011. "Enlightenment." In *The Stanford Encyclopedia of Philosophy*, ed. Edward N. Zalta. http://plato.stanford.edu/archives/sum2011/entries/enlightenment/.

Buck-Morss, S. 1992. "Aesthetics and Anaesthetics: Walter Benjamin's Artwork Essay Reconsidered." *October* 62:3–41.

Chakrabarty, D. 2009. "The Climate of History." *Critical Inquiry* 35 (2): 197–222.

——. 2012. "Postcolonial Studies and the Challenge of Climate Change." *New Literary History* 43 (1): 1–18.

Chertow, M. 2009. "The Ecology of Recycling." *UN Chronicle* 3 (4): 56–60.

Clark, N. 2005. "Ex-orbitant Globality." *Theory, Culture, and Society* 22 (5): 165–185.

——. 2010. "Ex-orbitant Generosity: Gifts of Love in a Cold Cosmos." *parallax* 16 (1): 133–143.

——. 2011. *Inhuman Nature: Sociable Life on a Dynamic Planet*. London: Sage.

Clark, N., and M. J. Hird. "Deep Shit." "Objects/Ecology," special issue, ed. E. Joy and L. Bryant, *O-Zone: A Journal of Object-Oriented Studies* 1 (2014): 44–52.

Collins, S. L., F. Micheli, and L. Hartt. 2000. "A Method to Determine Rates and Patterns of Variability in Ecological Communities." *Oikos* 91:285–293.

Crutzen, P., and E. Stoermer. 2000. "The Anthropocene." *IGBP Newsletter* 41 (May): 17–18.

Davis, M. 1996. "Cosmic Dancers on History's Stage? The Permanent Revolution in the Earth Sciences." *New Left Review* 217:48–84.

De Landa, M. 1997. *A Thousand Years of Nonlinear History*. Cambridge, Mass.: MIT Press.

Deleuze, G., and F. Guattari. 1987. *A Thousand Plateaus: Capitalism and Schizophrenia*. Trans. B. Massumi. Minneapolis: University of Minnesota Press.

Diamond, J. 1999. *Guns, Germs, and Steel*. New York: Norton.

Egan, M. 2013. "Commentary." In *The Future of Nature*, ed. L. Robin et al. New Haven, Conn.: Yale University Press.

Ehrlich, P. 1968. *The Population Bomb*. New York: Ballantine.

Elabroudy, S., et al. 2008. "Waste Settlement in Bioreactor Landfill Models." *Waste Management* 28:2366–2374.

Elton, C. 1958. *The Ecology of Invasions by Animals and Plants*. In *The Future of Nature*, ed. L. Robin et al., 367–377. New Haven, Conn.: Yale University Press.

Esposito, R. 2008. *Bios: Biopolitics and Philosophy*. Minnesota: University of Minnesota Press.

Foucault, M. 1976. *The History of Sexuality*. Vol. 1: *The Will to Knowledge*. Paris: Gallimard.

——. 1984a. *The History of Sexuality*. Vol. 2: *The Use of Pleasure*. Paris: Gallimard.

——. 1984b. *The History of Sexuality*. Vol. 3: *The Care of the Self*. Paris: Gallimard.

——. 2007. *Security, Territory, Population: Lectures at the College de France 1977–1978*. Trans. G. Burchell. Houndmills: Palgrave.

——. 2008. *The Birth of Biopolitics: Lectures at the College de France 1978–1979*. Trans. G. Burchell. Houndmills: Palgrave.

Freud, Sigmund. 1891. "The Psychotherapy of Hysteria." *In Studies on Hysteria* [2009], trans. J. Strachey and A. Freud, 253–306. New York: Basic Books.

Haraway, D. 2014. "Anthropocene, Capitalocene, Chthulucene [*sic*]: Staying with the Trouble." Lecture. http://vimeo.com/97663518.

Hardt, M., and A. Negri. 2000. *Empire*. Cambridge, Mass.: Harvard University Press.

Harman, G. 2010. "Asymmetrical Causation: Influence Without Recompense." *parallax* 16:96–109.

Hazen, T. C., et. al. 2010. "Deep-Sea Oil Plume Enriches Indigenous Oil-Degrading Bacteria." *Science* 330 (8): 204–208.

Hird, M. J. 2002. "Unidentified Pleasures: Gender Identity and its Failure." *Body and Society* 8 (2): 39–54.

——. 2009. *The Origins of Sociable Life: Evolution After Science Studies*. New York: Palgrave Macmillan.

——. 2010. "Indifferent Globality." *Theory, Culture, and Society* 27 (2/3): 54–72.

——2012. "Knowing Waste: Toward an Inhuman Epistemology." *Social Epistemology* 26 (3/4): 453–469.

——. 2013. "Digesting Difference: Metabolism and the Question of Sexual Difference." *Configurations* 20:213–238.

——. 2015. "Waste, Environmental Politics, and Dis/Engaged Publics." In "Geo-Social Formations," special issue of *Theory, Culture, and Society*, ed. N. Clark and K. Yusoff. Available online first.

——. 2016. "The Phenomenon of Waste World Making." In *Rhizomes: Cultural Studies in Emerging Knowledge* 30. http://www.rhizomes.net/issue30/.

——. Forthcoming. "Burial and Resurrection in the Anthropocene: Infrastructures of Waste." In *Infrastructures and Social Complexity: A Routledge Companion*, ed. P. Harvey et al. New York: Routledge.

Horne, R. H. 1850. "Dust, or Ugliness Redeemed." *Household Words*, July 13. http://omf.ucsc.edu/london-1865/dust-heaps/ugliness-redeemed.html.

Hulme, M. 2011. "Reducing the Future to Climate." *Osiris* 26:245–266.

Latour, B. 1988. *The Pasteurization of France/Irreductions*. Cambridge, Mass.: Harvard University Press.

Lehman, J. 2014. "After the Anthropocene: Politics and Geographic Inquiry for a New Epoch." *Progress in Human Geography* (February 7): 1–18.

Levis, J. W., et al. 2010. "Assessment of the State of Food Waste Treatment in the United States and Canada." *Waste Management* 30:1486–1494.

Margulis, L. 1998. *The Symbiotic Planet: A New Look at Evolution*. London: Phoenix.

Mirzoeff, N. 2014. "Visualizing the Anthropocene." *Public Culture* 26 (2): 213–232.

National Wildlife Federation. 2014. "What We Do to Protect Endangered Species." http://www.nwf.org/Home/What-We-Do/Protect-Wildlife/Endangered-Species.aspx.

Novacek, M. 2007. *Terra: Our 100-Million-Year-Old Ecosystem—and the Threats That Now Put It at Risk*. New York: Farrar, Straus and Giroux.

Olson, P. 2013. "Knowing 'Necro-waste.'" *Social Epistemology Review and Reply Collective* 2 (7): 59–63.

PlasticsEurope. 2009. http://www.plasticseurope.org/.

Raffles, H. 2007. "Jews, Lice, and History." *Public Culture*.

Ramachandra, G. 1997. "Radical American Environmentalism and Wilderness Preservation: A Third World Critique." In *Varieties of Environmentalism: Essays North and South*, ed. G. Ramachandra and J. Martinez Alier, 92–216. London: Earthscan.

Robin, L. 2013. "Commentary." In *The Future of Nature*, ed. L. Robin et al., 405–408. New Haven, Conn.: Yale University Press.

Robin, L., S. Sörlin, and P. Warde, eds. 2013. *The Future of Nature*. New Haven, Conn.: Yale University Press.

Rockström, J., et al. 2009. "A Safe Operating Space for Humanity." *Nature* 461:472–475.

Serres, M. 2011. *Malfeasance: Appropriation Through Pollution.* Stanford, Calif.: Stanford University Press.

Stengers, I. 2000. *The Invention of Modern Science.* Minneapolis: University of Minnesota Press.

Shah, Anup. 2014. "Loss of Biodiversity and Extinctions." *Global Issues.* http://www.globalissues.org/article/171/loss-of-biodiversity-and-extinctions.

Steffen, W., et al. 2004. *Global Change and the Earth System: A Planet Under Pressure.* IGBP Book Series. Berlin: Springer-Verlag.

Turner, C. 2011. "Dirt—an Exhibition of Our Fear of Filth." *The Guardian*, March 18. http://www.theguardian.com/artanddesign/2011/mar/18/dirt-paintings-posters-exhibition.

Van Wyck, P., and M. J. Hird. Forthcoming. "What Was the Anthropocene?" In *Assembling the Planet: The Postwar Politics of Globality*, ed. R. van Munster and C. Sylvest.

Wallsten, B. 2013. *Underneath Norrköping: An Urban Mine of Hibernating Infrastructure.* Linköping: Linköping Electronic Press.

Widger, W. R., et al. 2011. "Longitudinal Metagenomic Analysis of the Water and Soil from Gulf of Mexico Beaches Affected by the Deepwater Horizon Oil Spill." *Nature Precedings.* http://precedings.nature.com/documents/5733/version/1/files/npre20115733-1.pdf.

World Wildlife Federation. 2014. "How Many Species Are We Losing?" http://wwf.panda.org/about_our_earth/biodiversity/biodiversity/.

Wynne, B. 1987. *Risk Management and Hazardous Waste: Implementation and Dialectics of Credibility.* Berlin: Springer.

Yildiz, E., and R. K. Rowe. 2004. "Modeling Leachate Quality and Quantity in Municipal Solid Waste Landfills." *Waste Management and Research* 22 (2): 78–92.

Yusoff, K. 2012. "Aesthetics of Loss: Biodiversity, Banal Violence, and Biotic Subjects." *Transactions of the Institute of British Geographers* 37:578–592.

Zahara, A., and M. J. Hird. 2015. "Raven, Dog, Human: Inhuman Colonialism and Unsettling Cosmologies." "Learning How to Inherit in Colonized and Ecologically Challenged Lifeworlds," special issue of *Environmental Humanities* 7:169–190.

12

SPECTRAL LIFE

The Uncanny Valley Is in Fact a Gigantic Plain,
Stretching as Far as the Eye Can See in Every Direction

TIMOTHY MORTON

They talk to me about civilization, I talk about
proletarianization and mystification.

—AIMÉ CÉSAIRE, "DISCOURSE ON COLONIALISM"

Specter: a word that itself is spectral, by its own definition, since it wavers between appearance and being. It could mean *apparition*, but it could also mean *horrifying object*, or it could mean *illusion*, or it could mean *the shadow of a thing* (OED 2015). "Specter" floats around, like a specter. This chapter summons them because, as it will show, such a convocation of specters will aid us in imagining something like an ecocommunism, a communism of humans and nonhumans alike.

A specter is haunting the specter of communism: the ghostly presence of beings not yet formed according to the kind of "nature" that Karl Marx is talking about—the human economic metabolism of things. Things-in-themselves haunt data: this is possibly the shortest way of describing the continental philosophical tradition since Immanuel Kant. Marx's version of it is that use-value is already on the human metabolic side of the equation: the spoon exists insofar as it becomes part of how I organize my enjoyment, which is just what economics is, when we take the capitalist theological blinkers off. But the spoon is also a landing pad for a fly, a surface on which dust collects, a bacterial bioregion. Human use is a certain way of relating to this ovoid, metallic being, which for the sake of that use we call *spoon*. As the old term *economy of nature* hints at, what is eventually called *ecology* names a host of human *and* nonhuman enjoyment modes, a multiplicity of ways of accessing things such as spoons. The restriction of the use-value concept is the anthropocentric elephant in the Marxist room.

Perhaps communism is *only* fully thinkable as a coexisting of humans and nonhumans—and as we will see, this does strange things to the thought of communism. What is required is to think a radical "being-with" that is now deanthropocentrized. This coexistence drastically lacks something that we hear in the fifteenth chapter of the first volume of *Capital*, with its imaginative architects and mechanical bees, namely the production of sharp distinctions between human being and everything else, distinctions that Marx inherits from Kant, where Kant is still haunted by the specter of Descartes and his substance ontology of purely extensional lumps connected mechanically and acted on by (human) thought. Such an inclusive communism would also need to correct the lopsided (though at least half-true) correlationism (also from Kant) in which a (human) decider—in Marx's case human economic relations—makes reality "real," or "realizes" it as one might realize a script in the form of a movie.

The more we think ecological beings—a human, a tree, an ecosystem, a cloud—the more we find ourselves obliged to think them not as alive or dead but as spectral. The more we think them, moreover, the more we discover that such beings are not solidly "real" nor completely "unreal"— in this sense, too, ecological beings are spectral. In particular, ecological beings provide insights into the weird way in which entities are riven from within between what they are and how they appear. Another way of putting this is that beings, as a possibility or condition for their existing at all, are specters.

So many specters, so little time. Here I shall outline a way of proceeding into a thinking of ecological beings as necessarily spectral. To do so, we need to consider what it is to encounter an ecological being. The encounter as such is a moment at which I encounter something that is not me, in a decisive way, such that even if this being is obviously part of me—say, my brain—I do not experience it as part of the supposed whole that makes up "me." Ecological thought is Theodor Adorno's (1973) ideal of thinking as the encounter with nonidentity (5).[1] When it is not simply a cartoon of itself, pushing preformatted pieces around, thought meets specters, which is to say, beings whose ontological status is profoundly ambiguous and perhaps irreducibly ambiguous—and how can we even tell?

Given the precedent of Jacques Derrida's *Specters of Marx* so clearly evoked in its title, one should perhaps not be surprised that this is this chapter's eventual conclusion. For if communism is a specter that is haunting Europe, then ecological awareness—not to mention theory and political praxis—is a specter in many senses that most definitely haunts communism. Either communism is capacious enough, or can grow capacious enough, to accommodate it, or communism is now up against an internal limit easy to identify in an era of global warming and mass extinction. Simply put, communism may have a space for nonhumans—or not. Ecological awareness is a specter that haunts communist theory with the possibility either of its undoing or of its expansion. And given the uncertainty between undoing and shifting change, the choice itself is spectral.

Furthermore, and this is the most general concluding point, if ecology names relations between beings, then these relations are spectral. And since I hold that relationality just is what we call causality, stripped of its metaphysical, pre-Humean junk, we had better accommodate ourselves to the fact that causality is spectral, or, better, what we call the spectral is in fact the causal, whether we like it or not.

To encounter an ecological entity, then, is to be *haunted*. And although I must use this term metaphorically for now, it will become clear that haunting is a very precise term for what happens in ecological thought. For now we can try for some kind of detail, nevertheless. Thus, in the spirit of a certain deconstruction, this chapter will proceed in a sidewinding fashion from something one might suppose is only marginal to ecological discourse: what is called too glibly *experience*.

Something is already there, before I think it. When we talk about haunting, what we are talking about is what phenomenology calls givenness. Givenness is the condition of possibility for data (the Latin for *what is given*). There is already a light in the refrigerator before I open the door to see whether the light is on. The light's givenness—it's a light, not an octopus—is not something I have planned, predicted, or formatted. I cannot reduce this givenness to something expected, predictable, planned, without omitting some vital element of givenness as such. Givenness is therefore always surprising—and surprising in surprising ways: *surprisingly surprising*, we might say. Yet also, in haunting the phenomenon of the disturbing, surprising given, whose surprise cannot be reduced, also repeats itself.

Each time givenness repeats, according to the logic I have sketched out here, there is no lessening of surprise, which is perhaps why givenness is (as I have said surprisingly) surprising. Repetition does not lead to boredom but rather to an uncanny sense of refreshment. It is as if I am tasting something familiar yet slightly disgusting, as if I were to find, upon putting it to my lips, that my favorite drink had a layer of mold growing on its surface. I am, as it were, stimulated by the very repetition itself: stimulated by boredom. Another term for this is the familiar—because Charles Baudelaire made it so—term *ennui*. Ennui is the sine qua non of the consumerist experience: I am stimulated by the boredom of being constantly stimulated. In ennui, then, I heighten the Kantian window shopping of the bohemian or romantic consumer.

The experience of vicarious experience—wondering what it would be like to be the kind of person who wears *that* shirt—itself becomes too familiar, slightly disgusting, distasteful. I cannot enjoy it "properly"; to wit, I am unable to achieve the familiar aesthetic distance from which to appreciate it as beautiful (or not). Disgust is the flip side of good taste in this respect: good taste is the ability to be appropriately disgusted by things that are in bad taste. I have had too many vicarious thrills, and now I find them slightly disgusting—but not disgusting enough to turn away from them altogether. I enjoy, a little bit, this disgust. This is ennui.

Since in an ecological age there is no appropriate scale on which to judge things (human? microbe? biosphere? DNA?), there can be no pure, unadulterated, totally tasteful beauty. Beauty is always a little weird, a little bit disgusting. Beauty always has a slightly nauseous taste of the kitsch about it, kitsch being the slightly (or very) disgusting enjoyment-object of the other, disgusting precisely because it is the other's enjoyment-thing and thus inexplicable to me. Moreover, since beauty is already a kind of enjoyment that isn't to do with my ego and is thus a kind of not-me, beauty is always haunted by its disgusting, spectral double, the kitsch. The kitsch precisely is the other's enjoyment-object: How can anyone in their right mind want to buy this snow globe of the Mona Lisa? Yet there they are, hundreds of them, in this tourist shop.

Since beauty involves me in organizing enjoyment, it is (as Derrida would again have argued, slightly differently) a profoundly *economic* phenomenon, in the interesting sense that its use-value has not yet been determined. Beauty, in other words, strangely gives us a way to think economics that crosses over the correlationist boundary a little bit, the boundary between things and data, between what things are and how things appear. Beauty, in that case, provides a channel through which the nonhuman specters haunting the specter of communism can enter. They do not have to be left out in the prefabricated "nature" of bourgeois ideology, with its unquestioned metaphysical assumptions. Ecology, in other words, does not have to be excluded in advance from the New Left project as "a hippie thing" unlike race and gender.

Beauty is the nonhuman footprint of a nonhuman. And ennui is when we allow beauty to begin to lose its anthropocentric equalization. Now in ennui I am not totally turning my back on this sickening world—where would I turn to anyway, since the ecological world is the whole world, three hundred and sixty degrees of it? Rather, ennui is, and this is as it were the Hegelian speculative judgment of this chapter (though I am far from a Hegelian), the correct ecological attunement!

The very consumerism that haunts environmentalism—the consumerism that environmentalism explicitly opposes and indeed finds disgusting—provides the model for how ecological awareness should proceed. A model that is, moreover, not dependent on the "right" or "proper" ecological being and thus not dependent on a necessarily metaphysical (and thus illegal in our age) pseudofact (or facts). Consumerism is the specter of ecology. When thought fully, ecological awareness includes the essence of consumerism, rather than shunning it. Ecological awareness must embrace its specter.

With ennui, I find myself surrounded, and indeed penetrated, by entities that I can't shake off. When I try to shake one off, another one attaches itself, or I find that another one is already attached, or I find that the very attempt to shake it off makes it tighten the grip of its suckers more strongly. Isn't this just the quintessence of ecological awareness, namely the abject feeling that I am surrounded and penetrated by other entities such as stomach bacteria, parasites, mitochondria—not to mention other humans, lemurs, and sea foam? I find it slightly disgusting yet fascinating. I am "bored" by it in the sense that I find it provocative to include all the beings that I try to ignore in my awareness all the time. Who hasn't become "bored" in this way by ecological discourse? Who really wants to know where their toilet waste goes all the time? And who really wants to know that in a world where we know exactly where it goes, there is no "away" to flush it to absolutely, so that our toilet waste phenomenologically sticks to us, even when we have flushed it?

Consider the infamous "Spleen" poems of Baudelaire, poet-consumerist par excellence, bohemian inventor of the *flâneur*, or, rather, the one who christened this quintessential "Kantian" mode of consumption, and furthermore the poet who originated the notion of ennui. I quote the poems in their entirety for various reasons. First because I am not so much interested in analyzing particular poems as I am in tracing a more general structure of feeling across the poems. Second, because the provocative titling—exactly the same ("Spleen") for four poems in sequence in *The Flowers of Evil*—compels us to read them together, as if the same affect were collapsing, or going to sleep, then queasily restarting each time. Third, because the very form of the poems as such—as individual poems, without doubt, but also as a sequence, as the point about restarting is beginning to make—suggests being haunted, in the sense of being frequented, of an event occurring more than once. We all know how repetition is intrinsically uncanny: In other words, there would be no such thing as the uncanny without the notion of repetition. Freud's essay on the uncanny, for this very reason, is a startling exegesis of the puzzle of repetition.

Here, in sequence, are the four "Spleen" poems:

SPLEEN [1]

Pluvius, this whole city on his nerves,
Spills from his urn great waves of chilling rain
On graveyards' pallid inmates, and he pours
Mortality in gloomy district streets.

My restless cat goes scratching on the tiles
To make a litter for his scabby hide.
Some poet's phantom roams the gutter-spouts,
Moaning and whimpering like a freezing soul.

A great bell wails-within, the smoking log
Pipes in falsetto to a wheezing clock,
And meanwhile, in a reeking deck of cards
Some dropsied crone's foreboding legacy
The dandy Jack of Hearts and Queen of Spades
Trade sinister accounts of wasted love.

SPLEEN [2]

More memories than if I'd lived a thousand years!

A giant chest of drawers, stuffed to the full
With balance sheets, love letters, lawsuits, verse
Romances, locks of hair rolled in receipts,
Hides fewer secrets than my sullen skull.
It is a pyramid, a giant vault
Holding more corpses than a common grave.
I am a graveyard hated by the moon
Where like remorse the long worms crawl, and turn
Attention to the dearest of my dead.
I am a dusty boudoir where are heaped
Yesterday's fashions, and where withered roses,
Pale pastels, and faded old Bouchers,
Alone, breathe perfume from an opened flask.

Nothing is longer than the limping days
When under heavy snowflakes of the years,
Ennui, the fruit of dulling lassitude,
Takes on the size of immortality.
Henceforth, o living flesh, you are no more!
You are of granite, wrapped in a vague dread,
Slumbering in some Sahara's hazy sands,
An ancient sphinx lost to a careless world,
Forgotten on the map, whose haughty mood
Sings only in the glow of setting sun.

SPLEEN [3]

I'm like some king in whose corrupted veins
Flows agèd blood; who rules a land of rains;
Who, young in years, is old in all distress;
Who flees good counsel to find weariness
Among his dogs and playthings, who is stirred
Neither by hunting-hound nor hunting-bird;
Whose weary face emotion moves no more
E'en when his people die before his door.
His favourite Jester's most fantastic wile
Upon that sick, cruel face can raise no smile;
The courtly dames, to whom all kings are good,
Can lighten this young skeleton's dull mood
No more with shameless toilets. In his gloom
Even his lilied bed becomes a tomb.
The sage who takes his gold essays in vain
To purge away the old corrupted strain,
His baths of blood, that in the days of old
The Romans used when their hot blood grew cold,
Will never warm this dead man's bloodless pains,
For green Lethean water fills his veins.

SPLEEN [4]

When low and heavy sky weighs like a lid
Upon the spirit moaning in ennui,
And when, spanning the circle of the world,
It pours a black day sadder than our nights;

When earth is changed into a sweaty cell,
In which Hope, captured, like a frantic bat,
Batters the walls with her enfeebled wing,
Striking her head against the rotting beams;

When steady rain trailing its giant train
Descends on us like heavy prison bars,

And when a silent multitude of spiders
Spins its disgusting threads deep in our brains,

Bells all at once jump out with all their force,
And hurl about a mad cacophony
As if they were those lost and homeless souls
Who send a dogged whining to the skies.

And long corteges minus drum or tone
Deploy morosely through my being: Hope
The conquered, moans, and tyrant Anguish gloats
In my bowed skull he fixed his black flag.

This is John Keats left in the refrigerator too long, accumulating mold. Everything slips into the uncanny valley, even the difference between consumerism and ecological awareness. The breakdown of well-ordered poetry into something like prose is the deliquescence of lineation, writing as plowing. The narrator tells of being surrounded, permeated, by other beings, "natural" and "unnatural" and "supernatural," willy-nilly. The narrator is an abject ecosystem. The Sphinx, that monstrous hybrid, returns from death. Living and dead things become confused and weigh on the narrator, depressing him. Isn't ecological awareness fundamentally depressing in precisely this way, insofar as it halts my anthropocentric mania to think myself otherwise than this body and its phenomenological being surrounded and permeated with other beings, not to mention made up of them?

Which is to say, isn't ecological awareness a kind of spectrality that consists of awareness of specters? One is unsure whether a specter is material or illusory, visible or invisible. What weighs on Baudelaire is the specter of his bohemian, romantic consumerism, his Kantian floating, enjoyment tinged with disgust tinged with enjoyment. The specter is called *ennui*, badly translated merely as *boredom*. It is being enveloped in things, like a mist. Being surrounded by the spectral presence of evacuated enjoyment.

Ecological awareness is not unlike being the narrator of Baudelaire's "Spleen" poems. When thinking becomes ecological, the beings it encounters cannot be established in advance as living or nonliving, sentient or nonsentient, real or epiphenomenal. What we encounter instead are spectral beings whose ontological status is uncertain precisely to the extent that we know them in detail as never before. And our experience of these spectral beings is itself spectral, just like ennui. Starting the engine of one's car isn't what it used to be, since one knows one is releasing greenhouse gases. Eating a fish means eating mercury and depleting a fragile ecosystem. Not eating a fish means eating vegetables, which may have relied on pesticides and other harmful agricultural logistics. Because of interconnectedness, it always seems as if there is a piece missing. Something just doesn't add up, in a disturbing way. We can't get compassion exactly right. Being nice to bunny rabbits means not being nice to bunny rabbit parasites. Giving up on sophisticated boredom is also an oppressive option.

Science does not do away with ghosts. Rather, it multiplies them. As the human-nonhuman boundary and the life-nonlife boundary collapse, more and more specters emerge.

Art since 1800—since the inception of the Anthropocene—has been about allowing specters in. With whom or with what are we coexisting? How can we prove that a *who* isn't in fact a *what*? Photography's emergence in the nineteenth century gave rise quite rapidly to a fascination with *photographing ghosts*. The desire arose to see, in the flesh, rather than in the mind's eye, some kind of material ghost, imagined as an ectoplasmic, anamorphic being. *Anamorphic* means not simply *distorted* but "unshaped." Not exactly without a shape altogether but in a process of liquefaction, such that any discernible, obvious shape is collapsing. Why this fascination with the anamorphic specter? It is significant, to say the least, that the ability to imprint paper with photons reflected from actual things immediately suggests the possibility of seeing the unseeable, as if one could glimpse oneself from beside oneself. But isn't this the basic drive of science—to see what cannot (or indeed should not) be seen?

The emergence of specters becomes very vivid in the transition from romantic to expressionist and other forms of atonal music. *Strukturklang* is the term for the sound of the whole structure of a musical work (Lachenmann 2003). Yet if we think about it, this "sound" is an impossible, inaudible sound, or rather it is not a sound at all if by *sound* we mean *musical note*. What if, in some bizarre and disturbing way, we were to hear this inaudible yet pervasive sound? It would be a disturbing dissonant ambience that haunts the entire work, heard, for instance, in the vast reverberation of the three pianos at the end of the Beatles' "A Day in the Life" (1967). *Strukturklang*, far from being an abstract pattern, is nothing other than the resonance of the actually existing physical entities in collaboration with which humans make music: pianos, sitars, gongs, voices.

The Western term is *timbre*. Its etymology has to do with drums, but the sound evokes "timber" and hence the Greek *hyle*, which means *matter* and also *wood*. The history of European and American music since the start of the Anthropocene has been the story of the gradual liberation of these physical entities from their slavish role in telling human-flavored stories about human-flavored emotions. Instruments have become noninstrumental insofar as they are left to vibrate, twang, or otherwise "express" their physicality without an obvious human story. Atonalism liberated the notes; serialism in turn liberated the musical form in which the notes were played. Students of Arnold Schoenberg liberated the instruments themselves, from John Cage with his prepared piano to his student La Monte Young and his use of whole-number tunings (just intonation), tunings that allow for maximum harmonic depth and lucidity and minimum storytelling. Human stories disappear, and the timbre of the instrument manifests, because modulating between keys to symbolize emotion is impossible unless you bind the instrument to equal temperament, slightly fudging the ratios between the notes to make them shallower and blunter, as if one were seeing an old sepia photograph. Young domesticates this sound within the conceptual apparatus of Hindu devotion and its arts of the drone. Without this conceptual frame, what is heard is a spectral resonance, precisely the fullest possible *spectrum* of an instrument's timbre.

The encounter with this strange reverberation is, once again, spectral, insofar as it's an encounter with something already present (givenness) that was nevertheless suppressed in European and American music: the physicality of things and the necessary (and therefore ecological) collaborations between humans and nonhumans such as flutes, silicon chips, and valves.

But the inception of the spectral in modern art surely occurs at a moment we could call the Baudelaire Moment. In the decades of the nineteenth century before this moment, there has occurred a mapping of almost every move within the possibility space of consumerism. In particular, the top level of consumerism has become the subject of art itself. This is "bohemian" or romantic consumerism, a Kantian version in which one doesn't consume anything, per se: What one consumes is the pure possibility of consuming, without purpose or purchase. This is reflected in the activity of the *flâneur*, later in the window shopper, later still the web surfer.

Romantic consumerism takes an already reflexive mode—the *-ism* of consumerism gives this away—and bends that reflexivity back on itself: the reflexivity of reflexivity.

What are the moves within bohemianism as such?

In the hall of mirrors of romantic consumerism—the reflexivity of reflexivity—a very strange thing happens. It is not simply the case that there is a dizzying spiral of pure ideality, or idealism, with its physical flipside, an "I can do anything to anything" sadistic joy. For in this hall of mirrors, the mirrors themselves become appreciable as entities in their own right. And this in turn allows for the appreciation of other entities in their own right.

Of all people, it was Charles Darwin who opened the gate to the spectral world. Reason, jumping to a dimension at which emergent forces such as evolution could be discerned, was the force that opened it. When one collapses the life-nonlife boundary and relaxes the human-nonhuman boundary, all kinds of spectral creatures start to be seen, nightmarish beings that scuttle about. They are not categorizable. Yet they exist. They look like nightmare beings because of the extreme pressure they exert on existing frames of reference, existing categorical boxes.

But when the boxes dissolve, are these beings intrinsically horrifying? Is the gothic view of these beings the only view, for the rest of time, or is it a temporary effect of the pressure that such beings place on categories such as life and nonlife? The spectral realm is an uncanny valley into which more and more beings begin to slide. Here I am using a concept in robotics that explains why people have negative reactions to lifelike robots and more pleasant reactions to less human-looking ones such as R2D2. The more lifelike the robot, the more it approaches the condition of a zombie, an animated human corpse, that is to say, a being exactly like us yet not alive—a being that suggests that we ourselves might also be some kind of zombie.

But what happens when everything is in the uncanny valley? In other words, when the valley is no longer a valley but has transmogrified itself into a gigantic charnel ground with no center and no edge?

The valley is uncanny because it is familiar and strange, and strangely familiar—and familiarly strange—all at the same time. Beings we recognize—a human, a fruit fly—start to flit around outside the categorial boxes of human versus nonhuman and of life versus nonlife.

The uncanny valley is precisely not a void. One of the things under pressure in the modern period is precisely the empty container, the void, in which things sit. Such empty spaces fill us with dread (Pascal); they are the objective correlative of Cartesian reason and are inferred by Newtonian laws. This kind of void is objectively present. If I could just put it in its most paradoxical form, this kind of void lacks nothingness. It is not void enough, in a sense.

A feature of beings in the uncanny valley is nothingness. We should draw a strong distinction between nothingness and void. This distinction emerged in the long history of the assimilation of Kant, that fragmentary distorted echo of the start of the Anthropocene. Kant opened up the Pandora's box of nothingness by discerning an irreducible gap between a thing and its phenomena.

Is horror truly the most adequate attunement to the discoveries of the Anthropocene? The Pascalian dread of the infinite void is perhaps, we might surmise, only a temporary reaction to the discovery of spectral beings and the uncanny valley. If we replace the void with nothingness, we may need to replace horror. Nothingness is not intrinsically horrible. It is intrinsically *weird*.

H. P. Lovecraft's insane octopoid god Cthulhu is evoked many times to describe the putative "horror" of speculative-realist philosophy. Why is horror the compulsory philosophical affect with which thought confronts reality, which is to say the spectral? The horror of the story of Cthulhu is an attempt to *contain* the nothingness. Cthulhu dwells in a non-Euclidean city under the ocean. This is the precise inverse of the universe according to relativity theory. In this universe, Euclidean spaces are small-enough (human-scaled) regions of spacetime that appear to lack the general, mollusk-like ripples and twists in the universal fabric. Yet they only appear to do so. It is more accurate to say that human space occupies a small region of Cthulhu's universe.

If the initial reaction to this fact is horror, then it is only because one's habitual conceptual frame has been pointed out in its narrowness and arbitrariness. Horror is the feeling of void giving way to nothingness—of premodernity giving way to modernity. But what happens after modernity?

If we act to reduce carbon emissions, we will never know whether global-warming science was totally accurate or not. This is by no means a claim that global warming isn't happening—that's the whole point. The key term in the sentence is "totally." The exploding range of types of global-warming science and the proliferation of research on these types in the humanities demonstrates the urgency and difficulty of thinking an entity that exists on a bewildering set of spatiotemporal scales, whose accuracy parameters are highly variegated. For instance, it is settled that global warming is real and that the long-term picture involves a rise in temperature within a certain (still quite wide) range, depending on human action. As we approach a smaller temporal scale—what happens one year from now?—the patterns in the data become more ambiguous. As a humanities scholar whose husband was working on global-warming science recently observed, unverifiability is built into the theory of global warming.[2] There is a weird dance between science and mystery, which we are not expecting. We were expecting science to demystify. But reason itself finds itself driven slightly mad.

(This chapter names *global-warming science* in the same way that one might name *imperialism* or *Canada* and for the same hermeneutic reasons as the ones given for complicating the notion of verifiability. Canada is a rich and complicated and ambiguous thing, but it isn't Papua New Guinea. Global-warming science isn't string theory. Humanities scholars are too often hamstrung by a guilty and politicized injunction to produce distinctions between specific and general, where what is being named is not a "generalization" but simply an entity at a rather large yet still accurate scale.)

What seems trivially the case—as the Muppets and Michael Jackson and Olympic opening ceremonies remind me daily—is the fact that I am human; is this fact not actually *the most phenomenologically distant thing* in the known universe, more distant than the supermassive black hole at the center of the Milky Way? And isn't it precisely this thing, the fact that I am human, the thing that I must reckon with if I am to understand my role as a geophysical force on a planetary scale? Yet to think my human being is the task of ecological awareness.

The category *human* is, like all other modern categories, spectral. The human inhabits the uncanny valley. When one thinks the human, the category collapses into all kinds of entities. Humans are made of nonhumans—gut bacteria, mitochondria, "left-over parts from the apes" (Crash Test Dummies 1993). They also derive from nonhumans—as Richard Dawkins is fond of saying, one's million-times great-grandfather was a fish. And closer to the arrival of humans, there were all kinds of hominids and hominins. Then there are other species of human, such as the Neanderthals. As we approach the human, these entities become more and more obviously uncanny.

Let us return to the idea that spectral beings emerge when the life-nonlife boundary collapses along with the human-nonhuman boundary—behind which is lurking the (human) subject-object boundary. The long history of the Anthropocene so far has been the history of the emergence of these specters, which can also be said this way: the uncanny valley into which beings slip begins to cover a wider and wider range, so that it ceases to be a valley.

Miljohn Ruperto's 2014 exhibition at the Whitney in New York exemplifies how the spectral is an intrinsic part of thinking ecology. In *Voynich Botanical Studies*, pages from the weird, indecipherable (but is it?) Voynich Manuscript (composed in fifteenth-century Italy) depict nonexistent (or are they?) plants. We might call them *ghost plants*. Ruperto reworks these illustrations into three-dimensional, black-and-white renderings that almost look like photographs. Recall that within the first few decades since the beginnings of photography, photographers became fascinated with capturing ghosts on film. Modernity is not about disillusionment, despite what it keeps telling itself.

Friedrich Nietzsche's Zarathustra declares that man is somewhere between a plant and a ghost. *Between a plant and a ghost* (1982, 125): a human flickers between what she is and how she appears. Doesn't the idea also suggest a kind of scale in which you could be 30 percent plant, 70 percent ghost, or 40 percent plant, 60 percent ghost, and so on? Doesn't this in turn suggest the possibility that a plant could be a ghost?

And also—there is something not human about being human. When one looks at a human, one is looking at a gap between or within a thing.

The ghostly images of plants on the right-hand sides of the Voynich Manuscript bring out something that is intrinsic to being a plant. A plant's DNA exploits its appearance to attract pollinating insects. A plant's appearance is not a plant but a plant-appearance for some other being, such as a bee. A plant emits a sort of ghost of itself, an uncanny double.

What is called Nature is a forgetting of this ghostliness, this necessary doubling, the way in which a plant is haunted by its appearance, or the way in which a physical system such as DNA is also a semiotic one, and that there is a strange disconnection between these levels. Digital images of plants, made with electrons and silicon: mineral monsters that are metaphors for humans, wavering between plant and ghost. In an era in which nonhumans press on our glass windows ever more insistently, get stuck in our machines, caught in our webs of fate—in this ecological age, such a strange peeling apart of the human into its ghostly nonhuman components is necessary.

Another work of Ruperto's at the exhibition is called *Elemental Aspirations*. Different minerals are molded to the shape of a gold nugget. *Elemental Aspirations* is about the conversation that takes place within a thing between the past and the future or, to put it another way, between phenomenon and thing. The form of a thing is what happened to it. The form of a thing is the past. The essence of a thing is what is not yet fully revealed. The form of a thing is the future. This not yet is irreducible, such that there is a fundamental, wavering unpredictability about being a thing.

A thing is a sort of train station in which past and future slide against each other, not touching. This sliding has sometimes been called nothingness. Nothingness is not absolutely nothing at all but rather a flickering, sparkling play of presence and absence, hiding and revealing. A thing is a platform where ghosts glide past. The gold nugget's form is the past: it provides the cast for the liquid metals, the alchemical beings we call lead, iron, copper, tin, silver, mercury. But what are these strange forms, these twists of metal? Every answer we give will not be them. If they could speak, what they would say would not be them. They haunt us like ghosts, ghosts of minerals future.

They are not just lumps, these little spools of metal. They look like candies in twisted wrappers. They look like drills. They look like heads. Or ducks. Or rabbits. They *aspire*: They are breathing, as if exhaling simulacra of themselves, into the realm of appearance.[3] *Elemental Aspirations* is a conversation between Aristotle and post-Humean science. *Mineral Monsters* talks about how the aesthetic has been sidelined in contemporary theories of physical things such as plants, rocks, and ducks or rabbits. Yet post-Humean (that is to say, contemporary) science *also* might be aesthetic in a strange and surprising way that is not acknowledged by scientism, which is a tactic to prevent just this thought. By illegally reducing the world to bland substances or bland extension, scientism covers over the anxiety lit up by Hume and the continuing conflagration called Kant, who explained the deep reason why Humean skepticism works: There is an irreducible gap between a thing and how it appears. This means that for Hume and his successors—the scientists in their white coats—we live in a world of phenomena, of data, without direct access to things in themselves. We live in a world not unlike the world of art, where things appear mysterious and withdrawn, where we are confused and bedeviled with the octopus ink of appearance.

To reenchant the world is to discover its necessarily aesthetic comportment, which I can't peel away from the world without damage. This is Miljohn Ruperto's project. How does one reenchant the world without just saying anything, without resorting to the kinds of statements that just ignore science, and without recourse to primitivism or to Nature?

How can one know, after Hume and Kant, a thing in itself? One knows data. One knows statistical correlations, not causes and effects. One knows ghosts, impressions of things: a pattern in a diffusion cloud chamber, a map of weather changes in a high-dimensional phase space, hundreds of monkeys who react the same way over and over again to some stimulus. And yet, and yet . . . this does not mean that *there is nothing*. This doesn't mean that *there are only atoms*. This doesn't mean that *only the subject decides on reality*. All these trusty positions are reactions against the explosiveness of the founding assertion, that when I try to find out whether the light is on in the refrigerator, I have to open the fridge and thus change the fridge. There is no way to get at a thing in itself. But we don't just have appearances. We don't just have stuff plus illusory appearances. We don't just have the subject's decision as to what counts (or history's, or economics', or whatever kind of Decider you want, George Bush included).

Materialism is a way to avoid the panic that ensues when the gap between what a thing is and how it appears becomes clear. Materialism comes in two basic flavors: regular and new. It is strange, this new materialism, this new solution to our modern anxieties. Aristotle had already refuted reductionism to sheer matter over two thousand years ago. Aristotle understood that things could not be reduced to sheer matter.

Aristotle understood that to be a thing is to have a certain specific and unique form (Greek, *morphē*). But for Aristotle a thing is substantially, constantly "there." Hume shows us that there is no way to assert this because all we have are data, not things themselves. We have sparkling rocks, we have twists of metal, we have the evidence of eyes and ears and measuring devices of all kinds. Yet this is precisely evidence of just this nugget of stone. It's not a popsicle. It isn't just any old thing. New materialism likes to think there is an underlying substance, something fluid rather than something solid, as if being fluid were the solution to all our problems. What is seen in modernity is not matter. It is specters.

The solution to our problems is to realize that there is no solution to the intrinsic weirdness of a thing. In what does this weirdness consist? Yet another part of Ruperto's exhibition is an animation of a duck morphing into a rabbit, a time-lapse version of the duck-rabbit illusion. As an old, cryptic joke puts it: *What is the difference between a duck? One of its legs is both the same.* The duck-rabbit is dying but not dead. Yet not fully alive, whatever that means. A spectral duck-rabbit with a strangely startled expression, weirded out by its own uncanniness.

To be alive just is to be a duck-rabbit. To maintain one's existence is to maintain a necessary rift between what one is and how one appears. In this respect, death is the reduction of a thing to consistency. To be alive without imposing a concept of life rigidly opposed to death is to be undead, to be a specter. This is already the afterlife, in the sense that what wriggles around here cannot be categorized according to the long tradition of concepts of life (Thacker 2010).

Death is the end of spectrality. When I die I become my appearances. There are some notes in a wastebasket, some memories in your mind, a corpse. The difference between me and these appearances has evaporated. Life is inconsistency. Existing is inconsistency. Being a thing is inconsistency. There is a fundamental, irreducible gap between being me and appearing to be me. I am a duck-rabbit. This in turn means that I am a sort of weird, true lie. It is perfectly logical to allow some things to tell the truth and lie at the same time. You can tell it's logical because if you try to eliminate true lies, you get worse true lies.

Consider the sentence "*This sentence is false*," which is both true and a lie at the very same time. It is true that it's a lie, so it's telling the truth, which means that it's false. And so on. A spectral sentence. Imagine you can make a rule that goes " *'This sentence is false' is not a sentence.*" But someone can make a new sentence that blows up your rule: *This is not a sentence.* How come they can do this? Because it is okay for some sentences to be contradictory. Because things are contradictory. To be a thing is to be a duck-rabbit. Trying to exorcise the specters that haunt logic only results in more specters. There is a simple conclusion: Something about reality is spectral, so that something about the structure of logic is spectral, incapable of being categorized as true or false.

A certain kind of person (not to mention a certain kind of social structure, a certain kind of ideology) wants to contain these double-truthed things, these *dialetheias*. This would be someone anxious about the spectral nothingness opened up by Hume and Kant, the flickering nothingness we call modernity. Someone like that wants to draw a line, to say, for instance, that there are can be no mineral monsters, to draw a line between the living and the nonliving, like Georges Canguilhem, for instance. But there is nothing in the data to prove that this thing, palpitating before me, is alive, or sentient, or conscious, or a person, even when that thing is my own reflection.

The default condition of being a person is being paranoid that one might not be a person: to be a puppet, a thing manipulated by some external, demonic force. One can stave off the anxiety of this thought by concluding that being a person is a pure illusion—there is only matter, or there are only relationships between systems, or what have you. This is scientism. Ruperto's conversation with medieval minerals and alchemy is a way to talk to scientism, not to regress from modernity to an older, safer age where things meant what they said they meant but rather to channel something from a future in which we have somewhat accommodated the anxiety that things are, and are not, as they appear. To bring back Aristotle and Ptolemy in this light is not to block one's ears to the modern but to try to open them a little bit wider.

A tiny piece of metal (thirty microns long) breathes. When isolated by cooling it close to absolute zero in a vacuum, it can be observed vibrating and not vibrating at the same time. This sliver of metal is far larger than an electron or a photon or other entities often associated with quantum-theoretical effects (O'Connell et al. 2010). What does this mean? It means that quantum effects (such as vibrating and not vibrating at the same time) are possible because to be a thing at all is to be like that. To be a thing is to be a breathing duck-rabbit, living-dying, moving-still.

Modernity has allowed us to open to the possibility of connecting to the nonhuman without a conceptual framework. Can I prove a plant is not sentient? Can I prove that there can be no mineral monsters? Isn't monstrosity installed at the mineral level, since to be a monster is to be on display (Latin, *monstrare*)? In other words, to be a mineral is to have an appearance that floats weirdly in front of a thing, lying and telling the truth at the same time. There can be mineral monsters, and the same modernity that generates the phobic reaction to its most strange discovery is also capable of thinking this thought.

Thus to be a monster is to be a distortion that is also true. The distorted metal nuggets, the ghostly plants, the dying duck-rabbit, all proclaim this distortion. To be is to be monstrous. This is why the experience of beauty is also an experience of something slightly disconcerting, to say the least, and monstrous, when its contours become more explicit. Beauty tells me that a thing cannot be grasped in its essence. A thing exceeds my capacity to grasp it. There is an inevitable gap between how it appears and what it is. A thing is not smooth. This was true in the Kantian age, but it is more explicit in an ecological one. There is no authority that orders me to like this and not that. I have to look for the rule in my inner space, but when I look there, I encounter this spectral not-me, something about which I am uncertain. Furthermore, since there is no accounting for taste for Kant, the idea of a single recognizable standard seems spurious from the start, and Kant's attempts to police what he means by beauty are symptoms of that. In an ecological age, that there is no one scale on which to judge anything becomes ominously clear. Which scale should we use? Microbe? Human? Biosphere? Planetary? In the absence of an authoritative scale, all art sinks into the uncanny valley called kitsch: the slightly (or hugely) disgusting enjoyment objects of the other, who for some bizarre reason likes this particular ceramic horse. Disgusting because of the other.

Thus a specter haunts the specter of communism: the specter of the nonhuman, or we might now even say, with Derrida (1994), the specter of spectrality as such—and I hope this chapter has shown how this is indeed the same thing as the nonhuman. No longer able to exclude them with a straight face, thought is confronted with its anthropocentrism. It simply cannot be proved, as Marx wants to, that the best of bees is never as good as the worst of (human) architects because the human uses imagination and the bee simply executes an algorithm (Derrida 1994, chap. 7). Prove that I am not executing an algorithm when I seem to be planning something. Prove that asserting that humans do not blindly follow algorithms is not the effect of some blind algorithm. The most we can say is that human architects pass our Turing Test for now, but that is no reason to say that they are better than bees. It is instead to assert that we are hamstrung as to whether humans are executions of algorithms or not, casting doubt on our certainty that bees really do execute algorithms blindly, since that certainty is only based on a metaphysical assertion about humans and is thus caught in fruitless circularity.

Bees and architects are important because for Marx, in the lineage of Kant, there is a Decider that makes things real. For Marx, the Decider is human economic relations. But ecological relations without doubt subtend human relations of all kinds (let alone economic), and ecological relations extend beyond them throughout the biosphere. Human economic relations are simply general ecological relations with arbitrary pieces missing—colossal amounts of them. Either Marxism can be thought in a way that includes this irreversible knowledge, or it cannot. If it can, then communism must involve greater and better relations with nonhumans. As Marx says in the chapter on machines in *Capital*, capitalism produces the misery of the worker and the depletion of the soil (1994, chap. 15). And soil is decomposing life forms and the bacteria whose extended phenotype these life forms are (Dawkins 1999). In short, Marx implies nonhumans while erasing them.

What is called Nature is also a way to blind and deafen oneself to this strangeness. Ecological awareness, now occurring to everyone on Earth, is a way to take one's hands away from one's ears, to hear a message that was transmitted loud and clear in the later eighteenth century, a message that not even its messengers wanted entirely to hear.

Kant blocked his ears, limiting the gap he had discovered to the gap between human beings and everything else. It is time to release the copyright control on this gap. The name of this release is ecological awareness. Ecological awareness is coexisting, in thought and in practice, with the ghostly host of nonhumans. Thinking itself is one modality of the convocation of specters we summoned at the beginning of this chapter. To this extent, one's "inner space" is a test tube for imagining a being-with that our metaphysical rigidity refuses to imagine, like a quaking peasant with a string of garlic, warding off the vampires. Like Adorno, we need to brave the encounter with nonidentity.

This would not necessarily be a bizarre stretch. Recall that commodity fetishism means that a table, a piece of fruit, a cloud of carbon dioxide, begin to operate like computer programs, chattering with one another about their exchange value and that this is *far stronger* than if we accepted that they could act in a paranormal way, which is to say, a "magical" way outside of normative modernity, by dancing around. That is precisely what Marx (1990) says about commodity fetishism:

> A commodity appears at first sight an extremely obvious, trivial thing. But its analysis brings out that it is a very strange thing, abounding in metaphysical subtleties and theological niceties. So far as it is a use-value, there is nothing mysterious about it, whether we consider it from the point of view that by its properties it satisfies human needs, or that it first takes on these properties as the product of human labour. It is absolutely clear that, by his activity, man changes the forms of the materials of nature in such a way as to make them useful to him. The form of wood, for instance, is altered if a table is made out of it. Nevertheless the table continues to be wood, an ordinary sensuous thing. But as soon as it emerges as a commodity, it changes into a thing that transcends sensuousness. It not only stands with its feet on the ground, but, in relation to all other commodities, it stands on its head, and evolves out of its wooden brain grotesque ideas, far more wonderful than if it were to begin dancing of its own free will.
>
> (163)

The future thought that Marx is unable quite to articulate himself is right there, not exactly in the argument but in the imagery. This future thought is strangely rather easy to decipher. In commodity fetishism, spoons and chickens do not have agency: They become the hardware platform for capitalist software. It is far *easier* to allow in to thinking the host of dancing daffodils that Wordsworth talks about, the dancing tables of Marx, let alone dancing chimpanzees. In allowing this spectral, paranormal supplement of modernity to enter the thought of communism, it is not that capitalism flirts with the spectral but that *capitalism is not spectral enough*. And in not being spectral enough, capitalism implies a substance ontology that sharply divides what things are, considered to be "normal" or "natural" fixed essences (extensional lumps without qualities), from how things appear, defanging the spectral and "demystifying" the thing, stripping it of qualities and erasing its data, nice blank sheets or empty hard drives. The extent to which any given form of Marxism retweets this metaphysics is the extent to which it cannot imagine an ecological future. But this requires accepting that some forms of mystery are not so bad.[4]

NOTES

1. See also: "Dialectics is the consistent sense of nonidentity" (Adorno 1973, 147–148, 149–150).
2. Wendy Chun, at The Nonhuman Turn, University of Wisconsin–Milwaukee, May 2012.
3. *Oxford English Dictionary*, "aspire," v.I.
4. This thought is strangely akin to what Aimé Césaire (1997) argues about the colonized person. She or he needs to be remystified: "They talk to me about civilization, I talk about proletarianization and mystification" (82).

WORKS CITED

Adorno, Theodor W. 1973. *Negative Dialectics*. Trans. E. B. Ashton. New York: Continuum.

Baudelaire, Charles. "Spleen." http://www.poetic-love.com/publicdomainpoetry/authors .php?pdaid=86&pagenum=3.

Césaire, Aimé. 1997. "Discourse on Colonialism." In *Postcolonial Criticism*, ed. Bart Moore-Gilbert, Gareth Stanton, and Willy Maley. New York: Routledge.

Derrida, Jacques. 1994. *Specters of Marx: The State of the Debt, the Work of Mourning, and the New International*. Trans. Peggy Kamuf. London: Routledge.

Dawkins, Richard. 1999. *The Extended Phenotype: The Long Reach of the Gene*. Oxford: Oxford University Press.

Lachenmann, Helmut. 2003. "Hearing [*Hören*] Is Defenseless—Without Listening [*Hören*]: On Possibilities and Difficulties." Trans. Derrick Calandrella. *Circuit: Musiques Contemporaines* 13 (2): 27–50.

Marx, Karl. 1990. *Capital*. Vol. 1. Trans. Ben Fowkes. New York: Penguin.

O'Connell, Aaron D., M. Hofheinz, M. Ansmann, et al. 2010. "Quantum Ground State and Single-Phonon Control of a Mechanical Ground Resonator." *Nature* 464:697–703.

Nietzsche, Friedrich. 1982. *Thus Spoke Zarathustra*. In *The Portable Nietzsche*. Trans. Walter Kaufmann. New York: Viking Penguin.

Thacker, Eugene. 2010. *After Life*. Chicago: University of Chicago Press.

MUSIC CITED

The Beatles. 1967. "A Day in the Life." *Sgt. Pepper's Lonely Hearts Club Band*. Parlophone.

Crash Test Dummies. 1993. "In the Days of the Caveman." *God Shuffled His Feet*. Arista.

13

DARKLIFE

Negation, Nothingness, and the Will-to-Life in Schopenhauer

EUGENE THACKER

CENTER WITHOUT PERIPHERY

Over the past few decades, the discourse surrounding the posthuman and
its avatars—the transhuman, the nonhuman, the inhuman, the unhuman—has provided several insights into the legacy of humanist thinking
and the ongoing mode of crisis that has come to define it. This is the case
whether the posthumanism being discussed is of the more critical, cultural posthumanism coming out of the humanities and social sciences
or whether it is the more egregious, technophilic variant often associated
with broad futurist proclamations of a science-fictional convergence or
singularity.[1]

The first insight is that "the human" in the term posthuman often
has an ambivalent status. On the one hand there is the desire somehow to bypass the human by reaching a kind of escape velocity, through
science, technology, culture, or even thought itself. But, on the other
hand, there is also an attendant conservationalism, in which selected
aspects of the human are retained, if only so that there would still be
a recognizable subject to apprehend, acknowledge, and appreciate this
escape from the human. On the one hand, a desire to move beyond the
confines of anthropomorphism, the literal form of the human; on the
other hand, an equal desire to retain the benefits of anthropocentrism,

the vantage point of the human, even as it divests itself of its human moorings.

This leads to a second insight to be gained from the discourse of the posthuman, which is that posthumanism is conceptually reliant on some distinction of "the human" from something else that is close to the human (perhaps, threateningly close) but also distinct from it. For discussions of posthumanism that highlight the role of technology, that operative distinction is between the human and the machine. Likewise, for those discussions of posthumanism that highlight the role of science (and, in particular, the life sciences), the operative distinction is between the human and the animal. In both cases, the posthuman viewpoint is one that seeks to transcend the division while retaining the distinction, usually in favor of some notion evoking hybridity, heterogeneity, assemblage, or mixture.

But these distinctions (human/machine, human/animal) themselves rely on a more fundamental conceptual distinction, one that undergirds them and one that has, in many ways, proven to be the most difficult to overcome theoretically, and that is the distinction between the living and the nonliving. This distinction is not only foundational for branches of the life sciences, but it is articulated early on in the Western tradition in the work of Aristotle, who distinguishes a general principle of life (*psukhē*) from the various manifestations of life, be they plant, animal, or human.[2]

The biases of the Aristotelian tradition of thinking about "life itself" are apparent. There is, first, a stratification of life according to increasing complexity, with human beings on the top of the pyramid. While the principle of life may be immanent and common to all living beings, this life principle is not manifest equally. This leads to a second bias, which is the assertion of a fecund notion of life as constantly productive, generative, and characterized by, in Aristotle's words, "coming-to-be and passing-away." What remains anathema to the Aristotelian tradition is a purely "flat" notion of human life with respect to other life forms and, more crucially, a notion of life that is not some beneficent gift of plentitude but something quite different, something both more obscure and more opaque to the varieties of posthumanism, many of which simply appear to be arguing for a return to the humanist project that prompted the term "posthuman" to begin with (Braidotti 2013).[3]

DARK NIGHTS OF THE SOUL

But where might one find such examples of nonanthropomorphic, non-anthropocentric life? Contemporary science provides one possible example, an example we might read as an allegory.

In the depths of labyrinthine caves, embedded in gigantic rocks, buried in the hottest geothermal vents, and in the cold stellar dust of space, life is stealthily creeping. In environments in which it was previously assumed that life could not exist, scientists have discovered a whole range of microscopic life forms that not only survive but actually flourish under conditions of extreme heat, cold, acidity, pressure, radioactivity, and darkness. Their very existence suggests to scientists scenarios that for us human beings can only ever be speculative: the emergence or the extinction of life on the Earth, the adaptation of life forms to extreme environmental and climate changes, the existence of life on other planets or in outer space.

Dubbed "extremophiles," such organisms have been recently discovered by scientists working in a range of fields, from microbiology and oceanography to lesser-known fields such as abiogenesis and astrobiology. Their discoveries have garnered attention both within and outside the scientific community, primarily because in many cases their findings end up questioning the basic premises of the life sciences.[4] A recent report, *Investigating Life in Extreme Environments*, provides the decidedly nonhuman setting for understanding extremophiles:

> That part of the Earth's biosphere permanently inhabited by human beings is rather small and most of the planet, its deep core or mantle, will clearly never see a living organism. In between these two zones (inhabited and uninhabited), a variety of environments exist where human beings cannot live permanently, or physically access, although other forms of life exist within them.[5]
>
> (ESF 2007, 13)

The report goes on to define an extreme environment as "a given environment, where one or more parameters show values permanently close to the lower or upper limits known for life" (15). Generally speaking, extremophiles are united by the fact that they constitute novel forms of life that exist in extreme conditions—conditions that would be unfavorable if not fatal to most life forms. In some cases scientists have discovered microbes that appear to live without either sunlight or oxygen: A group of bacteria called autolithotropes, for example, live deep within rock formations and derive all their nutrients entirely from granite, while the bacterium *Desulfotomaculum* thrives in the darkness of radioactive rocks: "The bacteria exist without the benefit of photosynthesis by harvesting the energy of natural radioactivity to create food for themselves" (MacKenzie 2006).

As living beings whose existence questions life, extremophiles pose interesting problems for philosophy—they serve as philosophical motifs, or philosophemes, that raise again the enigmatic question: "What is life?" On the one hand, extremophiles are forms of life living in conditions antagonistic to life. Microbes existing in the conditions of the absence of light—indeed, feeding off of the absence of light—are an anomaly for biological science. And, in their anomalous existence as scientific objects, they also serve as reminders of the anomaly that is the concept of "life itself"—everywhere in general yet nowhere in particular.

In the science of extremophiles, two factors—hyperbole and contradiction—intersect to produce a concept of life that ends up questioning the very idea of life itself. The hyperbolic nature of extremophiles highlights the relativistic character of the organism-environment relation. The boundary between an organism living in conditions of low light and one living in conditions of no light becomes blurry. Either one chooses to recuperate the extremophile back within the ambit of traditional science, in which "no light" really means "very very low light," or one chooses to accept the anomalous condition of "no light" as is, with the implication that light is in no way essential for life.

Tied to this hyperbolic factor is another one, namely the contradictory nature of the extremophiles—or of any example of biological life that fundamentally challenges the premises of the life sciences. Extremophiles are anomalous not simply because they live without light but because their living-without-light sets them apart from the existing epistemological qualifiers that ground the ability of human beings to identify and know what life is. In the case of the extremophiles, the hyperbole is the contradiction, or it is the paradoxical ground of the contradiction; light and darkness define any environmental condition to some degree, and it is the hyperbolic nature of the environment that leads to the contradictory nature of the extremophile. Biological science, in so far as it is rooted in systematic description and classification, relies on its own principle of sufficient reason, namely that life and logic bear some basic relation to each other—in other words, the principle that all that can be identified and known as life is ordered or organized in such as way that it can live. The extremophiles are, in a way, examples of living contradictions, a living instance of the inverse relationship between logic and life.

ON ABSOLUTE LIFE

However, we need not delve into the deepest caves to discover hyperbole and contradiction at the heart of life. It is a core part of philosophical reflection on life, from Aristotle to Kant to contemporary biophilosophy. For instance, Aristotle's enigmatic *De Anima* (2000), rendered more enigmatic by generations of commentary and translation, is perhaps the first systematic ontology of life that hinges on contradiction—namely, one between a general life principle or life force (*psukhē*) and the manifold instances of the living, so exhaustively catalogued in works such as *Historia Animalium* (2002). The former (Life) is never present in itself, only manifest in the diverse concretions of the latter (the living); the latter serve as the only conceptual guarantee of the former. But it is Kant's (2001) treatment of the teleology or purposiveness of life (*zweckmäßigkeit*) that not only revives Aristotle's problematic but adds another dimension to it—any instance of life is always split between its purposiveness in itself and its purposiveness for us, the beings who think life. In the Kantian paradigm, the possibility of knowing the former always compromised by the presumptions of the latter.

On the one hand life is phenomenal, since we as subjects are also living subjects. Life is amenable to the manifold of sensation, is given as an object of the understanding, and results in a synthetic knowledge of the nature of life. Life is an object for a subject. On the other hand, the Aristotelian problem—what is the life-in-itself that is common to all instances of the living?—returns again in Kant's critical philosophy. Life-in-itself is neither the knowledge nor the experience of the living (be it biological classification or the subjective phenomenon of living), and life-in-itself is also not the living being considered as such (for example, the object given to science as an object of observation).

In short, it would seem that the life common to all living beings is ultimately enigmatic and inaccessible to thought, since any given instance of the living (as subject or object) is not life-in-itself but only one manifestation of life. It seems there is some residual zone of inaccessibility that at once guarantees that there is a life-in-itself for all instances of the living while also remaining, in itself, utterly obscure.[6] It is precisely as living subjects, with life given as objects for us as subjects, that we are cut off from yet enmeshed within life in itself.

Schopenhauer (1969) once noted that Kant's greatest philosophic contribution was the division between phenomena and noumena, the world as it appears to us and the enigmatic and inaccessible world in itself. Whereas for Kant this division served a critical or regulatory function, providing philosophy with a ground to stand on, for idealism this division is an impasse to be overcome—by and through philosophy itself.

We know that, for the generation immediately following on the coattails of Kant, the important task was to identify this split as the key impasse in Kantianism and to provide ways of overcoming it.[7] This is a significant project because for Kant the critical philosophy was not, of course, a problem but rather a solution to a whole host of metaphysical quandaries that pitted empiricists against rationalists, materialists against idealists, and so on. In a sense, German idealism's first and most important gesture is to restate Kantianism as a problem to be overcome. The concept of the Absolute, and the various avatars of the Absolute proposed by Fichte, Schelling, and Hegel (Spirit, the Infinite, the World-Soul), have to be understood as the outcome of this initial gesture.

But this split between phenomena and noumena can only be overcome if it is in some way collapsed—or rendered continuous. Since we, as thinking subjects, cannot have access to noumena, we must begin from phenomena, and in particular the phenomena of thought. Hence thought must not be taken as split from the world in itself but as somehow continuous with it. But this itself is a difficult thought precisely because thought is presumed to be specific to living, rational, human subjects—thought is internalized, rendered proprietary, owned and instrumentalized. Kant's split implicitly relies on an internalist model of thought, one that begins and ends with the philosophical decision of anthropocentrism.

The key move that idealism makes is to externalize thought, to render it ontologically prior to the individual thinking subject that thinks it. Only if thought is understood to be ontologically prior to the human, only if thought is ontologically exterior to the human, can it then become that continuum between the "for us" and the "in itself," phenomena and noumena. The idealist operation is, in a sense, to subtract the noumenal from the Kantian split, leaving only a continuum that stretches without demarcations between the world-for-us and the world-in-itself.

In place of the phenomena-noumena split, then, one has a new totality, which raises thought above its Kantian, anthropocentric bias and establishes it as that which enables the split between phenomena and noumena—as well as the split within phenomena between thought and world and subject and object. Thought is raised to the Absolute, and, in this continuum, the thought of a subject and the world in itself are both manifestations of a single Absolute. If this is the case, then Kant's epistemological framework is not just a reflection or representation of the world but is itself a manifestation of the Absolute. What results is a new kind of philosophical drama, a drama of the Real and the Rational (and their comingling), or, in Schelling's terms, a drama of Nature as the manifestation of the Absolute.

This continuum is neither a transcendent, static category of Being, nor is it simply an affirmation of an unbounded, immanent multiplicity of beings; it attempts to play the role of both an inaccessible noumena "outside" us and a manifest field of phenomena that constitutes us from within. It is for this reason that idealism turns to the concept of life-in-itself. For post-Kantian idealism, the concept of life-in-itself establishes a continuum between phenomena and noumena, but without reducing itself to either biology or theology. For the early Hegel, the Absolute is inherently dynamic, "the life of the Absolute," moving, flowing, and becoming through the structured phases of the dialectic, with the living organism its exemplar.[8] For Schelling, with his longstanding interest in natural philosophy, it is in and through the process and expressive forces of nature that the Absolute manifests itself—thus Nature is "manifest Spirit" and Spirit is "invisible Nature."[9] Even Fichte, otherwise a logician, attempts to account for the movement between the I and not-I, the Ego and non-Ego, by resorting to the vitalist language of life itself, commenting on the "Absolute Life" through which the I/Ego spontaneously manifests itself.[10]

In their attempts to overcome the Kantian problematic, idealist thinkers exhibit a conceptual shift from a static to a dynamic ontology, or from being to becoming; they also effect a shift from a transcendent to an immanent concept of life, in which the framework of source/manifestation supersedes that of essence/existence. They turn their attention to conceptual models borrowed from natural philosophy and the philosophy of the organism, which has the broad impact of shifting the philosophy of life from a mechanist framework to a vitalist one.

Within idealism "life" becomes an ontological problematic and in the process becomes a metonym for the Absolute, resulting in what we can simply call, following Fichte, *Absolute Life*. This Absolute Life is monistic; it is a metaphysical totality that underlies all reality but that is not separate from it. It is a totality that exists beyond any part-whole relation but that is also only ever manifest in the particular. This Absolute Life is also immanence; it is an infinite process of becoming, flux, and flow; an infinite manifestation in finite Nature; an infinite expression of the living in an organic whole called Life. Finally, Absolute Life is paradoxical. It harbors a conceptual duplicity in which Absolute Life is at once omnipresent and omniabsent, accessible and inaccessible to the senses, thinkable and an outer limit for thought. Absolute Life, while not a pure thing-in-itself, is only ever manifest in Nature (and thus indirectly knowable). At its core, Absolute Life must necessarily have the conceptual structure of negative theology.

THE ONTOLOGY OF GENEROSITY

If we had to give a name to this kind of thinking, in which life itself is ontologized beyond its regional discourses (natural philosophy, biology, zoology) and comes to serve as a metonym for the Absolute, we could call it the *ontology of generosity*. The ontology of generosity states, first, that the precondition of the intelligibility of life lies in its innate propensity for continually asserting itself in the living. This propensity applies as much to the upscale processes of growth and development as it does to the downscale processes of decay and decomposition; indeed, as Schelling often notes, life is never so strongly asserted than in the process of decay.

Life, then, is generous not simply because it always gives itself forth but because it always asserts and affirms itself, even as it withdraws, withers, and returns to its inorganic foundations—where another life then continues. In the ontology of generosity, life is not simply present but *overpresent*.

In post-Kantian idealism, the ontology of generosity begins from these premises: the overpresence of life-in-itself and the split between Life (as superlative to the living) and the living (as always in excess of Life). However, the generosity of life does not flow forth in a single, homogeneous manner. In a number of the *Naturphilosophie* works of Schelling and Hegel, one can detect several variations to the generosity of Life. Each variation is defined by a basic philosophical question that serves as its principle of sufficient reason. There is, first, *Life as genesis* (also generation, production). Life is generous because it is defined by an ontology of becoming, process, and genesis. Here the question is: "Why is there something new?"[11] This mode is especially evident in Schelling's work in natural philosophy, where a "speculative physics" aims to account for the flux and flows of the Absolute in and as Nature. Life is ontologically prior to the living, but Life is also only ever explained in the living. When Schelling discusses the "potencies" (*Potenz*) of Nature—forces of attraction and repulsion, dynamics of electricity and magnetism, organismic physiology—he is evoking the generosity of Life in terms of its geneses.

In addition to this, there is a second mode, in which *Life is givenness* (also gift, donation). In this case, Life is generous because it is defined by its being given, its giving forth, its being already-there, its affirmation prior to all being. Here the central question is: "Why is there already something?" The idea of givenness is the spectral backdrop to the concept of Absolute Life. It enables the thought of Life to pass beyond the regional philosophies of nature and obtain a superlative ontological status. That this or that particular instance of the living is given is no great statement; it only points to the need for a concept of Life to account for all possible instances of the living. That Life (as opposed to the living) is given is another issue altogether; it means that the Absolute is not only an intelligible totality but that it is such within an ongoing process, an ordered flux and flow that is consonant with Absolute Life. In Hegel's epic schema, Spirit can only realize itself through its successive stages (Idea, Nature, Spirit) by virtue of this "life of the Absolute." Givenness is the necessary precondition for thinking Absolute Life.

The problem is that while idealism provides a solution to the Kantian problematic, that solution often ends up being compromised by the Kantian framework itself. There is, to begin with, the problem of genesis—generosity demands genesis, if only as its minimal condition. Within the ontology of generosity, one must still posit a source of life, even if this source is self-caused or self-generating, even if genesis remains immanent to itself. There is also the rather nagging problem of teleology. The positing of a life-source necessitates the positing of an end or purpose to organization, in order to qualify and to justify the organization inherent in life—order demands an end. This is true even if the end one posits is the process of becoming itself, without end. The positing of a source and end dovetail into the need to accept a minimally causal distinction between source and end, and this remains the case even if one asserts an immanent relation between source and end, in which source and end persist in a kind of tautology.

The result is that the ontology of generosity inherited from German idealism looks to be a compromised Kantianism, at once inculcated within the requirements of the Kantian framework and, at the same time, claiming to have absolved Kantianism of its own antinomies. In terms of the concept of life, the ontology of generosity must make do with a source that is self-caused, a process that is its own end, and an immanent distinction between essence and existence. The idealist resolution of Absolute Life comes to resemble an ouroboros—a split that is rendered continuous, only to have the split swallow its own tail and be recapitulated at a higher level.

With post-Kantian idealism, then, we see the concept of life raised up, as it were, beyond the regional discourses of natural philosophy, such that it can serve as a continuum bridging the Kantian gulf between phenomena and noumena. But this requires that one think not just of this or that living being but of Absolute Life—that which is not reducible to yet not separate from the fluxes and flows of life as we know it. Idealism's ambition is to put forth a concept of Absolute Life via an ontology of generosity, in which Life is conditioned affirmatively and positively by its overpresence. Absolute Life is thus overpresent in several ways—as genesis or as givenness. Note that these two paths—genesis and givenness—also form the two major channels through which flow contemporary biophilosophies, with Life-as-genesis constituting the vitalist ontologies of Henri Bergson

(1998), Alfred North Whitehead (1979), and Gilles Deleuze (1986, 1989, 1990, 1994) and Life-as-givenness constituting the phenomenological approaches of thinkers such as Edmund Husserl (1970), Jean-Luc Marion (2002), and Michel Henry (2003–2004).

SCHOPENHAUER'S ANTAGONISMS

At this point, the question is whether there is a post-Kantian response that does not adopt the ontology of generosity, and this is linked to a related question: whether there is a post-Kantian response that refuses refuge in a renewed concept of Being. When life is thought as life-in-itself, we seem to be driven to a fork in the road: either the framework of Being/beings or the framework of Being/becoming. There is, possibly, another approach, one that would think life-in-itself *meontologically*, as "nothing," though it too has its own limitations. The best exemplar of this approach is found in the work of Arthur Schopenhauer.

Schopenhauer's sentiments regarding German idealist thinkers are well known. He despised them.[12] Certain passages in *The World as Will and Representation* betray a profound personal distaste toward Fichte, Schelling, and above all Hegel, for whom Schopenhauer reserves his most vitriolic phrases:

The greatest disadvantage of Kant's occasionally obscure exposition is that . . . what was senseless and without meaning at once took refuge in obscure exposition and language. Fichte was the first to grasp and make vigorous use of this privilege; Schelling at least equaled him in this, and a host of hungry scribblers without intellect or honesty soon surpassed them both. But the greatest effrontery in serving up sheer nonsense, in scribbling together senseless and maddening webs of words, such as had previously been heard only in madhouses, finally appeared in Hegel. It became the instrument of the most ponderous and general mystification that had ever existed, with a result that will seem incredible to posterity, and be a lasting monument of German stupidity.

(1969, 429)

Metaphysical rants like these occur throughout Schopenhauer's writings, and there is an argument to be made for a certain charm behind Schopenhauer's curmudgeonly dismissals. Indeed, for many readers "obscure exposition" and "ponderous mystification" have come to define philosophy itself. Certainly Schopenhauer himself appears to be no stranger to the crime of obscurity, as demonstrated by his frequent uses of terms like *qualitas occulta* and *principum individuationis*.

Despite this, there is also a sense of clearing the air in Schopenhauer's writings, and no doubt "untimely" followers of Schopenhauer such as Nietzsche found inspiration in this tone. Not so fast, quips Schopenhauer: We have not even begun to address the problems put forth by Kant's antinomies. In the opening of *WWR*, Schopenhauer's first step is to brush aside the entirety of post-Kantian dogmatism and return to Kant's problematic—the split between phenomena and noumena. "Kant's principal merit that he distinguished the phenomenon from the thing-in-itself, declared this whole visible world to be phenomenon, and therefore denied to its laws all validity beyond the phenomenon" (1969, 434). This is, notes Schopenhauer, an acceptable constraint to our metaphysical thinking. But why stop there? It is "remarkable that he [Kant] did not trace that merely relative existence of the phenomenon from the simple, undeniable truth which lay so near to him, namely 'No object without a subject'" (434). Had Kant pushed his philosophy a few steps further, he would have arrived at the notion that "the object, because it always exists only in relation to a subject, is dependent thereon, is conditioned thereby, and is therefore mere phenomenon that does not exist in itself, does not exist unconditionally" (434, italics removed). Though inaccessible, noumena remain related—or correlated—to phenomena, with the former tending to become increasingly subsumed within the latter. When pushed a bit further, one ends up with something that looks a lot like idealism, with a metaphysical continuum between phenomena and noumena that promises to collapse Kant's split between them.

For Schopenhauer, idealism can only overcome this split by dropping out one of the two terms—the noumena—thereby allowing a phenomenal monism to fill the gap. "All previous systems started either from the object or from the subject, and therefore sought to explain the one from the other, and this according to the principle of sufficient reason" (25–26). Idealism attempts to think a continuum between phenomena and noumena that is not reducible to either. But what it really ends up doing—in Schopenhauer's opinion—is adopting a partial view (that of subject and object) and universalizing this in the Absolute.

What, then, does Schopenhauer propose? One must reexamine not only the Kantian framework but the basic presuppositions of the idealist response to Kant. For Schopenhauer, the principle of sufficient reason is primary among the presuppositions that must be reexamined. For the pessimist philosopher, that "everything that exists, must exist for a reason" must not be taken for granted. But this leaves a great deal open, too much perhaps: What if there is no reason for the world's existence, either as phenomena or as noumena? What if the world-in-itself is not ordered, let alone ordered "for us"? What if the world-as-it-is, let alone the world-in-itself, is unintelligible not in a relative way but in an absolute way? Once one dispenses with the principle of sufficient reason, what is left—*except a philosophy that can only be a nonphilosophy*? It would appear that two paths are left open—materialism or idealism, nihilism or mysticism, the hard facts and the great beyond, "it is what it is" and "there is something more . . ." For Schopenhauer, pessimism is the only viable philosophical response to such an abandonment of the principle of sufficient reason.

Schopenhauer dismisses the idealist response to Kant's phenomena-noumena split as inadequate. In its place he proposes a simple move—that Kant's split be recast in a way that allows for a collapse between them to take place. There is, first, the world as phenomena: "Everything that in any way belongs and can belong to the world is inevitably associated with this being-conditioned by the subject, and it exists only for the subject" (3). This includes the subject-object correlation as well as the finer distinction that Schopenhauer later makes between the representation and the object of representation, both of which are contained within the world of phenomenon. Put simply, "the world is my representation."

Then there is, on the other side, the world as noumena, which is a pure limit that at once conditions thought and remains inaccessible to thought—"something to which no ground can ever be assigned, for which no explanation is possible, and no further cause is to be sought" (124). The concept of noumena can only ever be an apophatic concept. Schopenhauer enters deep waters here, not least because any attempt to conceptualize the noumenal world is doomed from the start. This never seems to deter the philosopher-curmudgeon, however. The challenge is how to think both the inaccessibility and the immanence of the world as noumena, and Schopenhauer glosses this via a concept of nothingness/emptiness that is at the same time not completely separate from the phenomenal world. One need not soar into the infinity of the cosmos or the inner depths of Spirit to discover such a concept. In *WWR* Schopenhauer discovers it in the mundane materiality of the body: "Thus it happens that to everyone the thing-in-itself is known immediately in so far as it appears as his own body, and only mediately in so far as it is objectified in the other objects of perception" (19). What results is a strange immanentism of noumena: The correlation of subject and object that constitutes phenomena is the world considered as representation (*Vorstellung*), and that which is absolutely inaccessible to this world-as-representation, but which is also inseparable from it, is the world considered as will (*Wille*). "The world is, on the one side, entirely representation, just as, on the other, it is entirely will" (4).

SCHOPENHAUER AND THE WILL-TO-LIFE

Like his idealist contemporaries, Schopenhauer (1969) agrees that "Kant's greatest merit is the distinction of the phenomenon from the thing-in-itself" (417). And, like his contemporaries Schopenhauer views this distinction as something to be overcome. But whereas the idealist response is to adopt an ontology of generosity to bridge this gap, Schopenhauer will adopt a different approach. Instead of asserting an Absolute Life (grounded by its own principle of sufficiency and driven by an ontology of overpresence), Schopenhauer will drop the bottom out of the ontology of generosity. What remains is, quite simply, nothing. No overflowing life force, no pantheistic becoming, no immanent principle of life running throughout all of Creation. Just nothing. But nothing is, of course, never simple; it is also nothingness, or emptiness, or the void, and it quickly becomes a paradoxical and enigmatic something. So while Schopenhauer does not definitively resolve the Kantian problematic, he does provide a way of shifting the entire orientation of thought on the problem.

The new problem Schopenhauer is confronted with is how to overcome the Kantian split between phenomena and noumena without being determined by the ontology of generosity. This can be stated in even briefer terms: how to think "life" such that it is not always determined by overpresence (that is, by generosity, genesis, and givenness), *how to think life in terms of negation.* Certainly one would not want to return to a metaphysics of life, in which life obtains the quality of pure being that one finds in the concept of "soul" common to both Aristotle and Aquinas. But Schopenhauer is equally skeptical of the diffuse theism in the idealist notions of the Absolute, in which Absolute Life always radiates and flows forth, often finding its culmination in the heights of human life in particular. Schopenhauer notes, with some sarcasm, "life is thus given as a gift, whence it is evident that anyone would have declined it with thanks had he looked at it and tested it beforehand" (579).

The remaining option for Schopenhauer is to consider the role that negation plays in relation to any ontology of life, especially any ontology of life that would attempt to overcome the Kantian split of phenomena and noumena. Life, then, is not simply subordinate to a metaphysics of presence (as in Kant), but neither is it consonant with an infinite over-presence of generosity (as with idealism). In contrast to the ontology of generosity, which posits life as always affirmative, Schopenhauer will put forth a negative ontology in which life is paradoxically grounded in nothingness (it is, perhaps, "underpresent"). In a striking turn of phrase, Schopenhauer refers to all these relations between negation and life as the Will-to-Life (*Wille zum Leben*):

> As the will is the thing-in-itself, the inner content, the essence of the world, but life, the visible world, the phenomenon, is only the mirror of the will, this world will accompany the will as inseparably as a body is accompanied by its shadow; and if will exists, then life, the world, will exist. Therefore life is certain to the will-to-live.[13]

(275)

Schopenhauer's concept of the Will-to-Life is a response to an old dilemma concerning the ontology of life. It is found in Aristotle, and then in natural philosophy, before its recapitulation in Kant. We have seen it at play in German idealism, in the ontology of generosity and its affirmative over-presence. Put simply, the dilemma is how to articulate a concept of life-in-itself that would account for all the instances of the living. If one is to avoid both the naïveté of epistemological classification as well as the rhetorical games of nominalism, what is required is a concept of life that is at once synonymous with the living yet transcendentally separated from it.

The Will-to-Life is, then, Schopenhauer's attempt to overcome the Kantian split by asserting a subtractive continuity, a continuity paradoxically driven by negation. At the same time, sentences such as those in the citation above demand some unpacking, since in order to arrive at his concept of the Will-to-Life, Schopenhauer must make a number of steps (steps many of his critics perceived as fallacious or untenable). With this in mind, we can briefly consider the three aspects of the Will-to-Life as presented by Schopenhauer in *WWR*.

THE RIDDLE OF LIFE

Early on in *WWR* Schopenhauer recasts the Kantian problematic through the example of the living body. His concern, however, is neither a "body" in the sense of physics, which would commit him to mechanism or atomism, nor "body" in the sense of biology, which would commit him to natural philosophy. Instead the body is for Schopenhauer a kind of crystallization of abstract anonymity, a "Will" that is at once energy and drive but that has no origin or end and leads to no goal. The body is that which is the most familiar yet the most foreign to us as subjects. We are bodies, and we have bodies.[14] For Schopenhauer these are simply two ways of knowing the body—immediately as a living subject consonant with a living body and mediately as a subject relating to or thinking about the body as object. Both of these are well within the domain of the phenomenal world that Kant describes.

But in the second book of *WWR* Schopenhauer will take Kant a step further. If the body, as both subject and object, is on the side of the world as phenomena (as representation), then what would the living body as a thing-in-itself be? If there is a phenomena of life, is there also a noumena of life, a life-in-itself? On the one hand, such a noumenal life could not be something completely divorced from life as phenomenal, for then there would be no point of connection between phenomena and noumena (a logical prerequisite for Kant). On the other hand, this noumenal life must retain a minimal equivocity with regard to phenomenal life, else we are simply back within the phenomenal domain of subject-object relations.

Hence Schopenhauer's riddle of life: What is that through which life is at once the nearest and the farthest, the most familiar and the most strange? As Schopenhauer notes, "the answer to the riddle is given to the subject of knowledge appearing as individual, and this answer is given in the word *Will*" (100). The Will is, in Schopenhauer's hands, that which is common to subject and object but not reducible to either. This Will is never present in itself, either as subjective experience or as objective knowledge; it necessarily remains a negative manifestation. Indeed, Schopenhauer will press this further, suggesting that "the whole body is nothing but objectified will, i.e., will that has become representation" (100). And again: "My body and will are one . . . or, My body is the *objectivity* of my will" (102–103).

In reply to the riddle "what is nearest and farthest?" Schopenhauer answers with the Will—that which is fully immanent yet absolutely inaccessible. As we noted, Schopenhauer's first step is to recast Kant's framework in new terms—for Kantian phenomena he will use the term Representation, and for Kantian noumena he will use the term Will. His next step is to describe the living body, and more specifically life, as the nexus where Will and Representation meet. Schopenhauer's reply is that to each instance of the world taken as Representation there is the world as Will, and to each instance of life as Representation (whether as subject or object), there is a correlative Will-to-Life:

> The will, considered purely in itself, is devoid of knowledge, and is only a blind, irresistible urge, as we see it appear in inorganic and vegetable nature . . . and as what the will wills is always life, just because this is nothing but the presentation of that willing for the representation, it is immaterial and a mere pleonasm if, instead of simply saying "the will," we say "the will-to-life."
>
> (275)

Certainly life obtains a duality within the domain of Representation—there is the subjective experience of living, just as there is the scientific knowledge of the living, both inscribed within the world as Representation or phenomena. Schopenhauer's controversial move here is to assert that there is life outside of and apart from the world as Representation, that there is a life that remains inaccessible to the phenomenon of life, and his phrase Will-to-Life designates this horizon.

LIFE NEGATING LIFE

However, at this point, the problem is that Schopenhauer appears to have only elevated the concept of life beyond ontology, to the realm of unthinkable noumena. There still remains a part of the riddle to be answered, which is how what is nearest can—*at the same time*—be what is farthest. For this the role of negation in the Will-to-Life becomes more important.

Schopenhauer notes that the Will is not simply a static, transcendent category of being but a dynamic, continuous principle that is much in line with idealist concepts of the Absolute. But, as we've seen, Schopenhauer distances himself from idealism by opposing the ontology of generosity that it puts forth.[15] As Schopenhauer comments, "everywhere we see contest, struggle, and the fluctuation of victory, and . . . we shall recognize in this more distinctly that variance with itself essential to the will" (146–147). Schopenhauer provides a compendium of examples from the sciences, though they read more like scenes from a monster movie: insects that lay their eggs in the bodies of other host insects, for whom birth is death; the internalized predator-prey relationship in the hydra; the ant whose head and tail fight each other if the body is cut in two; invasive species such as ivy; giant oak trees whose branches become so intertwined that the tree suffocates. His examples continue, up through the cosmic negation of black holes and down to the basic chemical decomposition of matter in the decay of corpses, where life is defined by the negation of life.

Yet Schopenhauer is neither a Hobbes nor a Darwin; his emphasis here is less on the universalizing of struggle and more on what it indicates for an ontology of life. If the Will is flow or a continuum, it is, for Schopenhauer, one driven by negation—or by a negative flow, a negative continuum. The Will asserts itself through contradictions, oppositions, and subtractions, and its limit is the self-negation of life, through life. Thus "the will-to-live (*Wille zum Leben*) generally feasts on itself, and is in different forms its own nourishment, until finally the human race, because it subdues all the others, regards nature as manufactured for its own use" (147).

For Schopenhauer, there is an "inner antagonism" to the Will, one that is antagonistic at the level of this or that living being, as well as in the domain of inorganic nature, on through to the level of cosmic life. The Will-to-Life is driven by this process of "life negating life," from the inorganic to the organic and beyond.

COSMIC PESSIMISM

In the inner antagonism of the Will-to-Life Schopenhauer comes upon what is perhaps his greatest insight, and that is the Will-to-Life's radically unhuman aspect. Schopenhauer here pulls apart the Kantian split, suggesting that all claims concerning noumena are necessarily compromised by concepts derived in some way from the phenomenal domain. And it is here that Schopenhauer most directly counters the furtive anthropocentrism in post-Kantian idealism. In the same way that the domain of noumena does not exist for phenomena, so the Will-to-Life is utterly indifferent to any concept of life, be it "for us" or "in itself." In the Will-to-Life "we see at the very lowest grade the will manifesting itself as a blind impulse, an obscure, dull urge, remote from all direct knowableness" (149).

In statements like these, Schopenhauer is actually making two separate claims. The first has to do with the principle of sufficient reason and Schopenhauer's critical treatment of it. In so far as the Will-to-Life is noumenal as well as phenomenal, all statements concerning its causality, its teleology, its relation to time and space, and its logical coherence or intelligibility must only apply within the phenomenal domain. In this sense "the will as thing-in-itself lies outside the province of the principle of sufficient reason in all its forms, and is consequently completely groundless, although each of its phenomena is entirely subject to that principle" (113). Schopenhauer admits that one can always recuperate any and all statements about the Will into the phenomenal domain, a recuperative move in which one is still able to articulate what is inarticulable, to think what is unthinkable. But in this paradoxical mode there is always something that, taken in itself, for which no sufficient reason can suffice or for which there is only a negation of sufficient reason. We might even say that Schopenhauer's concept of the Will-to-Life ultimately points to a principle of *insufficient* reason at its core.

If the Will-to-Life, considered in itself, has no sufficient reason because it lies outside the phenomenal domain, neither can the Will-to-Life be granted any anthropocentric conceits, least of all that life exists "for us" as human beings or that it reaches its pinnacle in the human life. Like his German contemporaries, Schopenhauer posits a principle of continuity that would collapse the Kantian split between phenomena and noumena, but unlike them, he refuses to grant the human being, or the human perspective, any priority with respect to this principle. Certainly, as Schopenhauer readily admits, there are gradations and differentiations within the natural world. What remains, however, is this Will-to-Life that indifferently cuts across them all. "For it is indeed one and the same will that objectifies itself in the whole world; it knows no time, for that form of the principle of sufficient reason does not belong to it, or to its original objectivity, namely the Ideas, but only to the way in which these are known by the individuals who are themselves transitory" (159–160).

Even as it is rendered hierarchical for Schopenhauer, the Will-to-Life maintains this cosmic indifference throughout the world. Indeed, Schopenhauer will go so far as to say that this constitutes the tragic-comic character of human life in particular: "The life of every individual, viewed as a whole and in general, and when only its most significant features are emphasized, is really a tragedy; but gone through in detail it has the character of a comedy" (322).

For Schopenhauer, pessimism is the only viable philosophical response to this radically unhuman condition. This pessimism is something for which Schopenhauer is popularly known (and often dismissed). The problem is that Schopenhauer's pessimism is often understood to be about human life, for it is only human beings that sense the senselessness and suffering of the world. It is true that Schopenhauer's pessimism has to do with a view of life as essentially "incurable suffering and endless misery," an ongoing cycle of suffering and boredom. But this is only the case from the perspective of the individual, living subject, toward which, for Schopenhauer, the world-in-itself is indifferent. As Schopenhauer evocatively notes, the very manifestation of the Will-to-Life is doubled by a kind of Will-lessness (*Willenslosigkeit*), every sense of the world-for-us doubled by a world-without-us. Pessimism for Schopenhauer is not so much an individual, personal attitude but really a cosmic one—an impersonal attitude. The indifference of the Will-to-Life thus stretches from the micro scale to the macro scale:

> Thus everyone in this twofold regard is the whole world itself, the microcosm; he finds its two sides whole and complete within himself. And what he thus recognizes as his own inner being also exhausts the inner being of the whole world, the macrocosm. Thus the whole world, like

man himself, is through and through will and through and through representation, and beyond this there is nothing.

(162)

In an enigmatic way, negation courses through Schopenhauer's notion of the Will-to-Life. Evocations of the Will-to-Life as "nothing" or "nothingness" recur throughout Schopenhauer's writings. Certainly Schopenhauer was influenced by his encounter with classical texts in the Buddhist traditions.[16] As we've noted, this type of cosmic pessimism stands in opposition to the ontology of generosity in post-Kantian idealism, with its emphasis on over-presence, flux and flow, and the becoming of the Absolute. In response to the Kantian split between Life and the living and in contrast to the post-Kantian ontology of generosity, Schopenhauer opts for a negative ontology of life.

However, that life is "nothing" can mean several things. The enigmatic last section of *WWR I* bears out some of these meanings. Here Schopenhauer makes use of Kant's distinction between two kinds of nothing: the *nihil privativum*, or privative nothing, and the *nihil negativum*, or negative nothing. The former is nothing defined as the absence of something (shadow as absence of light, death as absence of life). For Schopenhauer the world is nothing in this privative sense as this interplay between Representation and Will; the world, with all its subject-object relations, as well as its ongoing suffering and boredom, is transitory and ephemeral. By contrast, the indifferent Will-to-Life courses through and cuts across it all, all the while remaining in itself inaccessible, and "nothing."

The problem is that, at best, we have a limited and indirect access to the world as a *nihil privativum*, and "so long as we ourselves are the will-to-live, this last, namely the nothing as that which exists, can be known and expressed by us only negatively" (410). For Schopenhauer the very fact that there is no getting outside the world of the *nihil privativum* hints at a further negation, one that is not a relative but an absolute nothingness:

In opposition to this *nihil privativum*, the *nihil negativum* has been set up, which would in every respect be nothing. . . . But considered more closely, an absolute nothing, a really proper *nihil negativum*, is not even conceivable, but everything of this kind, considered from a higher standpoint or subsumed under a wider concept, is always only a *nihil privativum*.

(409)

At this point it seems that one must say—or think—nothing more. It is as if philosophy ultimately leads to its own negation, to Wittgenstein's claim that what cannot be thought must be passed over in silence. That *WWR* closes with an enigmatic affirmation of life as nothingness is indicative of the limits of Schopenhauer's negative ontology. On the one hand the Will-to-Life is nothingness because, considered as the interplay between Life and the living, the Will-to-Life in itself is never something in an affirmative or positive sense. But Schopenhauer suggests that the Will-to-Life is nothingness for a further reason, which is that, in itself, the Will-to-Life indicates that which is never manifest, that which is never an objectification of the Will, that which is never a Will for a Representation. To the relative nothingness of the *nihil privativum* there is the absolute nothingness (*absolutes Nichts*) of the *nihil negativum*. While Schopenhauer is himself opposed to the post-Kantian idealists, he is united with them in his interest in the concept of the Absolute, albeit one paradoxically grounded in nothingness. His contribution is to have thought the Absolute without resorting to the ontology of generosity and its undue reliance on romantic conceptions of Life, Nature, and the human. To the negative ontology of life, it would seem, therefore, that there is a kind of *meontology* of life. It is for this reason that Schopenhauer can close *WWR I* by stating that "this very real world of ours with all its suns and galaxies, is—nothing" (412).

CODA (THE SPECTER OF ELIMINATIVISM)

At a recent conference given at the New School in New York, Steven Shaviro characterized contemporary speculative philosophy as polarized between what he terms "panpsychism" and "eliminativism." Such a polarization relies on a number of presuppositions. If one accepts that philosophy is broadly conditioned by the correlation between self and world (but also subject and object, or thought and the intentional object of thought), and if one accepts that this "correlationism" is a central problematic within philosophy (insofar as philosophy is by definition unable to think outside of correlationism), then for Shaviro this leaves one of two extremes open for philosophy. Either one must opt for a kind of diffuse immanence, in which some quasi-monist entity (thought, affect, object, life, etc.) is already everywhere—the view of panpsychism—or one must opt for an equally diffuse reductionism, in which all claims about existing entities are in themselves groundless, masking a potential void within everything—the view of eliminativism. In Shaviro's presentation, current speculative philosophy is being polarized between, on the one hand, a view of everything-already-everywhere and, on the other hand, a view of nothing-ultimately-nowhere.

As I read them, Shaviro's comments are meant more as a provocation than a proof. In his talk he also notes alternatives that avoid moving toward either pole: "I should also note though . . . that there is also the alternative of abrogating both eliminativism and panpsychism at the same time" (2011).[17] Shaviro cites the work of Reza Negarestani (2008), Ben Woodard (2013), and yours truly (Thacker 2010, 2011) as examples, noting that "these thinkers have a very negative view of the efficacy of thought, and in that sense they're eliminativists. And yet they couldn't find the universe as horrible as they find it, in this Lovecraftian way, without being kinds of inverted panpsychists."[18] However, what remains an open question is the way in which the work of Negarestani, Woodard, and myself arbitrates between eliminativism and panpsychism—whether it is in the form of a synthesis, an implosion, a double negation, or something else altogether. But it is worth noting how this alternative described by Shaviro, which would avoid both the plenum of panpsychism and the reductionism of eliminativism, results in a paradoxical plenum of nothing or, better, a notion of immanence that is indissociable from nothingness. In short, the implosion of becoming and unbecoming into Schopenhauer's "will-to-nothing" or Will-lessness.

Eliminativism is more commonly understood as a branch of analytical philosophy that also goes by the name of "eliminative materialism." Often associated with thinkers such as Paul Churchland and Daniel Dennett, eliminative materialism questions the existence of "qualia" such as mental states, psychological behaviors, or subjective affects. At its most extreme, it challenges any claims for an independently existing mind beyond a neurological and biological basis. As fields such as cognitive science progress, many commonly held notions such as "belief" or "desire" will be discovered to have no viable scientific basis and may even be relegated to the dust heap of folk psychology. Eliminativism also has a broader significance, especially in the philosophy of science, where it questions the existence of any entity beyond its material basis (be it of the vitalist "soul" or the "luminiferous ether").

Shaviro's comments are, of course, meant to evoke a different type of eliminativism, one that would take up its fundamental challenge to philosophy's principle of sufficient reason while also departing from eliminative materialism's fidelity to biological, neurological, or physical "baseline" concepts. In a way, traditional eliminative materialism doesn't go far enough; put differently, given its critical questioning of basic philosophical premises, eliminative materialism's reliance on positivist science can only seem as an arbitrary stopping point. Why claim that subjective states or psychological categories like "faith," "joy," "despondency," or "dereliction" can only be assessed to the degree that they reduce to the biological or neurological level and then not continue on to "eliminate" that biological or neurological basis as well? It would seem that, for eliminative materialism, philosophy once again reinstates its Kantian, juridical capacity to regulate boundaries and reestablish grounds, precisely at the moment that it questions the concept of "ground" altogether. In short, this more ambiguous, "dark" eliminativism would suggest that any eliminative materialism must ultimately eliminate matter itself.[19]

The trials and tribulations (mostly tragic) of "life" as a philosophical concept readily lend themselves to the eliminativist approach. Surely no other concept has been so vociferously asserted and questioned, from historical debates over vitalism in the philosophy of science to contemporary evocations of "vibrant matter" and "the life of things." Eliminativism haunts the ontology of life, constantly questioning its theological pretentions, while also maintaining a minimal baseline or ground that would enable fields like neuroscience to make scientifically sound claims about what is or isn't living. At its extreme, the search for a material basis for life (be it in a molecule or even, ironically, in biological "information") ends up reducing life to its material constituents—at which point there is no life at all . . . or there is nothing but life. Interestingly, eliminativist approaches to the ontology of life tend to split it along the lines that Shaviro describes: Either everything is alive, or nothing is alive; either everything is pulsating flux and flow, autoaffecting and self-transforming, or everything is silence, stillness, and the enigmatic, vacuous hum of nothingness.

For both Aristotle and Kant, the proliferating, generous, and over-present manifestations of life are always shadowed by a concept of life-in-itself that must, by necessity, enter the eliminativist abyss. Nowhere is the awareness of this duplicity more evident than in post-Kantian idealism. In Schelling's *Naturphilosohpie* and his concept of the World-Soul, in Hegel's meditations on the organicist flows of Spirit, and in Fichte's lectures on "Absolute Life," one sees in idealism a concerted attempt to ameliorate this shadowy aspect of life itself while also refusing the options of either mechanistic science or a return to Scholastic theology.

In contrast to this tradition, one finds thinkers like Schopenhauer, the misanthrope from Danzig who, again and again, rails against his contemporaries for not having adequately grasped the nothingness at the heart of life itself. But if there is nothingness at the heart of life, then how does one account for its prodigious generosity and overpresence? How does one think the negation at the heart of life when life is commonly understood to be the concept of affirmation par excellence? Despite the animosities between them, this is the question that concerns both the idealists as well as Schopenhauer, and it is a problem first fully articulated by Kant.

Post-Kantian idealism did not end with Fichte, Schelling, or Hegel. In a way, its conceptual contours are resurrected by subsequent generations. A thread runs from the notion of life-as-generation to philosophical vital-ism and biophilosophies inspired by Deleuze or Bergson, just as another thread runs from the notion of life-as-givenness to the phenomenology of life, the life-world, or the flesh, as found in Husserl (1970), Merleau-Ponty (2013), or Michel Henry (2003–2004). Are we, for example, witness to a contemporary post-Kantian idealism today, in the correlationism of a neo-Fichteanism, in the transcendental geology of a neo-Schellingianism, or in the metamorphic plasticity of neo-Hegelianism?[20]

One of Schopenhauer's most contentious propositions is that all life is dark life, and thus even contemporary scientific fields such as those that study extremophiles recapitulate, through the methods of empirical science, this shift from life-in-itself as a regional problem of epistemology to a fundamental fissure within ontology. Its limit is one that Schopenhauer characterizes as life-as-nothing, life thought in terms of negation, ultimately leading Schopenhauer from a negative ontology to something that we can only call an *affirmative meontology* of life.

NOTES

Originally published as Eugene Thacker, "Darklife: Negation, Nothingness, and the Will-to-Life in Schopenhauer," *Parrhesia* 12 (2011): 12–27.

1. I am, admittedly, clustering a wide range of thinking on this topic under a single rubric. Under the heading of a critical posthumanism see Braidotti (2013), Colebrook (2014), Hayles (1999), Lyotard (1992), Kroker (2014), and Wolfe (2009), not to mention the numerous anthologies, readers, and journal articles on the topic. Under the heading of extropian posthumanism one might cite the work of Kurzweil (2006), Moravec (2000), More (2013), and work associated with the World Transhumanist Association.

2. For more on the role of Aristotle's *De anima* in the development of an ontology of "life itself," see Thacker (2010).

3. This is a move made at the end of *The Posthuman*, which looks to developments in climate change and sustainability to argue for a "neo-humanism."

4. The widespread coverage of extremophile research is evidenced by pop-science books (Taylor 1999) as well as a number of science documentaries (*Journey Into Amazing Caves*). There is a college-level textbook (Gerday and Glansdorff 2007), and there even exist a number of professional organizations, such as the International Society for Extremophiles.

5. See also the recent NASA press release announcing the discovery of an arsenic-based bacterium (2010).

6. Kant never says so, but one is tempted to state it: Life is noumenal.

7. I will be using the phrases "post-Kantian idealism" and "German idealism" interchangeably, though arguably there are reasons for treating them as separate terms.

8. The most frequently referenced example is in the opening sections of the *Phenomenology of Spirit* (1977), though the *Philosophy of Nature*, part of Hegel's *Encyclopedia of the Philosophical Sciences* (2004), also revisits these themes, from the perspective of Nature as manifest Spirit.

9. Schelling returned again and again to this relationship between Nature and the Absolute, from earlier works (2004) to his later work (2000).

10. This phrase plays a key role in Fichte's lectures (2005).

11. Here I borrow Steven Shaviro's paraphrase of Whitehead's process philosophy (2012), though used here in a different context.

12. There is an anecdote often told about Schopenhauer: While lecturing in Berlin in 1820, he intentionally chose the same time for his lectures as that of Hegel. Needless to say, the latter continued to draw huge crowds, while the former was faced with an empty hall.

13. I have chosen to translate Schopenhauer's *Wille zum Leben* as Will-to-Life. However, the Payne translation uses "will-to-live."

14. This is one of the greatest lessons of Cartesianism prior to Schopenhauer and of phenomenology after Schopenhauer.

15. Schopenhauer's negative approach is a position that is as much about being a curmudgeon as it is about critique—indeed the stylistic innovation in Schopenhauer's writings is to have rendered the two inseparable, culminating in a form of philosophical pessimism.

16. On Schopenhauer's complex relation to Eastern philosophy, see Abelsen (1993), Cross (2013), and Nicholls (1999).

17. Some of the material from the lecture is taken from Shaviro's work in progress, a science-fiction novel entitled *Noosphere*.

18. We might also add Brassier (2007) and Ligotti (2010).

19. As contemporary philosophy seems to be particularly fond of branding, one could coin new terms for this type of eliminativism: "dark eliminativism," "black eliminativism," "eliminative eliminativism," "the Abhuman Eliminativism of the Watching Mists," and so on.

20. This is in no way meant as a dismissal, simply a provocation. Consider the following as case studies: the role of Fichte in Meillassoux (2010), the role of Schelling in Hamilton Grant (2006), the edited volume *The New Schelling* (Norman and Welchman 2004), and the role of that greatest of resurrected corpses, Hegel, in works such as Nancy (2002).

WORKS CITED

Abelsen, Peter. 1993. "Schopenhauer and Buddhism." *Philosophy East and West* 43 (2): 255–278.

Aristotle. 2000. *On the Soul/Parva Naturalia/On Breath.* Trans. W. S. Hett. Cambridge, Mass.: Loeb Classical Library/Harvard University Press.

——. 2002. *Aristotle: "Historia Animalium": Volume 1, Books I-X.* Ed. D. M. Balme and Allan Gotthelf. Cambridge: Cambridge University Press.

Bergson, Henri. 1998. *Creative Evolution.* Trans. Arthur Mitchell and Henry Holt. Minneola, N.Y.: Dover.

Braidotti, Rosi. 2013. *The Posthuman.* Cambridge: Polity.

Brassier, Ray. 2007. *Nihil Unbound: Enlightenment and Extinction.* New York: Palgrave Macmillan.

Colebrook, Claire. 2014. *Death of the PostHuman: Essays on Extinction.* Vol. 1. Open Humanities Press.

Cross, Stephen. 2013. *Schopenhauer's Encounter with Indian Thought.* Honolulu: University of Hawai'i Press.

Deleuze, Gilles. 1986. *Cinema 1: The Movement-Image.* Trans. Barbara Habberjam and Hugh Tomlinson. Minneapolis: University of Minnesota Press.

——. 1989. *Cinema 2: The Time-Image.* Trans. Hugh Tomlinson and Robert Galeta. Minneapolis: University of Minnesota Press.

——. 1990. *Bergsonism.* Trans. Hugh Tomlinson and Barbara Habberjam. New York: Zone.

——. 1994. *Difference and Repetition.* Trans. Paul Patton. New York: Columbia University Press.

European Science Foundation (ESF). 2007. *Investigating Life in Extreme Environments—a European Perspective.* Ed. Nicolas Walter. Strasbourg: European Science Foundation.

Fichte, J. G. 2005. *The Science of Knowing—Fichte's 1804 Lectures on the Wissenshaftslehre.* Trans. Walter E. Wright. Albany: SUNY Press.

Gerday, Charles, and Nicolas Glansdorff. 2007. *Physiology and Biochemistry of Extremophiles.* ASM.

Hamilton Grant, Iain. 2006. *On an Artificial Earth: Philosophies of Nature After Schelling*. London: Continuum.

Hayles, N. Katherine. 1999. *How We Became Posthuman*. Chicago: University of Chicago Press.

Hegel, G. W. F. 1977. *Phenomenology of Spirit*. Trans. A. V. Miller. Oxford: Oxford University Press.

——. 2004. *Philosophy of Nature: Encyclopedia of the Philosophical Sciences, Part II*. Trans. A. V. Miller. Oxford: Clarendon.

Henry, Michel. 2003–2004. *Phénoménologie de la vie: "Tome I. De la phenomenology," "Tome II. De la subjectivité," "Tome III. De l'art et du politique," "Tome IV. Sur l'éthique et la religion."* Paris: PUF.

Husserl, Edmund. 1970. *The Crisis of European Sciences and Transcendental Phenomenology*. Trans. David Carr. Evanston, Ill.: Northwestern University Press.

Kant, Immanuel. 2001. *Critique of the Power of Judgment*. Trans. Paul Guyer and Eric Matthews. Cambridge: Cambridge University Press.

Kroker, Arthur. 2014. *Exits to the Posthuman Future*. Cambridge: Polity.

Kurzweil, Ray. 2006. *The Singularity Is Near: When Humans Transcend Biology*. New York: Penguin.

Ligotti, Thomas. 2010. *The Conspiracy Against the Human Race*. New York: Hippocampus.

Lyotard, Jean-François. 1992. *The Inhuman*. Stanford, Calif.: Stanford University Press.

MacKenzie, Debora. 2006. "Gold Mine Holds Life Untouched by Sun." *New Scientist* (19 October). http://www.newscientist.com/article/dn10336-gold-mine-holds-life-untouched-by-the-sun.html.

Malabou, Catherine. 2005. *The Future of Hegel: Plasticity, Time, and the Dialectic*. Trans. Lisabeth During. London: Routledge.

Marion, Jean-Luc. 2002. *Being Given: Towards a Phenomenology of Givenness*. Stanford, Calif.: Stanford University Press.

Meillassoux, Quentin. 2010. *After Finitude: An Essay on the Necessity of Contingency*. Trans. Ray Brassier. London: Bloomsbury Academic.

Merleau-Ponty, Maurice. 2013. *The Phenomenology of Perception*. Trans. Donald Landes. New York: Routledge.

Moravec, Hans. 2000. *Robot: Mere Machine to Transcendent Mind*. Oxford: Oxford University Press.

More, Max, and Natasha Vita-More, eds. 2013. *The Transhumanist Reader*. London: Wiley-Blackwell.

Nancy, Jean-Luc. 2002. *Hegel: The Restlessness of the Negative*. Trans. Jason E. Smith and Steven Miller. Minneapolis: University of Minnesota Press.

NASA. 2010. "Press Release: NASA-Funded Research Discovers Life Built with Toxic Chemical." http://www.nasa.gov/home/hqnews/2010/dec/HQ_10-320_Toxic_Life.html.

Negarestani, Reza. 2008. *Cyclonopedia: Complicity with Anonymous Materials*. Melbourne: re:press.

Nicholls, Moira. 1999. "The Influences of Eastern Thought on Schopenhauer's Doctrine of the Thing-in-Itself." In *The Cambridge Companion to Schopenhauer*, ed. Christopher Janaway, 171–212. Cambridge: Cambridge University Press.

Norman, Judith, and Alistair Welchman, eds. 2004. *The New Schelling*. London: Continuum.

Shaviro, Steven. 2011. "Panpsychism and/or Eliminativism." Paper delivered at the Third Object Oriented Ontology conference, 9 September. The New School, New York. http://www.ustream.tv/recorded/17269234.

——. 2012. *Without Criteria*. Cambridge, Mass.: MIT Press.

Schelling, F. W. J. 2000. *The Ages of the World*. Trans. Jason M. Wirth. Albany: SUNY Press.

——2004. *First Outline of a System of the Philosophy of Nature*. Trans. Keith R. Peterson. Albany: SUNY Press.

Schopenhauer, Arthur. 1969. *The World as Will and Representation*. Vol. 1. Trans. E. F. J. Payne. New York: Dover.

Taylor, Michael Ray. 1999. *Dark Life: Martian Nanobacteria, Rock-Eating Cave Bugs, and Other Extreme Organisms of Inner Earth and Outer Space*. New York: Scribner.

Thacker, Eugene. 2010. *After Life*. Chicago: University of Chicago Press.

——. 2011. *Horror of Philosophy*, vol. 1: *In the Dust of This Planet*. New York: Zero.

Whitehead, Alfred North. 1979. *Process and Reality*. New York: Free Press.

Wolfe, Cary. 2009. *What is Posthumanism?* Minneapolis: University of Minnesota Press.

Woodard, Ben. 2013. *On an Ungrounded Earth: Towards a New Geophilosophy*. New York: Punctum.

FILM CITED

Journey Into Amazing Caves. Directed by Steven Judson. 2001. Laguna Beach, Calif.: MacGillivray Freeman Films.

14

THINKING LIFE

The Problem Has Changed

ISABELLE STENGERS

I t is difficult to confer a unified sense upon the question of life in the work of Gilles Deleuze. The question, as such, is always insistent, but it appears in different modes, notably with authors like Gilbert Simondon and Raymond Ruyer but also, beginning with *Anti-Oedipus*, with the themes of desiring machines, the Body without Organs, and the difference in nature between the molar and molecular. In *A Thousand Plateaus*, all the components—and more—are present, yet they are not unified into a synthesis. A radicalized process-thinking, what had started with *Anti-Oedipus*, excludes the set of synthetic judgments that allow the "I understand" or "So that is it." One could argue that in the interval of eight years that elapses between the two books, a mutation occurred: There is certainly a question of tone, but not exclusively. The problem has changed.

At *very first* impression, *Anti-Oedipus* gives rise to a warlike reading: a (joyous) declaration of war against the organism and against the organs understood as means to the ends of the organism. In a rather Bergsonian mode (but a disheveled Bergson, intoxicated by his famous cup of tea) or perhaps a Spinozan one (but a Spinoza who no longer polishes his lenses but instead builds Tinguely machines), the stratified duality of what would refer to "knowledge" and to life "itself" is swept up in the very operativity of the text. The battle is transversal, and in a sense it continues the great Bergsonian theme of the incapacity of intellectual categories, dominated by the disjunction "either/or," to grasp the creative unfolding of life. But the fluid continuity of a becoming-musical is replaced by *productions*, of flux as well as of cuts, transfers, connections, and couplings against nature. And the battle is especially against the "judgment of God," spewed forth by Artaud, which assigns relations that authorize a permanent cross-reference between healthy body and healthy thought under the double sign of nature and logic.

Even if Bergsonian "health" has nothing to do with the operation of a well-oiled mechanism, the screeching, hissing, clashing, and shearing of the "machines" contest the mechanism altogether differently. There is no happy marriage between the supple and penetrating finesse of intuition and the immanent, undivided sense of which the analyzable, living order is but the trace or negative expression. Machines work, albeit in an aberrant mode, by breaking down, and if a whole is produced, it is like a part peripheral to other parts, incapable of unifying them.

> We no longer believe in a primordial totality that once existed, or in a final totality that awaits us at some future date. We no longer believe in the dull gray outlines of a dreary, colorless dialectic of evolution, aimed at forming a harmonious whole out of heterogeneous bits by rounding off their rough edges. We believe only in totalities that are *peripheral*.
>
> (Deleuze and Guattari 1983, 42, italics from French).

However, the very possibility of talking about "the very first impression" in regard to *Anti-Oedipus* draws attention to the proximity between the line of flight and the line of death that will haunt *A Thousand Plateaus*. The possibility of this impression signals the eventuality of both a new dualism and an idea of recognition and, with that, a new generation of judges, presiding in the name of the molecular, desiring machines, and the categorical imperative of destratification—feeding on the effects of terror they arouse. In *A Thousand Plateaus* there will no longer be a first impression, and as such, the plateau called "The Geology of Morals" forecasts a general disorder [*débandade*], which could easily include listeners-readers ready to identify with a "Body without Organs," even if it indeed kills them, but unable to understand what Challenger "is driving at."

Challenger "discombobulates." He gets lost in digression and ensures "that there was no possible way to distinguish the digressive from the non -digressive" (Deleuze and Guattari 1987, 49). As he digresses and loses his listeners, he himself loses his human appearance. Jumping between the present and imperfect tenses, the text no longer addresses anyone, and "Challenger was addressing himself to memory only" (57). Here, mind could be understood as a "perplication" (Deleuze 2001, 187), a coexistence without the confusion of ideational contractions, factual similarities, inseparable variations, connections between heterogeneous elements, folds on folds, and an innumerable multiplicity without the confusion of records for the same event. A Bergsonian cone twisted by ahuman convulsions.

"It is the brain that thinks, not Man" (Deleuze and Guattari 1994, 210), Deleuze and Guattari affirm later in *What Is Philosophy?*, and as such, the "Geology of Morals" plateau does not "critique" what it designates— without stopping there, for it never stops—as "the illusion constitutive of man . . . [which] derives from the overcoding immanent to language itself" (Deleuze and Guattari 1987, 63). It makes one feel—physically, cere- brally—the genealogy of a morality that begins with questions that, here, are rightly not asked: "But what do I mean? All this must be organized a little! To be able to recount it to myself . . . " When thought is made moral, Man comes *after*, in the sense that morality is populated by the set of interlocutors to whom it explains itself and for whom it rarifies, organizes, justifies, and traces a path that might be shared.

A Thousand Plateaus is a book that thinks on its own [*un livre-cerveau*], not a book written by authors [*un livre d'auteurs*]. It is two brains func- tioning in discordant accords, in proximities never stabilized into reliable convergences, in divergences never transformed into oppositions—pro- ducing by seizures, captures, and relays something that not only does not belong to either of the two (small success) but also transforms the 2 into *n* partial brains. Bursting at the seams, it constructed its line of flight in rela- tion to the sort of polemic that constitutes a vulnerability of *Anti-Oedipus*: The polemic is the kind that lends itself to the distinction between ends and means. "Vulgar words" ("it shits and fucks," starting from the second line [Deleuze and Guattari 1983, 1]) can become means for an end that will always be the same—a placard or banner to be displayed by those who are "in" against those whose indignation denounces them. In contrast, the geology of morals makes the very thing it recounts exist for the reader: Disarray [*débandade*] and panic arise during Challenger's "lecture." It is not that "bands" are denounced: Banding, packing, and hoarding are all events, even cerebral ones. Rather, *A Thousand Plateaus* has something of an immanent test, but it is not about sorting; rather it is a learning to pay attention and generate knowledge about the difference of nature between the mode of existence of bands—with their passwords, their secrecy, the unexpectedness of their rapid, swirling, and nonvectoral reorientations— and the "'individual' group effects spinning in circles, as in the case of chaffinches that have been isolated too early, whose impoverished, simplified song expresses nothing more than the resonance of the black hole in which they are trapped" (Deleuze and Guattari 1987, 334).

Everything becomes complicated in *A Thousand Plateaus* with the knowledge of the perils that surround any experiment, whether it is conducted under premature or brutal conditions, producers of "the emptied or cancerous doubles" (166), or whether it generates other repugnant double types, "as if each effort and each creation faced a possible infamy" (379). The worst is never far away,

> the worst is the way the texts of Kleist and Artaud themselves have ended up becoming monuments, inspiring a model to be copied—a model far more insidious than the others [that requires finding the method to express well what one believes oneself entitled to say] for the artificial stammerings and innumerable tracings that claim to be their equal.
>
> (378)

It is not enough to say, "I am an animal, a negro," and thus reject the image of the universal thinking subject. One must *do* it—and the infamy is to transform the animal or negro into monuments or models, in the name of which one will scorn majoritarian thought and wait for the repression that will confirm membership among the oppressed people. This constitutes a new infamous and logical marriage: of the truth and of the posters and placards that feed on the perception "it is unbearable" that they arouse.

There is something unbearable in *A Thousand Plateaus*, but without the slightest "monumental" announcement effect: an immanent production that defies all models because its trade secret is the best guarded in the world. This book is irreducible, as mathematicians say, which is to say that it cannot be formulated in a more economical mode (what do you mean?), as it is strictly coextensive with its process of production. There is not a hidden code but rather a decoding process undertaken by the very brain of the reader—it is literally "demoralized." "Philosophy is no longer synthetic judgment; it is like a thought synthesizer functioning to make thought travel, make it mobile, make it a force of the Cosmos" (343).

The question can no longer be, then, one of commentary, rendering explicit what would have remained implicit, clarifying or elucidating. Rather it is about "consolidating" just a little more—always a little more—which is to say, forming relays. As it happens, in regard to thinking about life, it will be a question of forming relays in the manner that *A Thousand Plateaus* struggles against the nearly irresistible slope that would transform the "voyage" of thought into the destination, into the position of its final definition, and simultaneously assign an end, in the double sense of the term, to thought.

The plateau "Of the Refrain" begins with a fragile little song, which at every moment runs the risk of disintegrating into the heart of chaos. Then, it is the drawing of a circle—of a home that does not preexist—that one must "do" around the fragile center: "The forces of chaos are kept outside as much as possible, and the interior space protects the germinal forces of a task to fulfill or a deed to do" (311).

The circle—a circle of witches who understood the need to protect themselves—is neither a timorous [*frileux*] enclosure nor a relation of forces. And neither is it is a membrane, which certain biologists identify with that which life requires: the distinction between an interior milieu and an exterior one. In *A Thousand Plateaus*, the question of life is associated with a matter of milieus and rhythms and the passing of milieus into one another. Each milieu is coded, and rhythm is the event of a transcoding, a coordination of heterogeneous space-times. This is why rhythm is not subject to regularity or a meter that would identify a milieu as such: "Meter is dogmatic"—it enslaves at once that to which it applies—"but rhythm is critical; it ties together critical moments, or ties itself together in passing from one milieu to another" (313).

However, coded milieus and rhythms are born of chaos, not from relations between forces. Did forces preexist the refrain? This is perhaps a "bad" question, in a Bergsonian sense. It seems that the problem of forces is no longer that of Nietzschean forces, in their irreducible plurality, always in relation with one another, either to obey or command: "Active and reactive are precisely the original qualities which express the relation of force with force" (Deleuze 2006, 40). The Nietzschean relation of force with force could suggest that to enter into relation was—as in mechanics—the very definition of forces. Here forces and relations take on meaning simultaneously. Force is no longer a subject, or, more precisely, it becomes a subject relative to an assemblage and the creation of a relation.

It could thus be said that "forces" emerge in reciprocal presupposition with the refrain, with the question of danger and protection. Perhaps even with the very sense of danger? It is not that forces are "dangerous"—the heroic figure of confrontation against the "forces of chaos"—perhaps instead chaos here becomes what it was not when codes and rhythms were emerging from it: what would overwhelm, "destroying both creator and creation" (Deleuze and Guattari 1987, 311). And when the circle opens, whether to permit entrance or exit, it is not in order to join the forces of chaos. "One launches forth, hazards an improvisation. But to improvise is to join with the World, or meld with it" (311). Here again, we should say that the World did not preexist the circle; it takes on meaning simultaneously with the circle or the germinal forces that the circle fosters [abrite].

The refrain is territorial, and the territory is not an effect—referring to anything that would precede it—but an "act" that affects milieus and rhythms. Act in this sense, of course, does not relate to the question of ethics or the subject in the sense of the subject-engaged-by-its-act, unlike the animal-defined-by-its-behavior. It is instead ethics that plunges into the ethos from the territorial question of distances, critical distances. There are no forces without both the danger of intrusion and the creation of that which makes distance. Remember, ethology was born from the refusal to consider the animal as isolated from "its" milieu and to impose upon it the full range of laboratory devices that were supposed to allow for its reduction to a function of manipulable variables. Ethology advocates for territory when it asserts the indissociability of the possessives—"its" behavior and "its" environment—and denounces the destruction in the laboratory of what we should be learning to identify.

However, Deleuze and Guattari insist that the expressive comes first in relation to the possessive. "There is a territory precisely when milieu components cease to be directional, becoming dimensional instead, when they cease to be functional to become expressive. There is a territory when the rhythm has expressiveness" (1987, 315). Mine is too close to me: my interests, my survival, my female. Would we say that the musician bird sings "its" rising sun? Ownership brings us too quickly to functionality, to "another way" of assuring functions in terms that can define selective advantages (Lorenz). The expressive raises the question of relation as such, as irreducible to any generality and to any dependence regarding words designed to be put in relation. Take the musician bird singing at sunrise. One could say that it goes from "sadness to joy" (interior milieu of impulses), "greets the rising sun" (exterior milieu of circumstances), and that "it endangers itself in order to sing" (drawn territory). But it is the melody itself that puts the sun in counterpoint, and "in the motif and the counterpoint, the sun, joy or sadness, danger, become sonorous, rhythmic, or melodic" (318–319).

Recall the ambiguity of emotion according to William James: Is the landscape moving because I am moved, or am I moved because it is moving? Is my disgust caused by a vile spectacle, or rather does the representation of the cause of my disgust follow after my body is repulsed? Or rather, as social constructivists will add, does emotion refer to what in a culture or given social class is "marked" as moving, as a distinction, as Bourdieu would say? However, in A Thousand Plateaus, this ambiguity begins with territory and, notably, the singing bird.

> Ethologists have a great advantage over ethnologists: they did not fall into the structural danger of dividing an undivided "terrain" into forms of kinship, politics, economics, myth, etc. The ethologists have retained the integrality of a certain undivided "terrain." But by orienting it along the axes of inhibition-release, innate-acquired, they risk reintroducing souls and centers at each locus and stage of linkage.
>
> (328)

They risk making the Sun a trigger, while the song demonstrates instead that the territory of the bird has opened up to the forces of the Earth and the terrestrial sun that bathes the territory. Where periodic changes in luminosity were transcoded into rhythms, the sun became a rhythmic figure to which the counterpoint of a melodic landscape responds. Before Akhenaton, and before the builders of Stonehenge, on another line, the singing bird confers upon the sun the power to affect it—the bird brings the sun into existence as a force.

Asserting rhythm against meter and the expressive against an appropriation reducible to the functional issues of selective survival, it is obviously
not a question here of "projecting" issues belonging to the human sciences onto the histories of the living; instead it is a question of defending
the life sciences against both anthropocentrism and the binary division
between the mode of explanation (causal, functional) that would be suit-
able for nonhumans and the symbolic issues that would give the human
sciences their (coarse) anthropological categories. Ethology, the behavioral science that affirms a territory, is crushed between these two tectonic
plates. Yet it is also the ground for a conceptual experiment that produces
a line of flight in relation to the stupid and nasty [*bêtes et méchants*] effects
of their confrontation: between those who represent "science," insisting
upon subjugating humans to the common lot, and those who will define
them by "rampart" categories, organizing a subjectivity that will be called
human—where Man is the focal point and majoritarian standard with the
power to dispense definitions by excess or default.

A pianist "composes a melody by trial and error that expresses the
vague sentiment he feels" (Ruyer 1958, 143). To give this case its required
conceptual extension, Raymond Ruyer, whose thought haunts the plateau
"Of the Refrain," realized he had to resist the functions that explain and
crush according to ready-made categories (here, it would be a process of
trial and error resulting in "hey, that is what I always wanted to express").
And he realized that resistance should start not with "man" but at the place
ethologists fear, anthropocentrism, attributing the animal with something
that would be on the order of representative consciousness. Start, for
example, with sexual display or the building of a nest. The territory itself
is certainly not "represented" as a matter of rivalry, but the question it
raises does not at all boil down to the issue of intraspecific aggressive-
ness established by Konrad Lorenz (2002). And what if "having a body,"
"being with," "being there"—the primordial givens that phenomenology
struggles to put into words—were to take on meaning with the intruder
who "knows" where he is but who hesitates between fight or flight? What
if, with the refrain, they were to take on meanings that we associate with
"reasons" rather than "causes": hesitating, venturing, crossing a border,
seeking admission into a group, being rejected?

Nietzsche-Zarathustra's eternal return is itself a little ditty, a refrain,
but one that "captures the mute and unthinkable forces of the Cosmos"

(Deleuze and Guattari 1987, 343). At the end of the plateau, the refrain has become a whirlwind, and simultaneously thought becomes force, and forces can henceforth be attributed only to the Cosmos—a matter of rendering them perceptible, visible, and audible. One could argue that this plateau was oriented toward what would appear as a true destiny, "illustrated" by the becoming of painting or music. We would be tempted to conclude then that the rupture with animality has been somewhat postponed, but, ultimately, it is precisely here, with the opening up on forces that, unlike the terrestrial sun, are "not visible" and are no longer those of the great expressive Form Earth but free instead the Earth as pure material. However, *A Thousand Plateaus* is a book struggling with modes of narrative haunted by the axis of progress, which leads to the definition of a "superiority" that, from its axial morality, would come to intensify [*redoubler*] a story that moves from relative deterritorializations to this absolute deterritorialization, a story that we would know henceforth as the only one worthwhile. Never borrow Zarathustra's ditty as if it were a ready-made truth. Never forget the young chaffinches prematurely isolated. Never disregard the power of the Natal: One must belong in order to take leave or venture out. Becoming offers no shortcuts, economic principle, or shoulders of giants upon which to climb. Lines of flight and cosmic deterritorialization have nothing to do with an emancipation that would free humans from what "still" imprisons them. "Sound invades us, impels us, drags us, transpierces us. It takes leave of the earth, as much in order to drop us into a black hole as to open us up to a cosmos. It makes us want to die" (1987, 348). The plane of consistency must be drawn—the relations of form and substance cannot be abolished in the name of a story or an evolution that would lead, like the irreversible arrow of time, from Earth to the Cosmos, with the infamy of the "we can no longer . . . we must henceforth . . ."

In order to complicate things a little, let us return to the refrain in its early stages and to the distinction of rhythm as transcoding. Has there been progress? It is not that simple; we read in *A Thousand Plateaus* that with certain transcodings—when a coded milieu "is not content to take or receive components that are coded differently and instead takes or receives fragments of a different code as such"—it is as though there is a Nature that becomes music: "It is as though the spider had a fly in its head, a fly 'motif,' a fly 'refrain'" (Deleuze and Guattari 1987, 314). Similar, perhaps, is the famous orchid, which "captivates" the wasp, presenting it with a sexual morphologic "motif." Yet the "it is as though" and the quotation marks are not unimportant. The question is not: Isn't there "already"—before the territorial assemblage—the beginnings of a refrain there? This is an evolutionist and progressive question. The refrain was certainly "already there," but here again, foremost, it is the problem that has changed. The fly is not a rhythmic figure *for* the spider; it is *in its head* that the spider has a fly. "Nature as music" excludes both the musician and the ear that listens. It excludes perhaps just as much the body itself that sings or listens. "Becoming Spider" would be perhaps to become a being who is not "on the alert" [*aux aguets*] but "is alertness" [*est aguet*].

One might recall, here, the great Bergsonian bifurcation in *Creative Evolution* between instinct and intelligence. Here too, nothing was pure and no separation was entrenched. It is no less true, affirmed Bergson, that there has been a hesitation concerning the two ways of posing the problem: hesitation "between two modes of psychical activity—one assured of immediate success, but limited in its effects; the other hazardous, but whose conquests, if it should reach independence, might be extended indefinitely" (Bergson 1970, 158). Bergsonian instinct (or von Uexküll's Nature as music) can be said to be "brilliant." It does not provoke the ethologist's hesitations ("how does one avoid anthropocentrism?") but instead arouses wonder. The Bergsonian contrast then passes between the perfection of an action that, if we were to associate it with the animal's representation of what it does, would imply a knowledge that is beyond us and the tentative clumsiness of intelligence, subject to error and hesitation. How to deal with this situation: This question, the question of Bergsonian intelligence, is foreign to the spider. But this does not signify that we are not also spiders, wasps, orchids, or even ticks. Incidentally, was not Bergson himself passionately interested in what was called in his time "metapsychic" [*métapsychique*]? What does the seer have "in his head?" What we know about seers is that when they are dishonest, it is in situations where the question "How do I deal with this situation?" is vital, the question Bergson assigns to intelligence. In contrast, what we do not know, what other practical traditions have apparently discovered how to cultivate, is how not to make clairvoyance a "big deal" in itself—the mark of a territory whose ownership should be authenticated.

A *Thousand Plateaus* departs from Bergsonian evolutionism, creator or not, because between the bird that builds its nest—or the chimpanzee engaged in the laborious task of cracking a nut—and intelligence that mobilizes, proceeds by equivalence, and submits to general categories, the result is not good and far too hasty—leading to the Bergsonian morality of final convergence: With intuition, instinct returns, but as disinterested, self-conscious, and capable of an understanding that espouses life. The wholly spiritual vocation of humanity is to be situated in this privileged point where the current of life "passes freely, dragging with it the obstacle which will weigh on its progress but will not stop it" (Bergson 1970, 323).

Bergson is too quick to summon a "confrontation at the summit," convoking the philosophers whom he considers as testament to this intelligence that can only understand life by stopping its course. Likewise he too quickly submits language to the principle of a stable view of the instability of things. Linguistic equivalence—different expressions to articulate the "same" eidetic content ("say what you mean!")—is not truth finally appearing from the regime of signs. Language was not made for communication but for translation between groups that did not speak the same language (Deleuze and Guattari 1987, 430) and for whom semiotics and material were articulated differently. The "static" categories of Bergsonian intelligence have the gravity of the apparatus of the State. Where are the itinerant peoples and the ambulant sciences—those that invent and follow a problem, in an "exploration by legwork" (373)? What has happened? What has "come over" us?

It is in the plateau "Apparatus of Capture," interrogating among other things the "origin" of the State, that the struggle against progressive evolutionism—which puts evolution under the sign of functions satisfied by increasingly effective means—is the most explicit. There is certainly a moment when the State appears; one can never be mistaken about that. But, as with the invention of territory, the invention of the State is not explained in a functional mode. The functions assured by the State are not invariants of every human society. They appear with the State and impose a reorganization of what they capture, like territory imposes a reorganization of milieus and rhythms.

"Once it has appeared, the State reacts back on the hunter-gatherers, imposing upon them agriculture, animal raising, an extensive division of labor, etc.," but it was acting "in a different form than that of its existence" already: "as the actual limit these primitive societies warded off, or as the point toward which they converged but could not reach without self-destructing" (431). The appearance of the State responds to a question of threshold. The series that converges does not lead to this threshold; rather it is inhabited by the evaluation of the limit beyond which "Stateless societies" will be annihilated, beyond which lies the threshold, the ultimate one, the tipping point where "the problem has changed."

The distinction between limit and threshold itself constitutes an incantation, counter to the thousand and one versions of an evolution leading triumphantly to us. Series do not lead somewhere. Therefore the refrain could have been "already in the series," but the series did not lead to territorial animals or to the mutation that accords its own necessity to the refrain as a concept. It may have been "already there," but it is the transition from the naturalist wonder (how is it possible?) to the perplexity of the ethologist (how can it be described properly?) that indicates that a threshold has been crossed. The same perhaps applies each time, that "life reconstitutes its stakes, confronts new obstacles, invents new paces, switches adversaries" (Deleuze and Guattari 1987, 500). It is not, however, a question of an inventive continuity that it would be possible to espouse. It is the term "new" that matters. Life, "inorganic, germinal, and intensive, a powerful life without organs" (499), dismantles every representation, even intuitive ones, and, therefore, also destroys every representative. "Absolute" deterritorialization may well "henceforth" be distinguished qualitatively from "relative" deterritorialization—the "smooth" space without assignable coordinates from a striated space and "stationary" whirlwind movement from vectorial movement—but then the infamy would be to transform this "henceforth" into a principle of judgment: "Your space is still striated, young man!" Any hastiness or imperative ("I must escape the striations") signals that we are not yet done with the morality of salvation and election.

Prudence, prudence, and restraint: "To become is to become more and more restrained, more and more simple, more and more deserted and for that very reason populated" (Deleuze and Parnet 1987, 29). When one reads—regarding the movement of deterritorialization sweeping away music as much as contemporary painting and philosophy—that "the problem is no longer that of the beginning, any more than it is that of a foundation-ground. It is now a problem of consistency or consolidation: how to consolidate the material, make it consistent, so that it can harness unthinkable, invisible, nonsonorous forces" (Deleuze and Guattari 1987, 343)—what must be thought, transcendental empiricism, is "the problem has changed," never that "we got the upper hand" with respect to anyone or anything. It is the same refrain in *What Is Philosophy?*: Let us not be proud of "no longer believing in God," or confuse the reference to a transcendent existence of God, or God as a standard [*étendard*], with "the infinite immanent possibilities brought by the one who believes that God exists" (Deleuze and Guattari 1994, 74–75)—Zarathustra leaves the cave, the gathering place of Overmen who are proud to believe no longer. And the problem of the "empiricist conversion," the problem of he who "believes in the world . . . in its possibilities of movements and intensities, so as once again to give birth to new modes of existence" (74), is not what we have conquered but what is the most difficult for us, "our" problem: "We have so many reasons not to believe in the human world; we have lost the world, worse than a fiancée or a god" (75).

Life reconstitutes its stakes: There is, in *A Thousand Plateaus*, a "trusting" [*faire confiance*], a trusting in the power of undecidable propositions, nondenumerable sets, and numbering numbers that do not stop straying from every organized scheme in which possibilities are formulated and where the absence of every "perspective of struggle" [*perspective de lutte*] can be deplored. But this trust has nothing to do with an adhesion—like when we allow undocumented migrants, suburban youths, or hackers to be transformed into guarantees for a new theorization or a slogan for a new mobilization. Trusting implies an empiricism that notions of progress abhor because it involves distinctions to make again and again in that which mixes them all: "Is a smooth space captured, enveloped by a striated space, or does a striated space dissolve into a smooth space, allow a smooth space to develop?" (Deleuze and Guattari 1987, 475). Even the most striated city gives rise to smooth spaces, and this fact can itself become an order-word [*mot d'ordre*]. Trusting [*faire confiance*], and above all not "being trustful" [*avoir confiance*], is reflected in the last sentence of *A Thousand Plateaus*: "never believe that a smooth space will suffice to save us" (500).

NOTE

Translated from the French by Jami Weinstein and Zachary R. Hagins. Originally published as Isabelle Stengers, "Penser la vie: problème a changé," *Revue Internationale de Philosophie* 3 (2007): 241, 323–335, http://www.cairn.info/revue-internationale-de-philosophie-2007-3-page-323.htm.

WORKS CITED

Bergson, Henri. 1944. *Creative Evolution*. Trans. Arthur Mitchell. New York: Random House. (1970. *L'évolution créatrice*. In *Oeuvres*. Paris: PUF.)

Deleuze, Gilles. 2001. *Difference and Repetition*. Trans. Paul Patton. London: Continuum. (1972. *Différence et repetition*. Paris: PUF.)

——. 2006. *Nietzsche and Philosophy*. Trans. Hugh Tomlinson. New York: Columbia University Press. (1962. *Nietzsche et la philosophie*. Paris: PUF.)

Deleuze, Gilles, and Félix Guattari. 1983. *Anti-Oedipus: Capitalism and Schizophrenia*. Trans. Robert Hurley, Mark Seem, and Helen R. Lane. Minneapolis: University of Minnesota Press. (1972. *L'anti-Oedipe*. Paris: Minuit.)

——. 1987. *A Thousand Plateaus: Capitalism and Schizophrenia*. Trans. Brian Massumi. Minneapolis: University of Minnesota Press. (1980. *Mille plateaux*. Paris: Minuit.)

——. 1994. *What Is Philosophy?* Trans. Hugh Tomlinson and Graham Burchell. New York: Columbia University Press. (1991. *Qu'est-ce que la philosophie?* Paris: Minuit.)

Deleuze, Gilles, and Claire Parnet. 1987. *Dialogues*. Trans. Hugh Tomlinson and Barbara Habberjam. New York: Columbia University Press. (1996. *Dialogues*, coll. "Champs." Paris: Flammarion.)

Lorentz, Konrad. 2002. *On Aggression*. Trans. Marjorie Kerr Wilson. New York: Routledge.

Ruyer, Raymond. 1958. *La genèse des formes vivantes*. Paris: Flammarion. Cited sentence translated by Jami Weinstein.

CONTRIBUTORS

NICOLE ANDERSON is professor of communication and cultural studies at Macquarie University. She is the cofounding editor of the *Derrida Today Journal* (Edinburgh University Press) and the director of the Derrida Today biannual conference. She is an international fellow of the London Graduate School and a member of the Biocultures Project at University of Illinois, Chicago. Her publications include *Derrida: Ethics Under Erasure* (Continuum 2012) and *Culture* (forthcoming, Routledge 2016).

FRIDA BECKMAN is associate professor of English at Stockholm University. She is the author of *Culture Control Critique: Allegories of Reading the Present* (Rowman and Littlefield, 2016) and *Between Desire and Pleasure: A Deleuzian Theory of Sexuality* (Edinburgh University Press, 2013) and editor of *Deleuze and Sex* (Edinburgh University Press, 2011).

CLAIRE COLEBROOK is Edwin Erle Sparks Professor of English at Penn State University. Among her many books and edited volumes, she has authored *Twilight of the Anthropocene Idols* (coauthored with Tom Cohen and J. Hillis Miller); *Death of the PostHuman: Essays on Extinction*, vol. 1, and *Sex After Life: Essays on Extinction*, vol. 2 (Open Humanities Press, 2015); and *Deleuze and the Meaning of Life* (Continuum Press, 2010). Her next book is *Fragility: Species, Planet, Archive* (forthcoming, Duke University Press).

ZACHARY R. HAGINS is assistant professor of French at University of Arkansas at Little Rock. He has published "Staging Masculinity for the Lens: Performing 'les jeunes de banlieues' in Mohamed Bourouissa's Directorial Photography" (*Contemporary French Civilization*, 2016), and "Fashioning the 'Born Criminal' on the Beat: Juridical Photography and the *Police municipale* in Fin-de-Siècle Paris" (*Modern and Contemporary France*, 2013).

SUSAN HEKMAN is professor of political science at University of Texas at Arlington. Among her many books are *The Feminine Subject* (Polity, 2014), *The Material of Knowledge: Feminist Disclosures* (Indiana University Press, 2010), *Material Feminisms* (coedited with Stacey Alaimo, Indiana University Press, 2008), and *Private Selves, Public Identities: Reconsidering Identity Politics* (Penn State Press, 2004).

MYRA J. HIRD is professor, Queen's National Scholar, and FRSC in the School of Environmental Studies, Queen's University. She is also director of both Canada's Waste Flow project and the genera Research Group (gRG). She has published many books and edited collections, including *The Origins of Sociable Life: Evolution After Science Studies* (Palgrave, 2009), *Queering the Non/Human* (coedited with Noreen Giffney, Routledge, 2008), and *Sex, Gender, and Science* (Palgrave Macmillan, 2004).

ALASTAIR HUNT is associate professor of English at Portland State University. Among his publications are *The Right to Have Rights* (coauthored with Stephanie DeGooyer, Werner Hamacher, Samuel Moyn, and Astra Taylor, forthcoming, Verso, 2017), *Against Life* (coedited with Stephanie Youngblood, Northwestern University Press, 2016), and "Romanticism and Biopolitics" (special issue of *Romantic Circles Praxis*, coedited with Matthias Rudolf, 2012).

AKIRA MIZUTA LIPPIT is professor of film and literature at the University of Southern California. His books include *Cinema Without Reflection: Jacques Derrida's Echopoiesis and Narcissism Adrift* (University of Minnesota Press, 2016); *Ex-Cinema: From a Theory of Experimental Film and Video* (University of California Press, 2012); *Atomic Light (Shadow Optics)* (University of Minnesota Press, 2005); and *Electric Animal: Toward a Rhetoric of Wildlife* (University of Minnesota Press, 2000).

TIMOTHY MORTON is the Rita Shea Guffey Chair in English at Rice University. Among his many books are *Dark Ecology: For a Logic of Future Coexistence* (Columbia, 2016); *Hyperobjects: Philosophy and Ecology After the End of the World* (Minnesota, 2013); *Realist Magic: Objects, Ontology, Causality* (Open Humanities, 2013); and *The Ecological Thought* (Harvard University Press, 2010).

JEFFREY T. NEALON is Edwin Erle Sparks Professor of English and Philosophy at Penn State University. His most recent books include *Plant Theory* (Stanford University Press, 2016); *Post-Postmodernism: Or, the Cultural Logic of Just-In-Time Capitalism* (Stanford University Press, 2012); and *Foucault Beyond Foucault: Power and Its Intensifications Since 1984* (Stanford University Press, 2008).

LUCIANA PARISI is reader in cultural theory and codirector of the Digital Culture Unit at Goldsmiths University of London. She has authored *Contagious Architecture: Computation, Aesthetics, and Space* (MIT Press, 2013), and *Abstract Sex: Philosophy, Biotechnology, and the Mutations of Desire* (Continuum Press, 2004).

JOHN PROTEVI is Phyllis M. Taylor Professor of French Studies at Louisiana State University. Among his books are *Life, War, Earth* (University of Minnesota Press, 2013); *Political Affect* (University of Minnesota Press, 2009); and *Political Physics: Deleuze, Derrida, and the Body Politic* (Bloomsbury, 2001).

ARUN SALDANHA is associate professor of geography, environment, and society at the University of Minnesota. He is the author of *Psychedelic White: Goa Trance and the Viscosity of Race* (2007) and the coeditor of *Deleuze and Race* (with Jason Michael Adams, Edinburgh University Press, 2013) and *Sexual Difference Between Psychoanalysis and Vitalism* (with Hoon Song, Routledge, 2013).

ISABELLE STENGERS is professor of philosophy at the Université Libre de Bruxelles. In addition to her many books published in French, her most recent English publications are *In Catastrophic Times* (Open Humanities Press, 2015); *Thinking with Whitehead: A Free and Wild Creation of Concepts* (Harvard University Press, 2014); and *Cosmopolitics I* and *Cosmopolitics II* (University of Minnesota Press 2010 and 2011).

EUGENE THACKER is professor at the New School in New York and the author of several books, including *Cosmic Pessimism* (Univocal Publishing, 2015), the three-volume series *Horror of Philosophy* (Zero Books, 2011–15), and *After Life* (University of Chicago Press, 2010).

JAMI WEINSTEIN is associate professor and founding director of the Critical Life Studies Research Group at Linköping University. She is currently working on the project *Vital Signs: Life, Theory, and Ethics in the Age of Global Crisis* (Swedish Research Council) and completing her monograph *Vital Ontologies*. She has coedited a number of volumes and special issues, including "Tranimalities" (with Eva Hayward, *TSQ*, 2015), "Anthropocene Feminism" (with Claire Colebrook, *philosSOPHIA*, 2015), and *Deleuze and Gender* (with Claire Colebrook, Edinburgh University Press, 2008).

CARY WOLFE is the Bruce and Elizabeth Dunlevie Chair in English and founding director of 3CT: The Center for Critical and Cultural Theory at Rice University. Among his many books and edited collections are *Before the Law: Humans and Other Animals in a Biopolitical Frame* (Chicago University Press, 2012); *What Is Posthumanism?* (Minnesota University Press, 2010); and *Animal Rites: American Culture, the Discourse of Species, and Posthumanist Theory* (Chicago University press, 2003). He is founding editor of the series Posthumanities at University of Minnesota Press.

INDEX

Abel, 106–7

Absolute, 300, 304; nature and, 301–3, 320n9; nothingness as, 315–16; Schopenhauer's skepticism of, 309

absolute life, 299–304; life-in-itself as, 305

action, 56, 62–63; intra-, 79; problem of, 334; time related to, 54–55

actors, 183–87, 191–93

Adams, John, 100, 190–91

Adorno, Theodor, 272–73

affirmative meontology, 12, 319

Agamben, Giorgio, xvi–xvii, 49, 107, 205n16

agency, 71, 79; in Anthropocene anti/ aesthetic, 264–65; constructionism and, 65–67; power and, 73–74. *See also* subjects

Alaimo, Stacy, 81

algorithms, 169–72, 290–91

Althusser, Louis, 112

American Revolution, 189–91, 194–95, 205n15

animal-human relations: on anthropomorphism, 34–38; deconstruction of, 31–38; formation of, 31–38; limits of, 17–38; literature on, 34–38; negative consequences of, 37–38; overview, 17–38; possibilities of, 17–38;

realms demonstration of, 25, 29–38. *See also* plants

animal life, 28–29; zoocentrism, 117

animals (*l'animot*), 8–9, 31, 121; Animal as, 28, 38n4, 53, 97; autobiographies related to, 88, 98–103; bees, 272, 290–91; behaviorism related to, 35, 39n11; being with, 93–94; cats: perspective of, 89–90, 92–93, 103n3; ethical relation with, 33–34; ethics and, 31; in *Glas*, 117–18, 122–29; human/ animal difference, xv, 110, 296; humans and, 110, 128, 140; identity and, 32–33; name for, 96–98; nonhuman animals, xiv–xvii, 25–26, 31, 57–58; as nonpersons, 203n3; personhood of, 181–82, 203n3; as persons, 180, 203n3, 204n6; plant life and, 106–9, 132n3; *vs.* plants, 117–18, 122–29; posthuman, 17–38; reason related to, 33–34, 39n9; rights of, 33, 180, 203n3, 204n6; speech and, 57–58; territory of, 332; wolves, 32; world and, 109, 132n4. *See also* dogs; *Glas*; prehuman animals

animal sacrifice, 106–7

animal studies, 87, 105–6

antagonism, 305–8, 312

information: difference in, 166–68; dynamic conception of, 166–67; dynamism of, 162–63, 177n2; energy of, 166–68, 174; entropy and, 163, 167–68; form in, 163–64; indetermination of, 164–66; individuation in, 164–66; measurement of, 165, 168; *negentropy* of, 163; randomness related to, 175; reduction in, 164
information theory, 156; technical objects and, 159–60
inhuman, the, xix, 5, 6, 9–10
inhuman rites: critical life studies and, 1–14; Inorganic Rites in, 9–10; Organic Rites in, 8–9; of posthuman, 6, 9–10; posthumous life associated with, 5
inhuman thought, 173–77
inorganic life, xvii, 13; Inorganic Rites, 8, 9–10
instinct, 335
integrated world capitalism, 241–42
intentionality, 44
intra-action, 79
Investigating Life in Extreme Environments, 297–98
ipseity, 17–22, 33, 38n2
Irigaray, Luce, xxiv
irreducibility of noncontradiction, 132n2
islands: mountains and, 138; world related to, 140–41

Jackson, Peter, 184–85
James, William, 331
Jefferson, Thomas, 197
Johnson, Barbara, 179, 201
Joyce, James, 91

Kant, Immanuel, 2, 203n5, 245, 271, 283; on bees, 272, 290–91; on life, 320n6; Schopenhauer on, 300, 306–11, 319
Kantianism, 300, 320n7. *See also* post-Kantian
Kantian wholes, 142–43
Kauffman, Stuart, 142–43, 147
Keeping Together in Time (McNeill), 215
King, Rodney, 82n7

kinship, 76–77, 80
Klein, Naomi, 242, 245
Kolbert, Elizabeth, 230–31
Korzybski, 151n11
Kristeva, Julia, 54
Kuhn, Thomas, 7
Kurzweil, Ray, xxiv, 18–19, 27

labor, 54–55, 190
Lacan, Jacques, 57, 91–92
Lambert, Gregg, 61
landfills, 259–62
Lange, Lisa, 181
language: cultural translation of, 76, 82n6; economy of, 97–98; Greek, 183, 204n11; humanism and, 46–47; Latin, 6, 101, 108, 186–87, 204n11; materiality and, 68–69, 81n2; personhood related to, 182; subjectivity and, 69. *See also* autobiographies; speech
l'animot. See animals
Laruelle, François, 3–4
Latin, 6, 101, 108, 186–87, 204n11
Latour, Bruno, xxiii, 3–4, 17, 81, 266n6
law, 204n10; nonpersons before, 183–88; performance related to, 186–88, 200; on personhood, 183–88
Laws (Plato), 220, 222
leachate, 260–61
Leas, James Marc, 203n1
Leavey, John, 119
Lebensztejn, Jean-Claude, 103n3
liberal personhood, ix–x
liberationist, 22
life: absolute, 299–305; the Absolute and, 302–4, 320n9; Aristotle on, 296; biases about, 296; critical related to, x, 2–4, 7, 13–14; Derrida on, 105–32; desire related to, 115–16; distance from, 2–4; eliminativism and, 318; as genesis, 304–5; as givenness, 303, 319; indifference and, 13; after life, 11–12; life negating, 312; meontology of, 12, 305, 316, 319; nature and, 117; negation

GPSR Authorized Representative: Easy Access System Europe, Mustamäe tee
50, 10621 Tallinn, Estonia, gpsr.requests@easproject.com